AGRI-FOOD CHAIN RELATIONSHIPS

AGRI-FOOD CHAIN RELATIONSHIPS

Edited by

Christian Fischer

Agribusiness, Logistics and Supply Chain Management
Institute of Food, Nutrition and Human Health
Massey University
Auckland, New Zealand

Monika Hartmann

Agricultural and Food Market Research
Institute for Food and Resource Economics
University of Bonn
Bonn, Germany

www.cabi.org

CABI is a trading name of CAB International

CABI Head Office	CABI North American Office
Nosworthy Way	875 Massachusetts Avenue
Wallingford	7th Floor
Oxfordshire OX10 8DE	Cambridge, MA 02139
UK	USA
Tel: +44 (0)1491 832111	Tel: +1 617 395 4056
Fax: +44 (0)1491 833508	Fax: +1 617 354 6875
E-mail: cabi@cabi.org	E-mail: cabi-nao@cabi.org
Website: www.cabi.org	

A catalogue record for this book is available from the British Library, London, UK.

Library of Congress Cataloging-in-Publication Data

Agri-food chain relationships / edited by Christian Fischer, Monika Hartmann.
     p. cm.
  Includes bibliographical references and index.
  ISBN 978-1-84593-642-6 (alk. paper)
  1. Food industry and trade. 2. Food supply. 3. Agricultural industries. I. Fischer, Christian, 1969- II. Hartmann, Monika. III. Title.

  HD9000.5.A3725 2010
  338.1--dc22

                    2010013799

ISBN: 978 1 84593 642 6

Commissioning editor: Meredith Carroll
Production editor: Tracy Head

Printed and bound in the UK from copy supplied by the authors by CPI Antony Rowe

Contents

Contributors 7

Preface and Acknowledgements 9

Introduction and Overview:
Analysing Inter-organizational Relationships in Agri-food Chains
Christian Fischer and Monika Hartmann 11

Part I Agri-food Chain Relationships – Context and Theoretical Foundations

1 Building Sustainable Relationships in Agri-food Chains:
  Challenges from Farm to Retail
  Monika Hartmann, Klaus Frohberg and Christian Fischer 25

2 Inter-organizational Relationships in Agri-food Systems:
  a Transaction Cost Economics Approach
  Fabio Chaddad and Maria E. Rodriguez-Alcalá 45

3 Behavioural Economics and the Theory of Social Structure:
  Relevance for Understanding Inter-organizational Relationships
  Monika Hartmann, Julia Hoffmann and Johannes Simons 61

4 Collaborative Advantage, Relational Risks and Sustainable Relationships:
  a Literature Review and Definition
  Christian Fischer and Nikolai Reynolds 74

Part II Empirical Evidence on Trust and Sustainable Relationships in Agri-food Chains

5 Trust and Relationships in Selected European Agri-food Chains
  Philip Leat, Maeve Henchion, Luis Miguel Albisu and Christian Fischer 91

6 A Review of the Trust Situation in Agri-food Chain Relationships in the Asia-Pacific
  with a Focus on the Philippines and Australia
  Peter J. Batt 105

7 Determinants of Sustainable Agri-food Chain Relationships in Europe
  Christian Fischer, Monika Hartmann, Nikolai Reynolds, Philip Leat,
  César Revoredo-Giha, Maeve Henchion, Azucena Gracia and Luis Miguel Albisu 119

8 Enhancing the Integration of Agri-food Chains: Challenges for UK Malting Barley
  César Revoredo-Giha and Philip Leat 135

9 From Transactions to Relationships: the Case of the Irish Beef Chain
  Maeve Henchion and Bridin McIntyre 150

10 Reviewing Relationship Sustainability in the Case of
the German Wheat-to-Bread Chain
Miroslava Bavorová and Heinrich Hockmann 164

11 Inter-organizational Relationships in the US Agri-food System:
the Role of Agricultural Cooperatives
Fabio Chaddad 177

12 Guanxi and Contracts in Chinese Vegetable Supply Chains:
an Empirical Investigation
Hualiang Lu, Jacques Trienekens, Onno Omta and Shuyi Feng 191

13 Inter-organizational Relationships as Determinants for Competitiveness in the
Agri-food Sector: the Spanish Wheat-to-Bread Chain
Azucena Gracia, Tiziana de Magistris and Luis Miguel Albisu 206

14 How Buyer–Supplier Relationships can Create Value: the Case of the Australian
Wine Industry
Lynlee Hobley and Peter J. Batt 220

Part III   Implications and Outlook

15 Best Practice in Relationship Management: Recommendations for Farmers,
Processors and Retailers
Hualiang Lu, Peter J. Batt and Christian Fischer 237

16 Improving Agri-food Chain Relationships in Europe: the Role of Public Policy
Luis Miguel Albisu, Maeve Henchion, Philip Leat and David Blandford 250

17 Lessons Learned: Recommendations for Future Research on
Agri-food Chain Relationships
Fabio Chaddad, Christian Fischer and Monika Hartmann 267

Index 281

Contributors

Luis Miguel Albisu is head of the Agri-food and Natural Resources Economics Unit, Agri-food Research and Technology Center of Aragón (CITA) in Zaragoza, Spain. lmalbisu@aragon.es

Peter J. Batt is an associate professor of food and agribusiness marketing at the School of Agriculture and Environment (Muresk Institute), Curtin University of Technology, Perth, Australia. p.batt@curtin.edu.au

Miroslava Bavorová is a senior researcher at the Institute for Farm Management at the Martin Luther University in Halle (Saale), Germany. miroslava.bavorova@landw.uni-halle.de

David Blandford is a professor of agricultural and environmental economics at the Pennsylvania State University, USA. dblandford@psu.edu

Fabio Chaddad is an assistant professor at the University of Missouri, Columbia, USA. chaddadf@missouri.edu

Shuyi Feng is an associate professor at the College of Public Administration, Nanjing Agricultural University, China. shuyifeng@njau.edu.cn

Christian Fischer, at the time of writing, was an associate professor of supply and value chain management at the Institute of Food, Nutrition and Human Health at Massey University, Auckland, New Zealand. He is now an associate professor of agricultural economics and management at the Free University of Bozen-Bolzano in Italy. christian.fischer@unibz.it

Klaus Frohberg is an emeritus professor at the University of Bonn, Germany. klaus.frohberg@uni-bonn.de

Azucena Gracia is a senior researcher at the Agri-food and Natural Resources Economics Unit, Agri-food Research and Technology Center of Aragón (CITA) in Zaragoza, Spain. agracia@aragon.es

Monika Hartmann is a professor at the Institute for Food and Resource Economics, University of Bonn, Germany. monika.hartmann@ilr.uni-bonn.de

Maeve Henchion is head of the food market research programme at the Ashtown Food Research Centre (AFRC), Teagasc in Dublin, Ireland. maeve.henchion@teagasc.ie

Lynlee Hobley is a graduate student and part-time lecturer in the School of Agriculture and Environment (Muresk Institute), Curtin University of Technology. l.hobley@curtin.edu.au

Heinrich Hockmann is a professor at the Leibniz Institute for Agricultural Development in Eastern and Central Europe (IAMO) in Halle (Saale), Germany. hockmann@iamo.de

Julia Hoffmann is a doctoral student at the Institute for Food and Resource Economics, University of Bonn, Germany. julia.hoffmann@ilr.uni-bonn.de

Philip Leat is a senior economist and former team leader of food marketing research in the Land Economy and Environment Research Group of the Scottish Agricultural College (SAC) in Aberdeen, UK. philip.leat@sac.ac.uk

Hualiang Lu is a senior lecturer at the School of Business Administration, Nanjing University of Finance and Economics, China. hualianglu@njue.edu.cn

Tiziana de Magistris is a researcher at the Agri-food and Natural Resources Economics Unit, Agri-food Research and Technology Center of Aragón (CITA) in Zaragoza, Spain. tmagistris@aragon.es

Bridin McIntyre is a senior researcher at the Ashtown Food Research Centre (AFRC), Teagasc in Dublin, Ireland. bridin.mcintyre@teagasc.ie

S.W.F. (Onno) Omta is a professor of food chain management at Wageningen University, the Netherlands. onno.omta@wur.nl

César Revoredo-Giha is a senior economist and team leader of food marketing research in the Land Economy and Environment Research Group of the Scottish Agricultural College (SAC) in Edinburgh, UK. cesar.revoredo@sac.ac.uk

Nikolai Reynolds is a senior research manager at Synovate in Frankfurt, Germany. nikolai.reynolds@synovate.com

Maria E. Rodriguez-Alcalá is an instructor of agricultural and applied economics at the University of Missouri, Columbia, USA. rodriguezalcalam@missouri.edu

Johannes Simons is a senior researcher at the Institute for Food and Resource Economics, University of Bonn, Germany. johannes.simons@ilr.uni-bonn.de

Jacques Trienekens is an associate professor of food chain management at Wageningen University, the Netherlands. jacques.trienekens@wur.nl

Preface and Acknowledgements

This book is the result of several years of research activity in an academic topic of growing importance: how to better link farmers, processors and retailers with each other in order to ensure and improve the supply of food products which meet consumer needs and wants.

The editors, and some of the book contributors, got more strongly involved in the topic in 2005 when they started the research project on 'Key factors influencing economic relationships and communication in European food chains' (FOODCOMM, SSPE-CT-2005-006458) which was funded by the European Commission as part of the Sixth Framework Programme. The collaborating laboratories were: Institute for Food and Research Economics, University of Bonn, Germany (coordinator); Land Economy and Environment Research Group, Scottish Agricultural College (SAC), Edinburgh and Aberdeen, UK; Institute for Agricultural Development in Central and Eastern Europe (IAMO), Germany; The Ashtown Food Research Centre (AFRC), Teagasc, Dublin, Ireland; Ruralia Institute, University of Helsinki, Finland; Institute of Agricultural and Food Economics (IAFE), Poland; and Unit of Agri-food Economics and Natural Resources, Agri-food Research and Technology Center of Aragón (CITA), Spain.

The main results of the research project have been published in several articles in scientific journals (all of which are referred to in this volume). However, when the research project finished it became evident that the results needed to be compared in a wider context than it was possible within the project. Simultaneously to the FOODCOMM work, complementary research has taken place nearly all over the world. This book attempts to integrate the experiences gained by a number of researchers in different countries but around the same topic.

We would like to express our sincere gratitude to Meredith Carroll of CAB International for publishing the book; to all contributors, and in particular to Maeve Henchion, Philip Leat and Luis Miguel Albisu, for reviewing many of the book contributions; and to Jan Schiefer, Julia Hoffmann and Stefan Hirsch of the University of Bonn and Bill Wang of Massey University for editorial assistance.

Christian Fischer
Monika Hartmann

March 2010

Introduction and Overview

Analysing Inter-organizational Relationships in Agri-food Chains

Christian Fischer[1] and Monika Hartmann[2]

[1] Massey University, Auckland, New Zealand
[2] University of Bonn, Germany

'Il faut créer des liens entre les hommes'

Antoine de Saint-Exupéry

Introduction

'One must create bonds between men', and – as we argue in this book – also between organizations, if a steady supply of food products is to be ensured and improved in an ever more demanding world. Against the background of global market liberalization, increasing consumer awareness and concerns, and the spreading of life-changing technologies, new ways to produce, distribute and consume food are evolving. As a result, the organization of production and distribution systems (including agriculture) has to adapt and one of the most fundamental drivers behind this process is the general recognition that, depending on the situation, collaboration can be more effective than competition as a general organizational mode to achieve economic efficiency.

Collaboration requires organizations to build and maintain effective relationships, hence the need for a clear understanding of what 'inter-organizational relationships' are, in particular within an agri-food chain context, when they are necessary, and how they can be improved. One of the main findings from the studies in this book is that while profit motives are at the core of agri-food chain relationships, gaining full satisfaction from them also requires the nurturing of personal bonds, in particular with farmers, if a relationship is to be sustained and the mutual benefits of collaboration are to be fully unlocked.

While the topic has already received considerable attention in the general economics, business, management and marketing literatures, it has not been extensively covered within the agricultural and food economics profession. This holds despite earlier work of Giovanni Galizzi and Luciano Venturini and their 1999 book *Vertical Relationships and*

Coordination in the Food System, which summarized the research that had been undertaken in the 1990s and before. Although over the last decade a number of articles have appeared on the topic, no comprehensive work has been published since then which covers theoretical and empirical investigations regarding agri-food chain relationships, adopting a global focus. This volume attempts to fill the gap.

The book is structured in three parts. Starting with an overview regarding main developments in the agri-food sector with relevance for chain relationships (Chapter 1), Part I is mainly concerned with providing the theoretical foundations for analysing agri-food chain relations (Chapters 2, 3 and 4). Building on this conceptual basis, the second part presents in-depth empirical evidence for different countries, food chains and chain stages regarding the issues of trust and sustainable relationships in agri-food chains (Chapters 5 to 14). The red meat industry (beef and pork) is the focus of Chapters 5, 7 and 9. Cereals (bread and malting barley) are analysed in Chapters 5, 7, 8, 10 and 13. Horticultural products (fresh produce and wine) are investigated in Chapters 6, 12 and 14. Regionally, the studies cover Europe, North America (the USA), China, Australia and the Philippines. While most studies were conducted in developed markets, Chapters 6 and 12 look at the particularities of transition or developing economies. As to individual agri-food chain stakeholders, a number of chapters (Chapters 5 to 12, 14 and 15) offer and discuss separate findings for farmers, food processors or retailers. Based on the theoretical and empirical findings in the first two parts of the book, recommendations for agribusiness managers (Chapter 15) and policy-makers (Chapter 16) are described in the third part. The final Chapter 17 discusses avenues for future research.

Three major topics are covered in this book. The first is about the use of suitable production and distribution system architectures. It addresses the question of whether members of agri-food chains should 'buy', 'make' or 'collaborate'. In other words, should agri-food products be produced by specialist organizations (farms and firms) and sold through spot market transactions; or should they be produced by large, multi-activity corporations that own farming, processing and retailing assets (or combinations hereof)? A potential third way covers hybrid arrangements, where legally independent and specialized organizations form collaborative partnerships and build sustainable relationships, thus working together in order to meet ever more sophisticated consumer requests. Within economics, this problem is analytically reduced to the appropriate choice of 'governance structures'. One of the main criteria used to facilitate decision-making about the best organizational architecture to use is the minimization of transaction costs. In this volume, three chapters are devoted to this topic. In Chapter 2, the reasoning behind transaction cost economics is reviewed and then expanded to not only include vertical (producer-to-consumer) but also horizontal (within the same industry) as well as cross-industry governance structures. Chapter 11 applies this approach to the US agricultural cooperative system. In Chapter 9, chain stakeholders' governance structure choices for selling beef in Ireland are empirically analysed. Using a number of social factors rather than exclusively relying on transaction cost minimization considerations, the authors explain the shift from primarily spot markets to repeated market transactions. Finally, Chapter 3 reviews the socio- and psycho-economic roots of human behaviour in the context of inter-organizational relationships. While the analysis in this chapter is somewhat different from the previous ones mentioned, conceptually this study belongs within this topic as it deals with the influence of social embeddedness of business relationships on governance structure choice.

The main focus of this book is the assessment of the 'sustainability' of inter-organizational relationships in agri-food chains. Relationship sustainability is defined as to include static 'quality' aspects of business relationships, as well as dynamic considerations which are summarized in a 'stability' component. The overall motivation behind the studies covering this topic is to identify the level of relationship sustainability and thus to reveal how good business relationships between agri-food chain stakeholders currently are. Another is to analyse determinants that influence relationship sustainability. Based on these findings, the studies suggest measures of how to improve business relationships in the agri-food sector. Eight chapters deal with relationship sustainability in one way or the other. Chapter 4 discusses underlying theories and formally defines the sustainable inter-organizational relationship (SIR) construct. Chapters 5 and 6 deal with trust, one of the most important components in SIRs. While using similar conceptual approaches, Chapter 5 covers European and Chapter 6 Asia-Pacific agri-food chains. Chapter 7 represents an empirical investigation into the main determinants of SIRs for six European countries, two commodities and both farmer–processor and processor–retailer relationships. Chapter 8 analyses the determinants affecting the sustainability of business relationships for the UK barley-to-beer and whisky supply chain. Chapter 10 uses a more qualitative approach to discuss the role of trust for sustainable business relationships in the German wheat-to-bread chain. Chapter 12 focuses on 'satisfaction', another particular component of sustainable relationships and empirically estimates the importance of major driving forces behind relationship satisfaction for fresh vegetable supply chains in China. Chapter 13 investigates determinants of relationship quality in Spanish wheat-to-bread chains but also analyses how good-quality relationships contribute to chain competitiveness.

The third main topic of this book covers the impact of sustainable relationships on chain performance. Although the book does not offer a systematic analysis of this matter, two chapters explore it in more detail. As already mentioned, Chapter 13 is concerned with chain competitiveness as a measure of performance while Chapter 14 uses the concept of 'relationship value' as a performance proxy, covering the Australian wine industry.

Chapter summaries

In the following we provide a synopsis of the contents of each chapter. The summaries were drafted by the respective chapter author(s) and reviewed by the editors.

Chapter 1 – Building sustainable relationships in agri-food chains: challenges from farm to retail

This chapter discusses the broader business context in which today's agri-food chains operate. It also describes the long-term trends which have shaped agri-food sectors and markets and which have contributed to the rise of vertical business systems such as supply or value chains. In particular, the changing economic, technological, social and policy environments are reviewed. These include globalization (i.e., the widespread trade and labour market liberalizations), increasing consumer quality demands and the linked rise in public and private standards, the ubiquity of information and communication technologies, and the increasing skills base and entrepreneurial capacities in today's knowledge-based societies which foster the development of highly networked economies.

Given the new reality, vertical business systems have emerged as highly competitive organizational formats which allow collaborating companies to exploit economies of size (scale and scope) while maintaining strategic flexibility, agility and managerial independence.

However, one requirement for establishing competitive vertical business systems is the development of sustainable inter-organizational relationships. Building and managing such relationships is in particular a challenge for farmers who traditionally have operated on the basis of spot market exchanges. Yet the decline of specialized food retailing and the rise of global supermarkets also pose problems. Supermarkets that deal with large numbers of (food and non-food) suppliers on a regular basis increasingly use procurement practices which build on private standards, formalized contracts and electronic auctions, therefore sidestepping individual supplier relationships.

While the transition from the transaction-based to a relationship-based agri-food economy will take time and be partial, the future seems to favour the extensively embedded, consumer-oriented agri-enterprise rather than the traditional owner-centred, price-driven farm business.

Chapter 2 – Inter-organizational relationships in agri-food systems: a transaction cost economics approach

This chapter analyses inter-organizational relationships in agri-food systems from a transaction cost economics perspective. Given the diversity of organizational and legal forms observed in inter-organizational relationships, the chapter focuses on understanding the nature and governance characteristics of such relationships.

The chapter first presents a typology of inter-organizational relationships treating them as intermediate forms along the market–hierarchy continuum of transaction costs economics. Subsequently, the chapter builds on recent theoretical advances and defines inter-organizational arrangements as 'true hybrids' combining governance instruments of 'pure' markets with instruments of 'pure' hierarchies. That is, governance of inter-organizational relationships is conceptualized as a multidimensional construct composed of several coordination and control mechanisms.

These governance mechanisms are designed and implemented by alliance partners to manage the basic tension between collaboration (or cooperative interdependence) and competition (or transactional interdependence) that is characteristic of every form of inter-organizational relationship.

Chapter 3 – Behavioural economics and the theory of social structure: relevance for understanding inter-organizational relationships

Neoclassical economists view economic agents as rational, atomized and purely self-interested individuals, maximizing utility in the presence of full information. Neither social relations on the one hand, nor limited resources to obtain and process all information on the other, are taken into account in this approach. However, a large body of evidence exists which indicates that people deviate from exclusively self-interested, efficiency-oriented behaviour.

The chapter describes the contribution that the theory of social structure and behavioural economics provides for a deeper understanding of business relationships.

Social structure theory, in the context of business relations, regards economic behaviour as embedded in and mediated by a complex and extensive web of social relations. It takes a comprehensive look at the different goals that actors in business relationships try to achieve and the rules they observe. Business relations are regarded as intertwined with social relations.

Behavioural economics is concerned with empirically testing neoclassical assumptions and explaining identified systematic deviations from those assumptions. Together, social structure theory and behavioural economics provide powerful insights regarding inter-organizational relationships, with the concepts of trust, reciprocity and reputation playing a crucial role. The chapter shows that social embeddedness of inter-organizational relations allows transactions to take place, even in cases where considerable vulnerabilities such as moral hazards exist that cannot (totally) be controlled by explicit contracts. Social ties between firms based on trust, reliability and reciprocity are able to create intangible, largely inimitable, resources which enhance firms' competitiveness. Fairness-driven reciprocity behaviour is a social norm with considerable implications for inter-organizational relationships, as is shown in behavioural economics. It can explain why actors get involved in or abstain from a business relationship, as well as why they are willing to punish or reward their partner even if this implies personal losses.

Numerous studies show that in markets with incomplete contracts, the interaction of fairness and reputation effects not only changes the outcome of a transaction, bringing it closer to the optimal efficiency level, but also influences the structure of relationships towards more long-term bilateral transactions.

Chapter 4 – Collaborative advantage, relational risks and sustainable relationships: a literature review and definition

Based on a review of the recent applied economics, marketing, business and management literature on inter-organizational relationships, a definition of sustainable inter-organizational relationships (SIRs) is derived.

Starting from the notion of 'collaborative advantage' (i.e., the competitive advantage arising from collaboration) it is argued that sustainability of inter-organizational relationships in vertical, buyer–seller partnerships can help to make business transactions more efficient, despite existing relational risks. In particular, it can contribute to: (i) reduce environmental uncertainty (e.g., by securing a more stable or higher inflow of orders); (ii) obtain better access to crucial resources (e.g., raw materials, capital, specialized skills or knowledge); and/or (iii) result in higher business productivity (e.g., by enhancing loyalty among suppliers).

SIRs can be defined as high-quality and stable business relationships which are responsive to changing environments and which collaborators continue as long as the benefits derived from a relationship outweigh the costs of maintaining it.

Chapter 5 – Trust and relationships in selected European agri-food chains

Trust is an integral part of agri-food supply chain operation, both for ensuring food safety and quality, and facilitating the exchange relationships between businesses. Through its implicit sharing of norms and values by transacting parties, it is central to good chain relationships.

This chapter explores the nature and role of trust in eight selected agri-food chains in four EU countries. Following a brief consideration of the concept of trust, the empirical analysis identifies key features of the business environments of the selected agri-food chains, and their possible influence on chain relationships. The assessment is based on 28 qualitative interviews with food chain experts. In particular, the connection between perceived trust levels and the prevailing type of vertical business relationships is explored.

The findings suggest that trust is more pronounced among small- and medium-sized enterprises (SMEs) whose representatives have personal relationships with each other. There is more trust upstream in the chain than downstream. However, there is also considerable mistrust at the farmer end in several of the studied chains. Two particular factors contribute to this situation. First, there may be a lack of chain transparency in relation to farm product quality and related prices. Second, where the general economic situation is difficult, the development of trust may be hampered, because all chain participants seek a share of a diminishing chain margin. Further downstream, where frequently larger enterprises are involved, the levels of trust between chain participants vary more substantially. Finally, if bargaining power is distributed unevenly, trust between chain partners may not fully develop.

Chapter 6 – A review of the trust situation in agri-food chain relationships in the Asia-Pacific with a focus on the Philippines and Australia

It is widely accepted that trust is the critical determinant of good buyer–seller relationships.

In the Philippines, the key antecedents to trust between growers and their preferred buyers is found to be the buyers' willingness to advise growers of supply problems, a close personal friendship and a perception that the buyer treats the grower fairly.

In Perth, Western Australia, where traditionally there has been a great deal of distrust between growers and the market agents who receive and distribute their produce, growers generally transact with more than one market agent. Trust is facilitated primarily by the market agent's good reputation. Preferred market agents are perceived to treat growers more equitably, to provide adequate rewards and to make available technical information that could assist or enable growers to better fulfil the needs of downstream buyers. Conversely, market agents' propensity to act opportunistically and to employ coercive influence strategies has a significant negative impact on trust.

Chapter 7 – Determinants of sustainable agri-food chain relationships in Europe

The role of inter-organizational relationships in European meat and cereals chains is analysed. Using survey data from 1442 farmers, processors and retailers in six countries (Germany, UK, Spain, Poland, Ireland, Finland), the empirical relevance of several factors potentially influencing the sustainability of agri-food chain relationships is tested.

Structural equation models are estimated for each country, each commodity chain (meat and cereals) and for two chain stages (i.e., the relationships between farmers and food processors, and the ones between processors and retailers).

Effective communication and the existence of personal bonds are consistently found to mostly affect relationship sustainability, followed by unequal power distribution between buyers and suppliers and the negative impact of key people leaving a company concerned.

The existence of personal bonds is particularly important in relationships with farmers, while effective communication is more crucial in relationships with retailers.

The empirical results are in line with the findings from other studies and confirm a number of theoretical expectations as discussed in Part I of this volume. In particular, the results reveal that inter-personal elements play an important role in commercial agri-food chain relationships.

Chapter 8 – Enhancing the integration of agri-food chains: challenges for UK malting barley

This chapter explores the theoretical and practical issues affecting the sustainability of business relationships and therefore the cohesion of an agri-food supply chain. This is important because improved supply chain coordination and cooperation amongst chain participants can improve its efficiency and effectiveness, and therefore, its competitiveness and long-term sustainability.

The chapter focuses on the UK barley-to-beer and whisky supply chain, drawing on two complementary analyses. First, a survey-based structural equation model is formulated to determine those factors that affect the sustainability of relationships in the chain. The second analysis consists of an in-depth case study based on an important malting barley-to-beer supply chain in eastern England.

The findings point to five factors affecting relationship sustainability: communication quality; compatibility of partners' aims; contractual relationships backed by professional regard and personal bonds; high levels of trust between chain participants and a willingness to resolve problems; and commercial benefit. Communication is found to be focused on creating awareness of customer requirements, facilitating logistical issues, problem resolution, maintaining the required quality of service, as well as the trust and friendship of participants. Relationships are assisted if the aims of the supply chain parties are strongly aligned and if contractual relationships are strengthened by professional regard, personal friendships and trust. Finally, each party has to derive clear benefit from a relationship if businesses are to retain a particular set of supply chain arrangements. Consequently, market power issues, that affect the distribution of rewards amongst the partners, require careful consideration in the maintenance of sustainable chain relationships.

Chapter 9 – From transactions to relationships: the case of the Irish beef chain

Pro-active management of inter-organizational relationships is a critical source of competitive advantage. However, this is easier said than done. Food chains across the world are characterized by a web of highly interdependent and interconnected associations that span a variety of relationship types. Furthermore, within any particular context, multiple exchange paradigms may be present.

This chapter seeks to illustrate how such pro-active management may be undertaken by attempting to provide further insight into how relationships operate within one particular sector, i.e., the beef sector in Ireland. Following an outline of important contextual factors within the sector, it examines the nature of business relationships that prevail within the sector, identifies factors that influence the selection of relationship type and investigates the link between relationships and performance, ultimately developing recommendations on how to optimize relationship type to enhance performance. Emphasis is also placed on the role of communication within the relationship. The focus is mainly on the farmer–processor relationship, however observations on the processor–retailer relationship are also provided where they illustrate a point regarding the farmer–processor relationship.

Chapter 10 – Reviewing relationship sustainability in the case of the German wheat-to-bread chain

The concept of relationship sustainability is used to evaluate business relationships in the German wheat-to-bread supply chain. Relationships are defined as sustainable when the returns gained from cooperation over time exceed the costs of their initiation and maintenance.

The survey results reveal that businesses in the chain rate the relationship with their main business partner as being of high quality and often long term. Results from additional interviews with buyers help us to isolate the effects of economic and behavioural dimensions on continuation and termination of relationships with suppliers. The economic dimension includes competitive price and quality, and supply continuity. The behavioural dimension is captured by trust as a proxy. These factors turn out to significantly govern the termination of long-lasting relationships. A lack of trust means that relationships are not developed or are terminated in the early phases. Furthermore, the results show that the actors in the German wheat-to-bread chain are free to choose the relationship type and length so as to maximize profit under given environmental conditions. Based on these findings we conclude that the analysed chain relationships can be considered as sustainable.

Chapter 11 – Inter-organizational relationships in the US agri-food system: the role of agricultural cooperatives

Despite fundamental changes in the business environment, agricultural cooperatives in the United States continue to perform a relevant economic role in agri-food chains as farmers' 'integrating agency.' The nation's 2594 agricultural cooperatives generated US$147 billion in sales and accumulated US$57 billion in assets in 2007. US agricultural cooperatives provide production inputs and services to farmers and process and market their commodities. In 2002, aggregate cooperative market shares for both farm commodity marketing and purchased inputs reached 27%. In addition, cooperatives are increasingly involved in upstream and downstream stages of the agri-food system.

This chapter describes and analyses the variety of organizational configurations adopted by US agricultural producers. Given the increased capital intensity and competitiveness in agri-food systems, cooperatives are forming alliances and joint ventures with other cooperatives, private firms and outside investors. A recent survey finds that 204 cooperatives had additional sales of US$14 billion in 2007 from other, non-cooperative business ventures. By means of several types of inter-organizational relationships, cooperatives have formed complex farmer-led supply chains and networks ('netchains').

Chapter 12 – Guanxi and contracts in Chinese vegetable supply chains: an empirical investigation

We develop and test an integrated analytical framework to investigate interactions among guanxi networks, buyer–seller relationships, contractual governance and relationship satisfaction in the Chinese vegetable industry.

The survey results show that guanxi networks positively contribute to trusting relationships and substantially enhance business satisfaction. However, farmers' approaches to achieving satisfying relationships with their buyers are different from those employed by vegetable processing and exporting firms. Relational governance substantially reduces transaction costs. Furthermore, the total effects of relational governance (based on guanxi networks) on contractual governance underline the importance of trust-building activities.

The complementary effects of guanxi networks on contractual governance imply that a combination of relational and contractual governance in vegetable supply chains may be the best way to improve business relationships and to enhance marketing performance.

Chapter 13 – Inter-organizational relationships as determinants for competitiveness in the agri-food sector: the Spanish wheat-to-bread chain

The objective of this chapter is to investigate whether the quality of inter-organizational relationships positively influences the competitiveness of small- and medium-sized enterprises (SMEs) in the wheat-to-bread chain in the Spanish region of Aragon. Relationship quality is composed of trust, satisfaction and commitment between buyers and sellers. A structural equation modelling approach is adopted using data from a survey that targeted farmers, processors and retailers.

The main finding of the study is that, as the quality of the relationship improves, stakeholder competitiveness increases. As relationship quality is strongly influenced by communication quality and, though to a lesser extent, by equal power distribution those two factors influence stakeholders' competitiveness in an indirect way. Moreover, personal bonds positively influence the quality of communication.

Chapter 14 – How buyer–supplier relationships can create value: the case of the Australian wine industry

This study examines customers' and suppliers' perceptions of the factors that are instrumental in optimizing relationship value in the Australian grape and wine industry.

A comprehensive examination of trading relationships between wine grape growers and the wineries to whom they supply grapes demonstrates that a willingness and ability to resolve conflict, communicate effectively, provide performance satisfaction and the propensity to trust and cooperate contributes towards relationship value. Conflict resolution is the core antecedent which directly drives communication and performance satisfaction to increase trust and cooperation, leading to superior relationship value for customers and suppliers.

Chapter 15 – Best practice in relationship management: recommendations for farmers, processors and retailers

Sustainable business relationships can create value for all chain partners. Using a practice-oriented approach, the chapter describes how to successfully build and maintain sustainable business relationships between agri-food chain stakeholders.

Exchange partners enter into a business relationship because they expect to receive net returns from it. Thus, among other factors, fair prices, the existence of trust, and the ability to deal with market power imbalances, form the foundations in the agri-food chain relationship management process. The successful management of business relationships should take into account multiple aspects, such as equitably sharing commercial risks, avoiding abuse of coercive market power, making relationship-specific investments, creating trusting partnerships, sharing commercial information, building reputation and fostering personal relationships.

Similar to growing agricultural products, there is a best-practice approach to managing agri-food chain relationships. The major ingredients for 'planting' business relationships are selecting the right partner, aligning business goals and procedures, and allocating appropriate resources. To 'grow' a relationship, effective communication, intensive collaboration, joint problem-solving and nurturing trust are pivotal. Following a successful 'harvest', rewards/benefits should be shared fairly and relationship outcomes be evaluated.

However, different actors in agri-food chains should follow different rules and routines to achieve successful business relationships. For farmers/growers, staying in regular contact with buyers, investing in relationship-specific assets, and building strong personal relationships are key elements. For processors, providing useful technology to upstream and downstream partners, offering fair contracts, providing relevant market information to suppliers, and investing in relationship-specific assets are the most important aspects. For distributors/retailers, maintaining effective communication with suppliers, using more professional management tools, and compensating power asymmetries by nurturing trust with suppliers are key recommendations.

Chapter 16 – Improving agri-food chain relationships in Europe: the role of public policy

This chapter deals with the role that government policies, and especially EU policies, can play in improving agri-food chain relationships.

The chapter starts with a review of the main factors that influence agri-food chain relationships, based on the previous chapters of this book, as well as other empirical work. The most important policies, with special emphasis on the Common Agricultural Policy, are analysed in relation to agri-food chain requirements. Other policies relating to food, small- and medium-sized enterprises (SMEs), consumers and related topics involving agri-food chains are examined to determine how they can affect relationships along the chain. The policies analysed can directly or indirectly affect agri-food chains and can have macroeconomic or microeconomic objectives. The main issues are separated, for any particular policy, between those directed towards agri-food chain stakeholders, with either horizontal or vertical implications for their relationships, and those that concern competitiveness in agri-food chains. Private and public implications are compared to understand the adequacy of current policies. Finally, propositions are offered for a more integrated policy approach to agri-food chain relationships.

Chapter 17 – Lessons learned: recommendations for future research on agri-food chain relationships

This chapter discusses potential research opportunities in the growing field of inter-organizational relationships that could build on the theories and empirical evidence discussed in this volume.

Several potential research avenues are identified regarding governance and chain collaboration architectures, the linkage between relationship governance and sustainability, and the definition and measurement of sustainable inter-organizational relationships. As to the latter issue, more detailed suggestions are proposed on how further progress could be achieved.

Extending, refining and combining existing theories and using additional sources of data and new research methods to confirm or disprove empirical evidence would be of significant value to researchers and practitioners interested in analysing and managing sustainable inter-organizational relationships in the global agri-food system.

References

Galizzi, G. and Venturini, L. (eds) (1999) *Vertical Relationships and Coordination in the Food System*. Springer, Heidelberg.

Saint-Exupéry de, A. (1943) *Le Petit Prince*. Gallimard Jeunesse, Paris.

Part I

Agri-food Chain Relationships – Context and Theoretical Foundations

Chapter 1

Building Sustainable Relationships in Agri-food Chains: Challenges from Farm to Retail

Monika Hartmann,[1] Klaus Frohberg[1] and Christian Fischer[2]

[1] University of Bonn, Germany
[2] Massey University, Auckland, New Zealand

Introduction

Over the last 20 years, the agri-food sector in industrialized countries has undergone substantial changes. It has become more complex and individual enterprise activities have been increasingly complemented by collective actions. Competition has been rising not only between retail stores but at all levels of the food chain. Locally produced goods increasingly compete with imported ones for consumers' spending.

Many elements shape the competitive and diverse environment of today's food chains such as, among others, globalization facilitated by trade liberalization, the increasing number of people travelling and the communication revolution, which has made it necessary for organizations to deal with more complex supply chains, and to cope with a large variety of product and process standards as well as with different consumer demands. Improvements in information and communication technologies (ICTs) have eased this process. Furthermore, ICTs allow the use of more sophisticated risk management tools, such as the tracking and tracing of products, which is a precondition for a higher level of food quality and safety.

In many countries, food is no longer primarily seen as a means to serve basic needs but is linked to lifestyles. Consumers expect not only high product quality but also that food is produced in a sustainable way, especially regarding animal welfare, environmental degradation and social conditions. The characteristics of modern food demand force retailers to acquire more information regarding food products along the supply chain to ensure that products are in accordance with food safety requirements and consumer preferences. Thus, the production and processing of food involves more and more information flows between various supply chain stages which requires collaborative efforts of the organizations concerned. With global sourcing gaining in relevance there is also an increasing need to build international business partnerships. Moreover, these developments are underway in less developed countries (LDCs) and emerging markets (EMs) as well, though differences in intensity still exist (Galizzi and Venturini, 1999; Boger, 2001; Ménard and Valceschini, 2005).

In this chapter, the primary developments shaping the agri-food sector are discussed and their impact on inter-organizational relationships is elaborated on. The second and third sections look at globalization and developments in consumer demand, respectively. The fourth section deals with food safety and quality standards, which is followed by a discussion of competition and concentration, ICTs, as well as human skills and entrepreneurial activities. The final section concludes and provides an outlook into the future.

Globalization

One of the determinants with an influence on the development of inter-organizational relationships is globalization; i.e., the trend towards increasing integration of the world's economies. This can be observed by many facts of which the expansion of trade in goods and services is one of them. In other words, the international division of labour is further on the rise.

Agri-food trade

Annual average growth rates of global trade (in volume terms) in all food products excluding fish varied substantially over different time periods between 1990 and 2006. In the first 6 years of the 1990s, global food imports grew at an annual average of about 3.0%, compared with approximately 6.2% per annum thereafter up to 2000 and they increased again from the latter year up to 2006 by approximately 5.9% per year on average (see Table 1). In 1996, imports reached an intermediate peak. The decline which followed in the late 1990s was caused by several factors. Among them were the financial crises which occurred in Asia and also in Russia during this time. On the other hand, the bursting of the bubble of the new economy in 2000 seems to have had no or only a slight effect on global food imports because this was also the year when trade figures rebounded and initiated strong growth until 2006. This average annual growth was about twice as strong as the one from 1990 to 1996. Global food exports exhibit a similar pattern as food imports.

Table 1. Average annual growth rates (in %) of global trade in food products excluding fish.

Imports			Exports		
1990–1996	1996–2000	2000–2006	1990–1996	1996–2000	2000–2006
3.0	−6.2	5.9	3.7	−7.0	6.2

Own calculations using the end years of each period from FAOSTAT (2010). FAO employs country-specific prices in dollars averaged over the period 1999 to 2001 to calculate dollar values from trade volume indices.

These variations in annual food trade changes are observed at all levels of food production and processing though the growth that took off in the year 2000 was more pronounced for consumer-oriented food items, and exceeded that of bulk food by about 50%, and of intermediate food items by roughly 3% (ERS, 2009). With global sourcing of agricultural and food products on the rise, the rapid emergence of global value chains can be seen. Processors buy their raw or intermediate products and retailers their consumer goods from all over the world, transforming the food sector in an interconnected system of complex relationships (e.g., Traill, 1997). This trend is caused by developments regarding the size and structure of the countries' populations as well as their incomes, changes in consumer preferences, public and especially private standards, advancements in transportation, procurement strategies due to foreign direct investments (FDIs, i.e., by investing in other countries with the intention of keeping control over the assets acquired) and adjustments in food trade policies.

Though protectionist policies have been eased in the past, there are still substantial tariff and non-tariff barriers to international food trade at play. While many countries lowered or even removed their tariffs altogether for a large number of commodities, they keep them high for sensitive food imports. Moreover, tariff reductions were usually smaller for fully processed food items than for semi-processed or primary products, as importing countries aim to protect their food processors (ERS, 2009). In addition, non-tariff barriers, mostly in the form of sanitary or phytosanitary measures, are widely applied in the agri-food sector. Though in general justified to protect consumers and society of the importing countries they also serve as a trade barrier. Thus, the assumption is that a further liberalization of international food policies in current trade negotiations will provide additional impetus for increasing agri-food trade.

FDIs in the food sector

Globalization is also reflected by FDIs, much of which take the form of acquisitions in the case of leading food companies. However, mergers, alliance formation and greenfield investments have played a role as well. Looking at the three largest food companies reveals their global span. In 2003 Nestlé, the world's number one food processing company was active in approximately 150 countries, Unilever in about 120 countries and Danone in roughly 70 countries (van Witteloostuijn, 2007). In the case of high import protection, penetrating foreign markets may be easier by acquiring large local companies, engaging in joint ventures or setting up new production facilities and selling the output in the country rather than by exporting from the home country. Another advantage of FDIs can be lower production costs. Economic conditions in the foreign country may allow cheaper production than is possible in the home country. Finally, FDIs may be used for buying foreign companies in order to achieve strategic objectives. In general, FDIs are seen as an important avenue in which food companies seek expansion in foreign markets. This holds especially for high-value food products. In the case of the USA, processed-food sales from foreign affiliates of US companies exceed the value of US processed-food exports by almost 400% (ERS, 2009; Gehlhar and Regmi, 2005). Moreover, many other OECD countries sell more food through their foreign affiliates than through exports (Bunte, 2007). In addition, the annual growth rates of FDIs over the period 1990 to 2004 have exceeded those of international trade in developed countries by 88% and in developing countries by 100% (Bunte, 2007).

Within the various stages of the agri-food value chain, FDIs have been dominant in the past in the food processing sector, however, the composition of FDIs is changing. While, for example, in the USA in 2003 the ratio of FDIs in food manufacturing to food retail and service was 3.5:1, it narrowed to 1.7:1 by 2007 (ERS, 2009).

Those countries which pursue stable and market-oriented policies and have regulations in place which protect foreign capital are more attractive for receiving FDIs. Thus, it is not surprising that most of FDIs in the food sector have taken place between the industrialized countries.[1] However, as over the last two decades many LDCs and also the former socialist countries have moved away from regulated and state-owned industries toward free-market enterprises and liberalized their investment regime, they have been increasingly attracting FDIs. This development has been promoted by additional pull and push factors. Regarding FDIs in the retail sector, push factors include fierce competition in this sector of the home country which limits potential profits as well as domestic planning restrictions with respect to the location of new retail outlets. Major pull factors are relative high returns and low levels of competition in foreign markets that at the same time are characterized by very high growth rate in demand (Reardon *et al.*, 2007; Malik *et al.*, 2009). In addition, the regulatory environment for modern retailing is more favourable in most LDCs and EMs, implying that much of the 'supermarket revolution' in developing countries has been pushed by FDIs.

FDIs can have considerable impacts on inter-organizational relationships. As shown in several studies, FDIs in the food processing sector of transition countries induced a move towards innovative vertical contracts. While hold-up problems usually in the form of excessively long delays in the payment for delivered products were common after the collapse of the socialist system, foreign owned enterprises secured immediate payments for delivered products (Gow and Swinnen, 1998). For the case of the Polish dairy sector, Dries and Swinnen (2004) show that after entering the Polish market, foreign investors strongly engaged in vertical coordination with suppliers. Though the dairies requested high product and process quality standards they provided at the same time assistance programmes to improve management at the level of the farmers and enhanced their access to technology, credit and other inputs, thereby linking input and output markets. This process induced considerable positive vertical spillovers promoting growth and in many cases even secured the survival of suppliers. Soon those strategies were copied by domestic dairy companies, thus the entrance of foreign firms changed business relationships beyond the companies' own suppliers and induced important horizontal spillover effects (Dries and Swinnen, 2004). Examples of proactive strategies by foreign companies to build up supply chains and provide assistance to upstream chain partners are also provided in other studies (see Reardon *et al.*, 2007).

However, FDIs can also threaten the existence of traditional enterprises: many LDCs and EMs received substantial FDIs in their food retail and food service sectors causing existing retailers to lose a considerable share of the market. In addition, FDIs in the retail sector often have considerable implications on local supply networks. As multinational retailers, such as Walmart, Carrefour or Ahold, have regional and/or global sourcing organizations they are able to shift from local suppliers to those of other countries (Schmid, 2007; Durand, 2007; Trienekens and Zuurbier, 2008).[2] In some countries such developments were further favoured by overvalued exchange rates and low import tariffs due to their membership in integration zones such as NAFTA in the case of Mexico (Durand, 2007). Thus, changes in trade become 'endogenous' to procurement strategies of major food retailers. With regional sourcing and regional procurement gaining in

---

[1] In food processing, 85% of FDIs take place between developed countries (van Witteloostuijn, 2007).

[2] However, it should be noted that this does not imply that products are primarily globally sourced. As Reardon *et al.* (2007) show, multiple combinations of local, regional and global sourcing can exist over time. In general, it seems that local and regional sourcing is of far greater importance than the global one. In fact, some case studies show that modern retailers such as Carrefour and Tesco source more than 90% of their food products from local sources (Reardon *et al.*, 2007).

importance, the supermarket revolution in developing countries stimulates south–south trade (Reardon *et al.*, 2007). However, not only imports (due to the increasing relevance of regional and/or global sourcing of foreign retailers) put pressure on local suppliers but also the whole modernization of procurement systems has led to profound changes for the upstream stages of supply chains. For example, foreign owned enterprises have centralized their purchases and integrated technological improvements. This development is often imitated by local retailers and has led to an increase in cross-regional competition between suppliers and thus has reduced their bargaining power vis-à-vis big retail chains (Durand, 2007; Reardon *et al.*, 2007). An additional interesting development that can be observed is 'follow sourcing'. Retailers that have invested in foreign countries encourage transnational logistics and wholesale firms with whom they cooperate in other markets, to follow them to the host country. This leads to a globalization of services to support the retail sector (Reardon *et al.*, 2007).

The growing pressure of imports as well as the increasing governance power of large retailers has pushed many traditional retailers out of business while leading at the same time to a concentration process in supply chains (Durand, 2007). For example, Frazão *et al.* (2008) found that in China money spent in western-style supermarkets[3] increased by approximately 500% and in fast-food services by about 120% over the period 1999–2005. In Indonesia, the respective numbers are 60% and 120% and in Morocco 120% and 20%. For the year 2007, it is estimated by ERS (2009) that supermarkets took a 50% share of all packaged-food sales globally. As the concentration of firms is concerned, the largest 15 supermarket chains worldwide accounted for more than 30% of the sales volume of the entire industry.[4] On another count, it is not unusual for these largest ranked transnational retail firms to have subsidiaries in more than 50 countries.

One reason for the expansion of supermarkets is the worldwide convergence of food consumption patterns and shopping habits (see below). This allows supermarkets to use standardized ways of retailing. Obviously, demand for such standardized services is high not only in developed countries but also in LDCs and EMs. To stimulate demand for their services, supermarkets provide incentives to shoppers such as competitive prices, large product varieties and convenient opening hours. In addition, the rapid growth in ownership of refrigerators, cheap motorized vehicles as well as better roads and bus infrastructures have enabled households to shift from daily shopping in traditional stores to a weekly one in supermarkets while the increase in ownership of microwave ovens has promoted the consumption of ready-to-eat foods available in supermarkets (Gehlhar and Regmi, 2005; Reardon *et al.*, 2007). As much as large retail stores gain from the new food consumption habits they also play a role in their convergence (Regmi *et al.*, 2008).

Consumer demand

Demographic, economic and social trends have altered the demands of consumers with respect to the quantity, variety, quality and safety of food with consequences also for inter-organizational relationships in the food chain.

---

[3] If not indicated otherwise the term supermarkets is used here to also include hypermarkets.

[4] The ranking is based on sales volume.

Food quantity

Demographic developments are a crucial determinant affecting food demand patterns. LDCs and some of the EMs experience high population growth while in developed countries the size of the population increases little, is stagnant or even declines with the consequence of ageing populations. This induces changes in the amount as in the composition of food consumed. In contrast, younger people account for a large share of the population in the LDCs and EMs, thus influencing food intake as well as shopping habits. According to a large ACNielsen consumer survey in Asia, younger consumers strongly favour modern supermarkets to the traditional wet markets for buying fresh produce. This preference will further accelerate the 'supermarket' revolution in those countries in the future (Reardon *et al.*, 2007). In addition, in those countries substantial migration from rural to urban areas takes place also inducing adjustments in diet and shopping practices. In countries such as China and Indonesia the rate of urban population growth outpaced that of total growth by a factor of three (World Bank, 2007). It is estimated that the urban population in China will increase by 300 million over the next 10 years (Coyle *et al.*, 2004).[5] The age structure as well as the level of urbanization influences the kind of food consumed and consequently food supply will need to adjust accordingly. In addition, feeding an ever increasing number of people in the metropolitan areas of LDCs and EMs will remain a challenge for the food supply chains.

Diet composition

Though the EU and US's joint global GDP share exceeds China's and India's together by a factor of six, the latter had a share in the global population in 2005 (37% compared to 12%) as well as an economic growth rate over the years from 1997 to 2006 (8.5% versus 2.8%) which is three times higher than that of the EU and the USA combined (UNdata, 2010). These numbers indicate that countries such as China and India will further gain relevance in shaping the world food system. Already today developing countries account for more than three-quarters of total global food consumers. Rising income in LDCs and EMs change not only the amount of food consumed but also the composition of peoples' diets: away from carbohydrate-rich staples to more expensive sources of calories, such as meat and dairy products as well as fresh fruit and vegetables (Regmi *et al.*, 2008; Bunte, 2007). Moreover, the increasing consumption of meat and dairy will lead to a rise in the demand and prices for cereals. This development implies not only negative repercussions especially for those who stay poor but also a challenge for food companies, as besides tightening regulations (e.g., environmental, food safety; see below) they have to cope with higher input prices (Bunte, 2007). In addition, the share of food expenditures spent on processed food rises with income. On a worldwide scale, convergence in food expenditure patterns can be observed (Frazão *et al.*, 2008; Regmi *et al.*, 2008). Besides income growth, this development in LDCs and EMs is due to a rapid modernization of food distribution mechanisms which themselves show convergence across high- and middle-income countries (see above). This trend is expected to continue as will the profound impacts this has on food supply and on coordinating agri-food supply chains in those countries.

---

[5] At the end of the last decade, the number of people living in urban areas worldwide outpaced those residing in rural regions for the first time in history.

Nevertheless, despite the movement towards global convergence of food demand, even consumers from different countries that belong to the same lifestyle segment may have different culture-based consumption patterns. Thus, diversified consumption patterns are likely to remain to some extent (Bunte, 2007; van Witteloostuijn, 2007; Traill, 1997).

Product quality and service

Rising incomes, lifestyle changes following urbanization, and changing family structures (e.g., a higher share of woman in the workforce) in countries of the western hemisphere but also in high-income segments of the population in LDCs and EMs also have effects on food consumption patterns (Gehlhar and Regmi, 2005). Convenience and pleasure (e.g., quality, taste and variety) receive more and more attention in decisions on how to spend money on food (Frazão *et al.*, 2008). Consumers increase attendance to outlays which provide complementary food services such as a high convenience level (pre-prepared food; extended shelf life) and smart packaging, eating away from home and on-the-go solutions (Schmid, 2007; Bunte, 2007). This can be illustrated with some numbers: on average, more than 50% of food expenditure in high-income countries is spent on packaged food while this share only amounts to about 14% in low-income countries (Gehlhar and Regmi, 2005). In developed countries increasingly high growth rates of 5–7% annually can be observed for retail sales of prepared food, such as ready-to-eat meals (Bunte, 2007). In the USA, 65% of the difference in spending for food between high-income groups and people in the low-income brackets is spent on eating away from home (Frazão *et al.*, 2008). Such differences mirror the higher purchasing power as well as the higher opportunity costs of time of the more affluent population groups (Gehlhar and Regmi, 2005). The taste of food is for many consumers a key determinant of their food choices (Grunert, 2005). In this respect, freshness and its influence on taste is an important consideration for consumers when they value overall food quality.

In LDCs, securing freshness and the continuous availability of products, such as fruits and vegetables but also fish and seafood products, which are prone to food safety risks, often is a considerable challenge for the retail sector given the poor institutional support, partly onerous regulations and an inadequate public infrastructure (Reardon *et al.*, 2005). To overcome such problems, supermarket chains build up private institutions as well as a whole new private supply infrastructure. Elements in this respect are the move away from a per-store procurement arrangement to a centralized distribution system. In addition, retailers sidestep or transform the traditional wholesale system by increasingly working with specialized wholesalers, some of whom were exclusively oriented in the past towards export markets. Finally, a rapid rise in the implementation of private standards can be observed to secure a high level of food safety and quality despite non-enforced or non-existent public standards. As a consequence of these developments suppliers are confronted with a whole series of new and additional requests. They have not only to adapt to best-practice logistical technology and make physical investments that allow a smooth interface with the supermarket warehouses but increasingly also have to adhere to higher food safety and quality standards (Reardon *et al.*, 2005, 2007). These changes have transformed inter-organizational relationships. For example, while in the past relationships between suppliers and retailers in Asia and Latin America were of a very informal personalized nature, operated under verbal informal contracts, this is increasingly replaced by formal contracts that ensure retailers of on-time delivery of products with desired quality attributes (Reardon *et al.*, 2005). This development offers an opportunity to those producers who are able to adapt to new and extended requests. These suppliers may even be successful in becoming 'preferred' suppliers of (specific) retail chains which implies 'assured' market access and in some cases also the provision of managerial and technical support. In addition, a supplier contract with a supermarket chain may serve as a collateral substitute and thus provide urgently needed capital for upgrading production. However, stricter standards and other requests that require investments imply the danger that many small farmers and firms might be excluded from modern supply chains as they lack the financial means to upgrade their production (Reardon *et al.*, 2005, 2007; Trienekens and Zuurbier, 2008; Henson and Reardon, 2005).

Adjustments in fresh product supply chain coordination are not confined to the LDCs and EMs. Consumer requests in developed countries or in high-income segments of LDCs and EMs for a large-variety, year-around delivery of fresh produce has also induced changes in inter-organizational relationships such as horizontal joint ventures or strategic alliances between firms in the northern and southern hemispheres (Malik *et al.*, 2009). The Global Berry Farms, a joint venture between Michigan Blueberry Growers Marketing, Hortifrut of Chile and Naturipe Berry Growers of California, is one example in this respect. Because the partners operate in a wide range of climatic regions in both the northern and southern hemisphere, Global Berry Farms is able to provide retail food outlets in the Americas and increasingly in Europe with berries year round (Reardon *et al.*, 2005).

Health and process quality

Alongside the evolutions in food demand, outlined above, have been trends in consumer requests that have forced enterprises in the agricultural and food sector to focus more strongly on issues of food safety and health (Codron *et al.*, 2005b). A variety of high-profile food scares in several industrialized countries has directed public attention to food safety issues increasing the concern of consumers and reducing their confidence in the safety of food supplied (Grunert, 2005). In a narrow sense, food safety refers to the absence of 'contracting a disease as a consequence of consuming a certain food' (Grunert, 2005, p. 381). However, given the increasing prevalence of diet-related chronic health problems, such as cardiovascular diseases (CVDs), cancer, obesity and diabetes, food safety in a broader perspective also covers issues such as the nutritional value of food products. Convergence in food systems implies that those problems associated with people's diet will rapidly become global issues (Regmi *et al.*, 2008). Yet, the trend towards healthy products is additionally motivated by a greater emphasis on healthy living in general and in particular in the ageing part of the population. This development has led to very high growth rates for health and wellness food and beverages. In the period 2002 to 2005 such products exceeded at a global level the overall sector's growth rate by about 12% (Malik *et al.*, 2009). Finally, existing wide ranging consumer concerns regarding production, processing and conservation practices such as genetically modified (GM) food or irradiation of food belong to that group (Grunert, 2005; Henson and Reardon, 2005).

People belonging to the high-income segments, more than others, consider additional, not necessarily health-related production processes as well as the way food is distributed also as important factors for their purchase decision. This development is revealed in trends towards an increased demand for local, organic, fair trade, animal-friendly and environmentally sound products (e.g., Grunert, 2005). Though social and environmental accountability of an enterprise is not always directly honoured by a higher demand for its products (Scordamaglia, 2008), not to adhere to those norms increasingly implies a risk of being attacked by the public and especially by non-governmental organizations (NGOs), such as Foodwatch and Greenpeace (Heyder and Theuvsen, 2008). Thus, the entire agribusiness environment is increasingly scrutinized by society. Though it may directly affect only some enterprises of an agri-food supply chain, the reputation and legitimacy of the entire chain may be in jeopardy. Coordination between all stages of the chain is often mandatory to secure a high level of production, processing and distribution standards. Corporate social responsibility (CSR) has been developed as a concept for gaining legitimacy and reputation. It implies integrating social and environmental values into corporate strategies and responsibilities and actively engaging in exchanges with society at large and with special interest groups about reaching these values (e.g., Regmi and Gehlhar, 2005). The main aspects of this concept are producing, processing and distributing food in a sustainable way, taking into consideration all dimensions of this term. Thus firms review not only their internal strategies and product portfolio but also re-evaluate their suppliers as well as all actors of the food chains they are involved in (Regmi and Gehlhar, 2005).

However, food safety as well as process attributes such as GM-free or environmentally sound activities is difficult or in many cases even impossible to measure at the time the food product is sold to the consumer, thus making the respective information asymmetrically allocated among supply chain participants (Starbird and Amonor-Boadu, 2007). The relevant attributes must be determined or verified and communicated throughout the entire agri-food supply chain. This requires adjustments in the traditional food supply chain to preserve the unique characteristics of the products and their processes and conveying information about those characteristics to consumers in a reliable way (King and Venturini, 2005). Vertical cooperation or integration and/or, as discussed below, standards controlled and enforced by third parties at the different chain stages reduce the measurement costs linked to verifying such attributes (e.g., Codron *et al.*, 2005b).

Food safety and quality standards

Private versus public standards

The previous section has discussed the profound changes in consumer requests that have taken place over the last decade regarding product and process quality of food products. This development has induced adjustments in public standards at national and multilateral level. But even more, it has led to an increasing prevalence of private food standards at different stages of the supply chain (Hammoudi *et al.*, 2009). The former are concerned with issues of liability and include requirements regarding the characteristics of the final product (e.g., maximum residue limits) and/or of processes in the food chain (e.g., HACCP; traceability) with a focus on food safety issues. The latter aim to secure the enforcement of public standards by providing more rigorous control measures and often are more stringent than public regulations for the same objective. In addition, private standards reach into areas such as environmental, social and ethical aspects of agri-food production, increasingly requested from consumers and so far not yet, or hardly covered, by public regulation (Fulponi, 2006). The central purpose of public as well as private standards in the agri-food sector is to communicate credible information on product and process characteristics, thereby easing coordination between actors in the chain across space and time (Henson and Humphrey, 2008).

Private standards are distinct from public ones in that no government force for compliance exists, thus making them from a legal perspective voluntary for businesses in the food chain. In addition, all key functions such as setting, assessing conformity and enforcing the standards are undertaken by private bodies (Henson and Humphrey, 2008). Despite this clear distinction, there exists a continuum between public and private modes of regulating the agri-food sector. Though legally mandatory standards fall in the domain of public institutions, several private standards have become *de facto* mandatory. This occurs if compliance with private standards becomes a precondition for market entrance of firms. Given the purchasing power especially of large retail chains in the food sector there is often no longer a freedom of choice whether or not to adopt private standards such as the British Retail Consortium Global Standard for Food Safety (BRC), the International Food Standard (IFS) or the Safe Quality Food (SQF). For actors who want to stay in the (global) market those standards are *de facto* compulsory despite their *de jure* voluntary nature (Fuchs *et al.*, 2009). At the same time, governments increasingly implement standards that allow 'freedom of choice', such as the 'Label Rouge' for chickens in France or standards referring to organic production in the EU (Henson and Humphrey, 2008). In addition, it is not uncommon for standards to change their domain. For example, private voluntary standards can develop to legally mandated ones if adopted by state actors and authorized with legal power. Moreover, standards, originally implemented as public voluntary standards, might need re-classification as private voluntary ones if they are subsequently acquired by private bodies (Henson and Humphrey, 2008).

There is, however, not only a continuum between public and private standards but also a close connection as will be shown below. As standards not only influence the attributes of the final product and the processes within firms but also affect the organization of supply chains and the distribution of profits along the supply chain, they can have a pronounced impact on inter-organizational relationships in the food sector (Hammoudi *et al.*, 2009).

Standards and inter-organizational relationships

Private standards in the agri-food sector follow two main objectives: risk management and product differentiation. Depending on the objective, standards have quite different implications for inter-organizational relationships in food supply chains.

Risk management standards

There are three main reasons for the implementation of private standards that are oriented at reducing risks. One is to close perceived gaps in the regulatory system, either because this is considered inadequate regarding its requirements (level of public minimum quality standards too low) or because the monitoring and enforcing system is too weak, or both. Accordingly, one important motivation for private standards is to 'top up' public standards that are considered to be insufficient (Codron *et al.*, 2005b; Henson and Reardon, 2005). With high-profile food scares having led to an erosion of consumers' faith in government safety controls in many industrialized countries, firms have implemented additional guarantees regarding the safety of food to regain consumer confidence. Companies that are successful in supplying products that consistently meet high food safety and quality requirements are able to maintain or increase their market reputation, 'the key asset for current and future earnings flows' (Fulponi, 2006, p. 6). In this sense, private risk management standards can be considered as a powerful tool for the protection of brand capital (Henson and Humphrey, 2008). Thus, the results of Fulponi (2006) based on a survey of 16 leading international retailers is not surprising. Her findings reveal that 85% of retailers require higher and 50% request even significantly higher food safety standards compared to governmental regulation. Besides regaining consumer confidence and increasing firm reputation, these higher standards aim at ensuring a margin of defence in court[6] in the case of a food hazard incidence (Codron *et al.*, 2005b; Henson and Reardon, 2005; Fulponi, 2006).

Even more than at the national level, private standards have gained importance in enforcing non-enforced or non-existent public standards in LDCs or in overcoming the problem of heterogeneous public standards even between the industrialized countries, thus easing international trade. In this respect, private standards help to coordinate supply chains by standardizing product as well as process requirements covering suppliers from many countries of different parts of the world, thereby enhancing efficiency and lowering transaction costs (Henson and Reardon, 2005; Reardon *et al.*, 2007). The trend towards global sourcing has increased the need for private standards as a tool to harmonize requirements (Fulponi, 2006). However, this development has considerable implications for inter-organizational relationships. For example, when looking at the implementation of private standards in developing countries this is in general linked to a shift from 'fragmented, decentralized procurement to a centralized supply system, from reliance on traditional wholesalers and spot markets to specialized/dedicated wholesalers and preferred suppliers operating under *de facto* contracts' (Henson and Reardon 2005, p. 245).

Food scares as well as the emergence of new sources of food anxiety have not only increased the relevance of private standards but, at the same time, resulted in an extension of the scope and rigour of the regulatory system over the last decade which also provided impetus to the evolution of private standards. Stricter public regulations are not only associated with tighter maximum residue limits (e.g., regarding pesticides) but they have induced a shift from more traditional product and process controls towards management-based approaches to secure food safety by making meta-systems such as hazard analysis and critical control point (HACCP), good agricultural practice (GAP) and traceability mandatory in the whole food chain in most of the developed world (Henson and Humphrey, 2008; Trienekens and Zuurbier, 2008; Codron *et al.*, 2005a).[7] Such systems provide 'codes of conduct' on how product and process attributes are to be achieved and increasingly play a central role in governing agri-food chains. To comply with those public standards farmers, processors and retailers had to adjust their production processes and improve coordination along the chain (Hammoudi *et al.*, 2009). Thus, a second central role of private risk management standards is to provide a level of assurance that the food fulfils minimum product and/or process requirements as requested by regulation. Hence, they serve as a tool to achieve and demonstrate compliance with regulation and/or to reduce

---

[6] Retailers want to secure an acceptable proof in a court of law. Thus, they set higher standards than those set by governmental regulation so that in court they can prove that they have undertaken all possible precautions to ensure food safety (Fulponi, 2006).

[7] Examples are the European regulation 178/2002 concerning food safety and traceability.

compliance costs linked to legal requirements (Hammoudi *et al.*, 2009; Henson and Humphrey, 2008).

In addition, governments have progressively shifted the primary legal responsibility for ensuring food safety to the private sector, making suppliers legally liable for the safety of the food they supply. For example, in the UK the only defence in the case of a food safety incidence specified in the 1990 implemented Food Safety Act is that of 'due diligence'. As private standards such as GlobalGAP and the BRC Global Standard provide such a due diligence defence, the shift in legal responsibility has been a third important motivator in increasing the prevalence of private risk management standards (Henson and Humphrey, 2008; Fulponi, 2006; Mondelaers and van Huylenbroeck, 2008). This development has further gained in importance with the increasing relevance of private retail labels as those place full liability for product quality and safety on the retailer even outside the UK (Codron *et al.*, 2005b).

Private risk management standards have not only induced adjustment in intra-organizational policies and in vertical chain management but also in horizontal coordination at various chain stages as the following example shows. While prior to 1998 British retailers had their own internal food safety schemes, the implementation of the British Retail Consortium Global Standard for Food Safety led to the establishment of a single food safety scheme for the UK. In the meantime, retailer alliances evolved from a national or regional to a global level. This resulted in the BRC becoming a leading global standard supported by major retailers throughout the world and adopted by over 10,000 food businesses in more than 96 countries. With the Global Food Safety Initiative (GFSI)[8] emerging in 2000 this development has reached a further step. Though the GFSI does not seek to create a single food safety standard at the global level, it nevertheless aims at benchmarking existing food safety schemes with GFSI requirements that are considered to be fundamental in managing food safety.[9] Besides the BRC Global Standard, the IFS, the SQF and the Dutch HACCP also have been approved to be in compliance with the GFSI requirements (Fuchs *et al.*, 2009). According to Fulponi (2006), almost all leading international food retailers are members of the GFSI. Those retailers taking part in her survey state that 75–99% of the food they sell is certified against the GFSI-benchmarked standard (Fulponi, 2006; Fuchs *et al.*, 2009). Setting 'global' private standards considerably influences the 'evolution of the global food system, determining how food is grown, processed and delivered' (Fulponi, 2006, p. 11). This development imposes new challenges for public governance at the local and global level (Fulponi, 2006).

The motivation for the emergence and dissemination of those collective standards is two-fold. As food safety crises have negative repercussions beyond the 'failing' firm, there is a *collective* rationality to cooperate with competitors to prevent or reduce the chance of the occurrence of such incidences (Hammoudi *et al.*, 2009). However, there is also an *individual* rationality as collective standards help retailers to reduce the costs of governing food safety along their supply chain (Henson and Reardon, 2005; Trienekens and Zuurbier, 2008; Mondelaers and van Huylenbroeck, 2008). Being of a collective nature they expand the population of suppliers from which retailers can procure. Since such standards make

---

[8] The Global Food Safety Initiative was created 2000. It is managed by the Consumer Goods Forum, an international food business association working with retailers and manufacturers from around the world.

[9] Another example is the Global Partnership for Good Agricultural Practice that was known until 2007 as EurepGAP, an initiative originally developed by European retailers.

suppliers interchangeable, they simultaneously reduce the need for vertical cooperation and thus the necessity to invest in long-term relationships (Schulze *et al.*, 2007). However, as suppliers are now certified to a standard that is also available to competing firms, buyers sacrifice product differentiation.

Product differentiation standards

In contrast to risk management standards, product differentiation standards, by their very nature, aim to differentiate the firm and/or its products from (those of) competitors with the aim to increase profit by gaining market share and/or secure price premiums (Henson and Reardon, 2005). While risk management standards are tightly interconnected with public (food safety) regulations, product differentiation standards primarily focus on product (e.g., added health and nutritional value of a product) and process attributes (e.g., worker rights, environmental protection and animal welfare) of interest to buyers that are not yet regulated. However, this division is far from being static. It can be expected that attributes that serve today as a means of differentiation evolve over time to basic product attributes while the focus of product differentiation standards will shift to other attributes where there is still scope for differentiation.[10] Overall, the evolution in product differentiation standards reflects the trend towards quality as a mode of competition in agri-food markets (Henson and Humphrey, 2008).

While most of the risk management standards are business-to-business approaches and thus not directly visible to consumers, product differentiation standards are used to communicate information to consumers via labels or emblems and thus help them to evaluate the quality of the product as well as the characteristics of the production process (Hammoudi *et al.*, 2009; Mondelaers and van Huylenbroeck, 2008; Schulze *et al.*, 2007). The main drivers of these standards are in general producers or producer groups. Being successful in attracting consumers' preferences for those labelled standards provides producers with some kind of countervailing market power, despite the leading role of retailers, forcing the latter to source those products. However, retailers also use individual company standards on their own private-label products (e.g., Tesco's Nature's Choice or Carrefour private-label beef in France) (Henson and Humphrey, 2008; Codron *et al.*, 2005a). Private product differentiation standards require the management of a trade-off between, on the one hand, the identification of attributes that allow distinguishing one's self from competitors in a way that is difficult and costly to imitate and, on the other hand, managing the implied transaction costs. In contrast to risk management standards which are increasingly of a collective nature, private differentiation standards require strong coordination between actors along the food chain thus fostering vertical cooperation and long-term inter-organizational relationships.

---

[10] For instance, while social standards serve at present as a means to differentiate from competitors, it might be that with child labour and workers' protection gaining in relevance, the further development of the Social Accountability 8000 (SA8000) standard and similar initiatives will lead over time to setting some kind of minimum social requirements (Schmid, 2007).

Competition and concentration

Cooperative relationships between organizations can create private benefits for the involved parties but may become collusive to the point where they undermine competition, thus resulting in a loss of public welfare (Palmer, 2001). Competition and collusion[11] have been characterized as opposite ends of a spectrum of economic organization systems, with cooperation/coordination/collaboration being in between these extremes (Palmer, 2001). While collusion has traditionally been a problem mainly of horizontal inter-organizational relationships such as cartels,[12] cases of vertical supply chain business partnerships have been investigated by anti-competition authorities in the EU, the UK and the USA in the recent past (Palmer, 2001). Collusive vertical relationships usually occur when one of the partners is in a dominant market position and the vertical business partnership serves to strengthen or extend this market power. Cooperation between organizations with significant mutual benefits may result in partners giving forbidden preferential treatment to each other. Such benefits can arise as a result of growing levels of trust leading to reduced dyadic contracting costs (see Chapters 3, 5 or 6), or when a partnership can unlock significant economies of size (Palmer, 2001). As long as supply chain alliances continue to compete against each other, thus providing public benefits such as low consumer prices or a wide range of products to choose from, the regulatory interest in the vertical partnership remains low. It is only in collaboration cases where competition is compromised and consumer benefits are eroded that governments typically intervene. Collusive cooperative relationships are more likely in markets characterized by high levels of concentration.

Rising levels of concentration in the food sector has been an issue of recent concern, in particular in Europe (Bukeviciute *et al.*, 2009; Commission of the European Communities, 2009; Palpacuer and Tozanli, 2008). Market concentration, which in Europe occurs mainly in the retail sector, can have negative effects on prices (Bukeviciute *et al.*, 2009), on food product quality (Palpacuer and Tonzanli, 2008), or on the value-added repartition (Commission of the European Communities, 2009). While there is no systematic evidence that European mechanisms of food price formation have been affected by rising downstream concentration in some countries (Bukeviciute *et al.*, 2009), there seems to be a perception that such concentration affects the sourcing decisions of food processors and manufacturers and thus results in more food products being made from low-cost international ingredients rather than European ones (Palpacuer and Tonzanli, 2008). Moreover, there is a trend that value-added is increasingly created and captured in the downwards parts of the European food chain rather than in the farming sector (Commission of the European Communities, 2009). This is a consequence of the development of more ingredients being sourced more cheaply outside Europe, thus reducing input costs and increasing value-added in the distribution sector, and the fact that the relative size of the agriculture sector versus the relative size of the processing and distribution sectors has been shrinking.

Overall, despite these developments, it needs to be remembered that the overwhelming share (96%) of food processing enterprises have less than 50 employees and thus fall in the group of small and micro-enterprises (Castenmiller *et al.*, 2008). According to data from the European Union's statistical agency Eurostat, in the EU-25 in 2005 there were still

---

[11] Collusion can be defined as unauthorized collaboration or working with others without permission.

[12] A cartel can be defined as a group of producers who act together to fix price, output or conditions of sale.

more than 12 million agricultural holdings (farms), more than 280,000 food processors and more than 500,000 food retailers, thus supporting Bukeviciute *et al.*'s (2009) statement that 'market power [in the food sector] in the euro area appears to be neither particularly high nor especially low' (p. 16). Hence, while rising concentration in particular at the retail level may limit collaboration choices in some cases, forcing agricultural producers and processors to engage in 'unsatisfying but stable' business relationships (Backhaus and Büschken, 1999), in general there still seems to be sufficient competition to allow most agri-food companies to build and maintain any number of non-collusive and sustainable chain relationships they need to.

Information and communication technologies

Technology improvements during the last decades and in particular advances in ICTs have enabled organizations to actually build and manage partnerships more effectively than was possible in the past (Dong *et al.*, 2009; Subramani, 2004; Sambamurthy *et al.*, 2003; Scott, 2000). This is especially true for international and global business partnerships. Yet, an economy-wide adoption of such new technologies also helps to intensify partnerships at the local and national level. New communication and electronic information exchange platforms allow keeping close contact and control around the clock and independent of business partners' physical locations.

Moreover, technology improvements make more sophisticated risk management tools such as the tracking and tracing of products possible, leading to more transparent agri-food supply chains (Banterle and Stranieri, 2009; van Dorp, 2002). Much of the data exchange in agri-food chains is done today in standardized ways on computer networks using electronic business-to-business data exchange interfaces (EDIs). There are two main issues involved. First, the development of cost effective and reliable data transmission links (wired or wireless) between organizations (the 'hardware'). In principle, this can be done by using private data exchange networks or public ones. During the last decade it has become increasingly common for most of the data exchange to occur through public cable networks such as the internet. The second issue is to develop suitable and safe software that allows the administration of differentiated data and information access rights. By keeping close control of commercially sensitive information, inter-organizational trust can be developed and maintained, thus enabling sustainable collaboration between independent organizations (Scott, 2000).

Nevertheless, there is wide agreement that technology is an enabler and not a driver (Spekman *et al.*, 1998). People, not machines, initiate, build and maintain partnerships. In the case of business partnerships (i.e., collaboration between independent organizations that are separated in space), effective coordination of transactions and interactions need to be provided in order to make a partnership successful. Building on Mintzberg's (1979) work, effective coordination involves the application of the following three mechanisms: (i) communication between people conducting interdependent work which leads to mutual adjustments; (ii) supervision by one or more individuals who take responsibility for the work as a whole; and (iii) standardization of work processes, outputs, skills, knowledge, and norms. It can be argued that improvements in ICTs during the last decades, and in particular the wide-spread proliferation and adoption of internet, e-mail and mobile phones, have enhanced all three coordination mechanisms. E-mail and mobile phones have allowed individuals to communicate from virtually any location, at any time and at low cost.

Through internet file sharing and video applications, supervision of work can be achieved in many instances inexpensively and around the clock. The worldwide availability of almost identical software (office as well as enterprise resource management applications) and the way people use it have helped to standardize work processes, outputs, skills and norms. As a consequence, the spreading of ICTs has indirectly contributed to the emergence and success of sophisticated vertically coordinated business systems such as supply chains.

The adoption of ICTs in particular in the farming sector has been slow (Gelb and Offer, 2005; Warren, 2004). However, there are significant age differences and the new generation of young farmers that is emerging is more willing and skilled to use ICTs both for private and professional purposes (Gelb *et al.*, 2008). It appears that in the future, ICTs will not be a significant constraint on partnership building and maintaining business relationships. Even in developing countries, mobile phones are already widely used and the eagerly expected introduction of the US$100-laptop will further improve internet access, even in remote rural areas (World Bank, 2007). Similar to cars, once a new technology is widely available and easy to use the challenge is to make the most out of it.

Human skills and entrepreneurial activities

In a networked economy, both organizations and individuals need to be highly skilled to remain competitive and to conduct business with others in a sustainable way. As organizations integrate into vertical or horizontal business systems, a refocusing on their core competencies takes place alongside the outsourcing of non-essential business functions. This opens up opportunities (or results in a necessity) for individuals to start their own companies (Global Entrepreneurship Monitor, 2010). In order to engage successfully in entrepreneurial activities, people require business skills. The higher people's formal education the more likely it is for them to possess such skills or to be in a position to acquire them quickly and successfully (e.g., by going through additional training). Moreover, it can be assumed that the more highly educated people are the higher is their willingness, propensity or desire to engage in entrepreneurial activities, or the lower their resistance to start up their own companies in cases where they need to do it (Global Entrepreneurship Monitor, 2010).

International data on long-term trends in both increasing skill levels and entrepreneurial activities are hard to find. However, the United Nations Educational, Scientific and Cultural Organization (UNESCO) education database shows that total tertiary graduates (i.e., those from all higher-education programmes) in those countries, out of the 215 included, for which data are available have increased from about 9 million in 2000 to more than 21 million in 2008. This is clearly more than overall population growth and suggests that the share of highly educated people in the total population has been increasing steadily. As to the number of total enterprises in the global economy, no consolidated information exists. However, data from the European Union's statistical agency Eurostat show that between 1999 and 2004 in the EU-25 the total number of enterprises across all economic activities (without agriculture and mining) has increased from 16.3 million to 18.4 million. This represents an increase of more than 13% while in the same period according to the same source, overall population has only grown by less than 2% (from 450.9 million to 459.2 million). In the USA, the number of enterprises that employ staff increased from 4.9 million in 1998 to 5.9 million in 2004, whereas the number of non-employing enterprises increased from 17.6 million in 2002 to 21.7 million in 2007 (US Census Bureau, 2010). Moreover, most of the growth of the former has come from small enterprises (fewer than 20 staff), representing more than 87% of the total number of employing firms. Although these data are not comprehensive and do not provide specific information on the food sector, they indicate the likely existence of a general trend to a structure of tomorrow's economy that, despite the existence of some very large enterprises in the food processing and retailing, can be characterized as a network of many and smaller, independent and interdependent organizations and individuals that are closely linked with each other by a web of tight business relationships.

Conclusions and outlook

This chapter has discussed the broader business context in which today's agri-food chains operate. It has also described the long-term trends which have shaped agri-food sectors and markets and which have contributed to the rise of vertical business systems such as supply or value chains. In particular, the changing economic, technological, social and policy environments have been reviewed. These include globalization (i.e., the widespread trade liberalizations), increasing consumer quality demands and the linked rise in public and private standards, the ubiquity of ICTs, and the increasing skills base and entrepreneurial capacities in today's knowledge-based societies which foster the development of highly networked economies.

The trends in agri-food chains outlined above will not stop any time soon. However, the speed of some of those developments discussed will vary between different parts of the world. For example, as economic integration of high-income countries has already reached relatively high levels, in the future the globalization process will primarily become evident by higher trade volumes and FDI flows between the developed and developing world as well as among LDCs. In the more sophisticated markets, convenience and pleasure aspect of food have become ever more important. Besides this, fulfilling public demands regarding safety, sustainability and social requirements in the production process provides nothing more than the 'license to operate' (Schmid, 2007). In the coming decade this trend will also reach nations with low per-capita income that are still lagging behind. This will lead to further convergence of consumption patterns, standards and distributions systems as well as technologies in the food sector at a global scale. Thus, some of the discussed implications for inter-organizational relationships that are today primarily of relevance in the developed world will likely gain in importance in LDCs and EMs in the future.

Given the new reality, vertical business systems have emerged as highly competitive organizational formats which allow collaborating companies to exploit economies of size (scale and scope) while maintaining strategic flexibility, agility and managerial independence.

However, one requirement for establishing competitive vertical business systems is the development of sustainable inter-organizational relationships. Building and managing such relationships is in particular a challenge for farmers who traditionally have operated on the basis of spot market exchanges. Yet the decline of specialized food retailing and the rise of global supermarkets also pose problems. Supermarkets that deal with large numbers of (food and non-food) suppliers on a regular basis increasingly use procurement practices which build on private standards, formalized contracts and electronic auctions, therefore sidestepping individual supplier relationships.

While the transition from the transaction- to a relationship-based agri-food economy will take time and be partial, the future seems to favour the extensively embedded, consumer-oriented agri-enterprise rather than the traditional owner-centred, price-driven farm business as the following chapters in this volume argue.

References

Backhaus, K. and Büschken, J. (1999) The paradox of unsatisfying but stable relationships – a look at German car suppliers. *Journal of Business Research* 46(3), 245–257.

Banterle, A. and Stranieri, S. (2009) The consequences of voluntary traceability system for supply chain relationships: an application of transaction cost economics. *Food Policy* 33(6), 560–569.

Boger, S. (2001) Quality and Contractual Choice: A Transaction Cost Approach to the Polish Hog Market. *European Review of Agricultural Economics* 28(3), 241–262.

Bukeviciute, L., Dierx, A. and Ilzkovitz, F. (2009) The functioning of the food supply chain and its effect on food prices in the European Union. Occasional Papers No 47. European Commission, Directorate-General for Economic and Financial Affairs, Brussels.

Bunte, F. (2007) The Food Economy: Global Issues and Challenges. www.foodeconomy2007.org/ UK/Papers+and+Presentations/.

Castenmiller, J., Fenwick, R., Lindsay, D. and Maat, J. (eds) (2008) European Technology Platform on Food for Life. Implementation Action Plan. Brussels.

Codron, J.-M., Giraud-Héraud, E. and Soler, L.-G. (2005a) Minimum quality standards, premium private labels, and European meat and fresh produce retailing. *Food Policy* 30, 270–283.

Codron, J.-M., Grunert, K., Giraud-Heraud, E., Soler, L.-G. and Regmi, A. (2005b) Retail Sector Responses to Changing Consumer Preferences. The European Experience. In: Regmi, A. and Gehlhar, M. (eds) New Directions in Global Food Markets. Agricultural Information Bulletin 794, 32–46. http://www.ers.usda.gov/publications/aib794/aib794.pdf.

Commission of the European Communities (2009) The evaluation of value-added repartition along the European food supply chain. Accompanying document to the Communication from the Commission to the European Parliament, the Council, the European Economic and Social Committee and the Committee of the Regions. Commission staff working document COM(2009) 591.

Coyle, W., Gilmour, B. and Armbruster, W. (2004) Where will demographics take the Asia-Pacific food system? *Amber Waves* 2(3), 14–21. USDA, Economic Research Service, Washington, DC. http://www.ers.usda.gov/amberwaves/june04/pdf/features_asiapacific.pdf.

Dong, S., Xu, S. and Zhu, K. (2009) Information technology in supply chains: the value of IT-enabled resources under competition. *Information Systems Research* 20(1), 18–32.

Dries, L. and Swinnen, J. (2004) Foreign Direct Investment, Vertical Integration and Local Suppliers: Evidence from the Polish Dairy Sector. *World Development* 32(9), 1525–1544.

Durand, C. (2007) Externalities from foreign direct investment in the Mexican retailing sector. *Cambridge Journal of Economics* 31(3), 393–411.

ERS (2009) Global Food Markets. http://www.ers.usda.gov/Briefing/GlobalFoodMarkets/.

Eurostat (2010) European Business – selected indicators for all activities (NACE divisions). http://epp.eurostat.ec.europa.eu/portal/page/portal/statistics/search_database.

FAOSTAT (2010) Trade Indices. http://faostat.fao.org/site/611/default.aspx#ancor.

Frazão, E., Meade, B. and Regmi, A. (2008) Converging patterns in global food consumption and delivery systems. *Amber Waves* 6(1), 22–29. USDA, Economic Research Service, Washington, DC. http://www.ers.usda.gov/AmberWaves/February08/Features/CovergingPatterns.htm

Fuchs, D., Kalfagianni, A. and Havinga, T. (2009) Actors in private food governance: the legitimacy of retail standards and multistakeholder initiatives with civil society participation. *Agriculture and Human Values*, Online First published. DOI: 10.1007/s10460-009-9236-3.

Fulponi, L. (2006) Private voluntary standards in the food system: the perspective of major food retailers in OECD countries. *Food Policy* 31, 1–13.

Galizzi, G. and Venturini, L. (eds) (1999) *Vertical Relationships and Coordination in the Food System*. Springer, Heidelberg.

Gehlhar, M. and Regmi, A. (2005) Factors shaping global food markets. In: Regmi, A. and Gehlhar, M. (eds) New Directions in Global Food Markets. Agricultural Information Bulletin 794, 5–17. *Economic Research Report 56.* USDA, Economic Research Service, Washington, DC. http://www.ers.usda.gov/publications/aib794/aib794.pdf.

Gelb, E. and Offer, A. (eds) (2005) ICT in agriculture: perspectives of technological innovation. Ebook: http://www.iol.ie/~harkin/ict_in_agriculture_book.htm.

Gelb, E., Maru, A., Brodgen, J., Dodsworth, E., Samii, R. and Pesce, V. (2008) Adoption of ICT Enabled Information Systems for Agricultural Development and Rural Viability. Pre-conference workshop summary. http://www.fao.org/docs/eims/upload/.../Workshop_Summary_final.pdf

Global Entrepreneurship Monitor (GEM) (2010) 2009 Global Report. http://www. gemconsortium.org/about.aspx?page=pub_gem_global_reports.

Gow, H. and Swinnen, J. (1998) Up- and downstream restructuring, foreign direct investment, and hold-up problems in agricultural transition. *European Review of Agricultural Economics* 25(3), 331–350.

Grunert, K. (2005) Food quality and safety: consumer perception and demand. *European Review of Agricultural Economics* 32(3), 369–391.

Hammoudi, A., Hoffmann, R. and Surry, Y. (2009) Food safety standards and agri-food supply chains: an introductory overview. *European Review of Agricultural Economics* 36(4), 469–478.

Henson, S. and Humphrey, J. (2008) Understanding the complexities of private standards in global agri-food chains. Paper presented at the workshop 'Globalization, Global Governance and Private Standards'. University of Leuven, November.

Henson, S. and Reardon, T. (2005) Private agri-food standards: implication for food policy and the agri-food system. *Food Policy* 30, 241–253.

Heyder, M. and Theuvsen, L. (2008) Legitimating business activities using corporate social responsibility: is there a need for CSR in agribusiness? Paper presented at the 110th EAAE Seminar 'System Dynamics and Innovation in Food Networks'. Innsbruck-Igls, Austria, February 18–22.

King, R. and Venturini, L. (2005) Demand for Quality Drives Changes in Food Supply Chains. In: Regmi, A. and Gehlhar, M. (eds) New Directions in Global Food Markets. Agricultural Information Bulletin 794, pp. 18–31. http://www.ers.usda.gov/publications/aib794/aib794.pdf.

Malik, M., Yawson, R. and Hensel, D. (2009) Destination 2025. Focus on the Future of the Food Industry. A collaboration between The BioBusiness Alliance of Minnesota and Deloitte Consulting LLP, Deloitte.

Ménard, C. and Valceschini, E. (2005) New institutions for governing the agri-food industry. *European Review of Agricultural Economics* 32(3), 421–440.

Mintzberg, H. (1979) *The Structuring of Organizations.* Prentice-Hall, Englewood Cliffs.

Mondelaers, K. and van Huylenbroeck, G. (2008) Dynamics of the retail driven higher end spot market in fresh food. *British Food Journal* 110(4/5), 474–492.

Palmer, A. (2001) Co-operation and collusion: making the distinction in marketing relationships. *Journal of Marketing Management* 17, 761–784.

Palpacuer, F. and Tozanli, S. (2008) Changing governance patterns in European food chains: the rise of a new devide between global players and regional producers. *Transnational Corporations* 17(1), 70–97.

Reardon, T., Timmer, C. and Berdegué, J. (2005) Supermarket Expansion in Latin America and Asia. In: Regmi, A. and Gehlhar, M. (eds) New Directions in Global Food Markets. Agricultural Information Bulletin 794, pp. 47–61. http://www.ers.usda.gov/publications/aib794/aib794.pdf.

Reardon, T., Henson, S. and Berdegué, J. (2007) 'Proactive fast-tracking' diffusion of supermarkets in developing countries: implications for market institutions and trade. *Journal of Economic Geography* 7, 399–431.

Regmi, A. and Gehlhar, M. (2005) Introduction. In: Regmi, A. and Gehlhar, M. (eds) New Directions in Global Food Markets. Agricultural Information Bulletin 794, pp. 1–4. *Economic Research Report 56.* USDA, Economic Research Service, Washington, DC. www.ers.usda.gov/publications/aib794/aib794.pdf.

Regmi, A., Takeshima, H. and Unnevehr, L. (2008) Convergence in global food demand and delivery. *Economic Research Report 56.* USDA, Economic Research Service, Washington, DC.

Sambamurthy, V., Bharadwaj, A. and Grover, V. (2003) Shaping agility through digital options: reconceptualising the role of information technology in contemporary firms. *MIS Quarterly* 27(2), 237–263.

Schmid, F. (2007) The Food Economy, Global Isssues and Challenges as seen by a food retailer. OECD Conference, The Hague. www.foodeconomy2007.org/UK/Papers+and+Presentations/.

Schulze, B., Spiller, A. and Theuvsen, L. (2007) A broader view on vertical coordination: lessons from German pork production. *Journal on Chain and Network Science* 7(1), 35–53.

Scordamaglia, L. (2008) The issues pertaining to the supply of meat. Presentation made at the '17th World Meat Congress'. Cape Town, South Africa, September 7–10.

Scott, J. (2000) Facilitating interorganizational learning with information technology. *Journal of Management Information Systems* 17(2), 81–113.

Spekman, R., Kamauff, J. and Myhr, N. (1998) An empirical investigation into supply chain management: a perspective on partnerships. *International Journal of Physical Distribution & Logistics Management* 28(8), 630–650.

Starbird, S. and Amonor-Boadu, V. (2007) Contract selectivity, food safety, and traceability. *Journal of Agricultural & Food Industrial Organization* 5(2), 23–128.

Subramani, M. (2004) How do suppliers benefit from information technology use in supply chain relationships? *MIS Quarterly* 28(1), 45–73.

Traill, B. (1997) Globalisation in the food industries. *European Review of Agricultural Economics* 24(3/4), 340–410.

Trienekens, J. and Zuurbier, P. (2008) Quality and safety standards in the food industry: developments and challenges. *International Journal of Production Economics* 113(1), 107–122.

UNdata (2010) http://data.un.org/Data.aspx?q=constant+GDP&d=CDB&f=srID%3a29918 and http://data.un.org/Data.aspx?q=Population&d=PopDiv&f=variableID%3a12

United Nations Educational, Scientific and Cultural Organization (UNESCO) (2010) Education database. http://www.uis.unesco.org.

US Census Bureau (2010) Statistics about Business Size (including Small Business) from the US Census Bureau. http://www.census.gov/epcd/www/smallbus.html.

van Dorp, K.-J. (2002) Tracking and tracing: a structure for development and contemporary practices. *Logistics Information Management* 15(1), 24–33.

van Witteloostuijn, A. (2007) Globalization in the food industry: the impact on market structure and firm strategies. University of Antwerp, Belgium. www.foodeconomy2007.org/NR/rdonlyres/6C1DDC4F-0FCF-4CCE-BD36-C5176CF6263A/50177/workshop12arjenvanwitteloostuijn.doc

Warren, M. (2004) Farmers online: drivers and impediments in adoption of internet in UK agricultural businesses. *Journal of Small Business and Enterprise Development* 11(3), 371–381.

World Bank (2007) *World Development Report 2008: Agriculture for Development.* The World Bank, Washington, DC.

Chapter 2

Inter-organizational Relationships in Agri-food Systems: a Transaction Cost Economics Approach

Fabio Chaddad and Maria E. Rodriguez-Alcalá

University of Missouri, Columbia, USA

Introduction

Since the seminal work of Ronald Coase (1937), scholars in transaction cost economics (TCE) have focused on the choice between market and hierarchy to govern transactions between economic agents (Klein *et al.*, 1978; Williamson, 1979). This dichotomy, however, overlooks the fact that transactions between firms are increasingly governed by organizational forms that are neither pure market nor pure hierarchy. More recently, scholars have recognized the existence of several intermediate or hybrid arrangements along a continuum of governance structures with market and hierarchy as the end points (Thorelli, 1986; Hennart, 1988; Williamson, 1991; Ménard, 2004). Such hybrid arrangements include (long-term) relational contracts, partnerships, cooperatives, strategic alliances and joint ventures.

Scholarly interest in understanding hybrid arrangements has coincided with the rise in the rate of alliance and network formation in many industries, including the agri-food sector. As agri-food systems industrialize – with increasing consolidation and the concomitant use of tighter forms of vertical coordination along supply chains – many interesting research problems emerge that are amenable to the 'get governance structure right' approach of TCE (Cook and Chaddad, 2000). In recent years, inter-organizational linkages among competitors in the same industry, and between firms in related industries, have proliferated to link farmers with consumers. Increasing vertical, horizontal and inter-industry coordination among firms continue to shape the competitive landscape and business strategies of agri-food firms. Accordingly, the capabilities and value creation potential of firms are increasingly dependent not only on their unique assets and resources but also on those of their alliance partners. Indeed, an increasing number of scholars and practitioners suggest that competition is shifting from 'firms versus firms' to 'supply chain versus supply chain' or to 'network versus network' (e.g., Gomes-Casseres, 1994; Zylbersztajn and Farina, 1999).

Notwithstanding growing interest in network forms of economic organization (e.g., Powell, 1990; Nohria, 1992; Lazzarini *et al.*, 2001), scholars analysing alliances or inter-organizational relationships in the agri-food system have in general adopted the supply chain approach as a conceptual framework. Supply chains are defined as a set of sequential, vertically organized transactions representing successive stages of value creation. The literature on supply chain analysis (SCA) suggests vertical interdependencies require a systemic understanding of resource allocation and information flow between firms engaged in sequential stages of production (Simchi-Levi *et al.*, 2000). Adopting a supply chain framework, agri-food scholars have applied transaction cost economics in conjunction with other new institutional theories of the firm to describe and analyse the emergence of inter-organizational arrangements between agri-food chain participants (for a recent review of this literature, see Cook *et al.*, 2008).

This chapter analyses inter-organizational relationships in agri-food systems from a TCE perspective. Given the diversity of organizational and legal forms observed in inter-organizational relationships, the chapter focuses on understanding the nature and governance characteristics of such relationships. The next section presents a typology of inter-organizational relationships treating them as intermediate forms along the market–hierarchy continuum of TCE.

Inter-organizational relationships in agri-food systems

In defining the unit of analysis and scope of this chapter, it is useful to understand the concept of 'inter-organizational relationship' and describe the many forms of such relationships that have emerged in the global agri-food system. Inter-organizational relationships are broadly defined as a set of organizations (and more specifically firms) and the linkages that connect them. These linkages can be horizontal, vertical or cross-industry. Horizontal linkages connect firms in the same market or industry; vertical linkages connect buyers and suppliers along a supply chain; and cross-industry linkages connect firms that operate in different industries or sectors.

Business and economics scholars, including agri-food specialists, are primarily interested in analysing these inter-organizational arrangements as modes of organizing economic activities or as 'institutional structures of production' (Coase, 1992; Madhok, 2002). The focus is on firms as the 'nodes' or 'differentiated units' to be coordinated and on contracts, chains and networks as 'nexuses of integration mechanisms' in addition to or in substitution for the market (Grandori and Soda, 1995). The unit of analysis in inter-organizational relationship studies can be a dyadic relationship between two firms or a set of relationships forming a chain, network or cluster of firms. For example, Dussauge and Garrette (1999, p. 12) define an alliance as a contract or set of contracts. The authors suggest that there is 'no legal form specific to alliances, and no legal definition of inter-firm cooperation.'

In reviewing this broad and amorphous literature, one comes across terms such as hybrids, intermediate forms, alliances, buyer–supplier relationships, inter-organizational collaboration, chains, channels, networks, clusters, among many others. Indeed, 'the set of arrangements that rely neither on markets nor on hierarchies for organizing transactions is broad and potentially confusing' (Ménard, 2004, p. 347). Perhaps not surprisingly, increased scholarly work on inter-organizational relationships has led to a 'messy situation' with a consequent lack of well-defined concepts and theories and research results that are difficult to reconcile (Oliver and Ebers, 1998, p. 550). This section of the chapter attempts to clarify some of the conceptual issues contributing to this 'messy situation' in the literature.

Types of inter-organizational relationships

Inter-organizational relationships or alliances do not refer to a specific type of contract between two or more firms, but rather to several forms of collaborative arrangements. Five types of alliances provide the basis for a generic classification of inter-organizational relationships. Two of these are based on the equity position of partners and the other three are based on the parent firms' relationship before the alliance takes place. Alliances based on equity position are simply classified as equity alliances or non-equity alliances. Additionally, the relationship of partners (i.e., competitors or non-competitors) is also useful in distinguishing alliance types. Firms that are industry competitors form horizontal alliances. Firms that are non-competing form either vertical alliances or cross-industry alliances.

Horizontal alliances are formed by two or more firms – which are competitors or potential competitors – at a similar level in the supply chain. Rival firms enter a horizontal arrangement to jointly produce or market a product or service at a given level of the marketing chain. An example is the joint processing alliance between two California-based prune packers (Sunsweet Growers and Shoei Foods) to process and market products and ingredients especially suited for the Japanese market.

A vertical alliance results between a buyer and a supplier in a supply chain. In a vertical arrangement one (upstream) partner supplies the other (downstream) partner with some input. The second partner then uses this input to manufacture and/or market products downstream in the supply chain. An example of a vertical relationship is the agreement between Ocean Spray Cranberries and PepsiCo. Under the terms of this contractual agreement, Pepsi distributes Ocean Spray single-serve beverage products to supermarkets, convenience stores, vending machines and other retail outlets.

Cross-industry alliances are cooperative agreements among firms operating in different industries or sectors. There are two types of cross-industry alliances. The first one is among firms in the same supply chain but positioned in non-subsequent stages. These firms, therefore, do not transact among themselves outside the inter-organizational relationship. An example is the alliance formed by Monsanto and Cargill to jointly develop biotechnology-based food products and ingredients. The second type of cross-industry inter-organizational relationship occurs among firms acting in different supply chains. Companies that form this type of alliance usually do so as a diversification strategy. The idea is to combine assets and competencies that are complementary in order to explore new business opportunities.

Non-equity alliances are usually referred to as contractual forms of governance, as the alliance is defined through a set of contracts and the parent firms do not share asset ownership and control (Doz and Hamel, 1998). Non-equity alliances can be vertical, horizontal, or cross-industry and can take many different forms of contractual arrangements. The arrangement between Ocean Spray and PepsiCo cited above is also an example of a non-equity alliance. The two parties involved in the contract have no equity invested in this alliance and no separate legal entity is formed between the two parent companies. It is solely a distribution contract between the two firms.

In an equity-based alliance parent companies share the residual claims and residual rights of control over assets. Equity alliances can also be vertical, horizontal, or cross-industry. Equity alliances usually, but not always, involve the formation of a separate legal entity. The joint venture known as Dairy Partners Americas (DPA) is an equity alliance involving Nestlé and Fonterra. The DPA joint venture is operated as a separate company with Nestlé and Fonterra as shareholders. Although the joint venture is owned by the two parent companies, which are the residual claimants with ultimate control over the joint venture assets, the day-to-day activities are managed independently from them.

Inter-organizational relationships as intermediate forms of governance

Vertical alliances are considered intermediate forms of governance in the transaction cost economics literature (Williamson, 1991). However, Williamson's (1991) conceptualization of intermediate forms does not include horizontal or cross-industry alliances. This section furthers Williamson's (1991) vertical continuum of governance structures to propose a generic classification of inter-organizational relationships. In this classification, the three generic types of alliances – horizontal, vertical and cross-industry – are characterized as intermediate forms.

Vertical relationships

Transaction cost economics (TCE) views vertical inter-organizational relationships as intermediate forms of governance between markets and hierarchies. The basic question addressed by TCE in analysing the vertical boundaries of the firm is 'to make, to buy or to cooperate.' Williamson (1991) classifies governance mechanisms into three generic forms: markets, hybrids and hierarchies. The market is defined as the arena in which transactions between autonomous buyers and sellers occur with the price system working as a coordinating mechanism. Intermediate forms (hybrids) are long-term relationships between autonomous parties that rely on added contractual safeguards that are specifically designed to sustain the transaction. Under internal governance, the transaction is organized under unified ownership – as the buyer and supplier are units of the same firm – and coordinated by administrative tools including authority and hierarchy (Williamson, 1991).

Markets and hierarchies are the two polar forms of governance, while intermediate forms (hybrids) stand in between. According to Barney and Hesterly (1996), there is a debate whether hybrids are discrete mechanisms of governance structures or consist of a continuum of plural forms combining characteristics of the pure market and the pure hierarchy. In this section of the chapter, the idea of positioning inter-organizational relationships as intermediate forms in a continuum of governance mechanisms is followed. In this continuum, inter-organizational relationships that do not involve equity capital are placed closer to the market, while equity-based alliances are closer to a hierarchy (see 'vertical' column in Table 1).

In the continuum of governance modes posited by TCE, as the degree of asset specificity increases, transactions tend to be internalized rather than occurring in the market. As we move away from the market and toward hierarchy, administrative controls, bureaucratic costs and vertical cooperation strengthen. On the other hand, incentive intensity is stronger or 'high powered' in markets. In addition, the market is more legalistic than a hierarchy as it operates under a rigid classical contract law regime. Along this continuum, there is a trade-off between coordination and flexibility. The closer one is to a hierarchy the easier it becomes to coordinate exchanges between subsequent vertical stages. However, the rigidity imposed by bureaucratic forms of organization decreases the flexibility to choose a different supplier or buyer. A strategic alliance, in this sense, is an intermediate form. When two separate firms form a vertical alliance, they are not transacting in the market nor internalizing the transaction, but adopting intermediate levels of coordination and control mechanisms.

Horizontal and cross-industry alliances

Horizontal alliances are formed among competitors or potential competitors and can be classified similarly to vertical alliances. However, in a horizontal alliance there are no vertical transactions involving partners. The firm boundary question is not 'to make, to buy, or to cooperate' but rather 'to compete, to unify, or to cooperate.' It is proposed here that horizontal alliances be placed within a governance continuum similar to the one offered by TCE. In this continuum, firms competing in the same industry are placed on one end, while unification is placed on the other end. The various forms of horizontal inter-organizational relationships and collaborations are placed along this continuum (see 'horizontal' column in Table 1). Inter-organizational relationships that are closer to competition are simple, formal or informal contracts and do not involve the sharing of equity among partners. Equity agreements, such as the classic joint venture, are closer to unification and involve more formal types of inter-firm cooperation. As we move along this continuum, bureaucratic costs increase and horizontal cooperation strengthens. Also, if the alliance is formed among actual competitors, industry structure becomes more concentrated.

Cross-industry alliances are not formed by a buyer and supplier or by competing firms. Instead, they are business linkages between firms that operate in different businesses or sectors. The main objectives in this type of inter-organizational relationship are to combine competencies and to provide a way to learn and/or diversify. If these inter-organizational relationships are placed in a continuum, on one end there are firms in different industries with no interaction at all, while on the other end there is unification under one single diversified firm (see 'cross-industry' column in Table 1). Inter-organizational arrangements between firms across industries or sectors are placed between these two polar cases, with non-equity agreements closer to the no interaction end and equity alliances closer to unification. As we move along this continuum, bureaucratic costs increase, cross-industry cooperation strengthens, and diversification as part of firm strategy becomes more important, as firms come closer to internalizing businesses operating in other industries or sectors.

Horizontal and cross-industry alliances are explained mainly by the resource-based view (RBV) of the firm, which suggests that the primary economic incentive for engaging in alliances or inter-organizational relationships is to exploit resource complementarity (Barney and Hesterly, 1996; Oxley and Silverman, 2008). Resources controlled by two or more firms are complementary when their combined economic value is greater than their separate economic value. When firms have complementary resources, value-creating economic synergies among these firms may be explored. A strategic alliance is one way that such synergies may be realized. The RBV not only explains the principal driving forces of horizontal and cross-industry strategic alliances but it also helps explain the formation of vertical alliances, since a buyer and a supplier engaging in some collaborative relationship must also share complementary resources and capabilities for the alliance to make economic sense. For this reason, as firms move closer to unification, resource complementarity increases in the three continuums shown in Table 1.

A proposed typology

Based on the above discussion, all types of inter-organizational relationships may be conceived as intermediate forms in the market–hierarchy (or unification) continuum. The three types of inter-firm alliances (vertical, horizontal, cross-industry) may be placed in a chart using three forms of relationships: market or arm's length relationship, unification (merger or acquisition), and some form of intermediate collaborative relationship between firms (Table 1). This typology should be regarded as an extended continuum in which the market and the hierarchy are seen as polar forms. This extended continuum is further categorized into vertical, horizontal, and cross-industry types of relationships.

Table 1. Typology of inter-organizational relationships.

MODE OF GOVERNANCE	INTER-ORGANIZATIONAL RELATIONSHIPS		
	VERTICAL	HORIZONTAL	CROSS-INDUSTRY
MARKET (Arm's Length Relationship)	SPOT TRANSACTIONS	COMPETITION	NO INTERACTION
	Repeated Transactions		
INTERMEDIATE FORMS	*NON-EQUITY CONTRACTS**	*NON-EQUITY CONTRACTS**	*NON-EQUITY CONTRACTS**
	*EQUITY-BASED CONTRACTS**	*EQUITY-BASED CONTRACTS**	*EQUITY-BASED CONTRACTS**
	Autonomous Equity Joint Venture*	Autonomous Equity Joint Venture*	Autonomous Equity Joint Venture*
HIERARCHY OR INTERNAL ORGANIZATION (Merger or Acquisition)	VERTICAL INTEGRATION	HORIZONTAL INTEGRATION	DIVERSIFICATION

* Settings in which there can exist inter-organizational relationships.

Understanding the distinction between vertical, horizontal and cross-industry relationships among firms is crucial to understanding this typology. Although the three types of inter-organizational relationships are illustrated by the same notion of a continuum, they represent very different situations. For example, in a vertical relationship transactions between firms occur in subsequent stages of a supply chain. This is one difference between vertical and horizontal relationships, where firms compete or can potentially compete at the same level of the supply chain, and cross-industry relationships, where firms do not transact or compete among themselves.

Additionally, in vertical relationships the issue of transfer pricing is always relevant as both sides of the transaction have strong incentives to bargain in order to appropriate the value created, which in turn might lead to disincentives ex-ante and haggling and maladaptation costs ex-post (Williamson, 2000; Joskow, 2008). This is very different from the rivalry faced by industry competitors that compete to increase their respective market shares. It is also different from a cross-industry relationship, where the possibility of an adversarial relationship is low. These differences imply that in the continuum from market to hierarchy, the analysis of vertical, horizontal and cross-industry inter-organizational relationships should be viewed differently.

Distinguishing between alliances and buyer–supplier vertical relationships

In general terms there are two broad types of vertical relationships in the agri-food system: alliances and buyer–supplier relationships. The conceptual boundary between alliances and buyer–supplier relationships is not precise. One useful distinction offered by Oxley and Silverman (2008) is that alliances involve inter-firm collaborations among peers suggesting the relationship is more symmetrical than in the case of buyer–supplier arrangements. Thus, even where an alliance relationship is vertical, in that output from one upstream partner is used as input by one downstream partner, the existence of an alliance indicates that each firm brings specialized resources to the relationship. As a result, the issue of partner selection is more relevant in alliance research than is commonly the case in the study of buyer–supplier relationships.

Another useful distinction is that an alliance is 'a reciprocal relationship with a shared strategy in common' (Harbison and Pekar, 1998, p. 18). In order to pursue a common strategy and achieve shared goals, alliance partners need to collaborate, share resources and invest in relationship-specific assets that increase the level of interdependency between them. Applying Thompson's (1967) classification of interdependencies, an alliance often deals with all types of interdependencies (pooled, sequential and reciprocal), whereas the organizational design of buyer–supplier relationships is primarily concerned with the governance of sequential interdependencies.

Oxley and Silverman (2008) identify four basic motivations for alliance formation: market-access, co-specialization, learning and co-option. Market-access alliances are very common in the agri-food sector as food and agribusiness firms form alliances to bypass trade and investment restrictions in foreign markets. For example, many cooperatives and multinational firms have recently formed alliances with partners in China to have access to this high-growth, yet restricted, market. Co-specialization alliances involve the combination of the output of independent partners and are thus characterized by low levels of operational integration and information and resource exchange. An example is the alliance between Dairy Farmers of America (DFA), Starbucks and PepsiCo in the production and distribution of a coffee-flavoured dairy beverage (known as Frapuccino), which is analysed in more depth in Chapter 11 of this volume. Firms form learning alliances with the purpose of acquiring knowledge or capabilities from partners. An example of a learning alliance in the dairy industry is DairiConcepts, a joint venture between DFA and Fonterra to develop and market dairy ingredients to food processing and pharmaceutical clients. More common in the agri-food sector are co-option alliances, which bring together current and potential rivals. Farmer-owned cooperatives, industry associations and trade groups are examples of alliances involving competitors that collaborate to achieve common objectives including market access, commodity promotion and lobbying.

Buyer–supplier relationships in the agri-food system entail all sorts of contractual arrangements between producers, cooperatives, processors and retailers. In the United States, for example, the share of total agricultural output value marketed through some contractual form between farmers and a downstream buyer has increased from 11% in 1969 to 41% in 2005 (MacDonald and Korb, 2008). The combination of dispersed family farm ownership and increasing concentration in the post-farm gate levels of agri-food supply chains, coupled with increasing market information and knowledge asymmetries and heightened consumer awareness about food quality and origin, suggests that farmer–processor and farmer–retailer relationships are rarely among equals. Rather, these buyer–supplier relationships are marked by a considerable degree of asymmetry, in size and market influence, between the parties but none the less can entail some form of collaboration and coordination. Examples include contract agriculture schemes between buyers and smallholders in developing countries, poultry and hog product-specification and resource-providing contracts, and farm supplier networks led by one large supply 'chain captain', in general a processor or a retailer.

Regularities and challenges in inter-organizational relationships

Despite their diversity and complexity, Ménard (2004) suggests that all intermediate modes of governance, including inter-organizational relationships, share three 'regularities' or common characteristics: resource pooling, contracting and competing. Irrespective of the organizational arrangement or form, hybrids organize economic activities by means of inter-firm coordination and cooperation and, as a result, important investment and strategic decisions are jointly made. On one hand, the search for rents provides the motivation and rationale for resource pooling and coordinated decision-making. On the other, sharing rents under incomplete contracts and risk of counter-party opportunism is a major source of conflicts and transaction costs – especially in the presence of relationship-specific investments – that might destabilize the inter-organizational relationship. The first common problem of such relationships then emerges – how can alliance partners secure cooperation to achieve efficient coordination without losing their independence?

This first problem is partially managed by means of some contractual arrangement providing a framework for the inter-organizational relationship. Given the incompleteness of contracts and the need to secure ongoing relationships, however, a governance structure must be designed and implemented to complement and secure formal and informal contracts between alliance partners. Following standard transaction cost economics, governance 'mechanisms must be designed that are aligned with the characteristics of the transaction they support, filling blanks left in contracts, monitoring the arrangement, and solving problems without repeated renegotiation' (Ménard, 2004, p. 352). The second problem facing all forms of inter-organizational relationships can now be posed as – how can such governance structures secure contracts while minimizing costly renegotiations?

A third common characteristic of inter-organizational relationships is that partners compete against each other since they remain independent residual claimants and pursue their own set of organization-specific goals. Since inter-organizational relationships develop in increasingly competitive markets, external competitive pressures and uncertainty make these arrangements highly unstable. Ménard (2004) then poses a third problem for inter-organizational relationships – what coordination and control instruments should be used as a basis for joint decision-making, preventing opportunistic behaviour and free-ridership, and conflict resolution?

In summary, inter-organizational relationships involve a basic tension between collaboration (e.g., resource pooling, joint decision-making, information sharing, coordination of operations, etc.) and competition (rent sharing in particular). Alliance parties need to collaborate and coordinate their activities to create value (cooperative interdependence). But at the same time they compete to divide the value created (transactional interdependence). Rent-seeking and opportunism are major sources of contractual inefficiencies that might jeopardize the success of the relationship. This tension is perhaps more acute in vertical relationships (and particularly in buyer–supplier relationships) as they involve a clear zero-sum game in dividing the value created. This basic tension calls forth some governance structure to protect the interests and relationship-specific investments of alliance partners and also to coordinate their interdependent activities. The next section discusses the transaction cost economics approach to inter-organizational relationship governance.

Governing inter-organizational relationships

This section addresses inter-organizational relationship governance from a transaction cost perspective focusing on the mechanisms or instruments of governance that are utilized to cope with the relationship challenges discussed in the previous section. Since Coase's pioneering work, the make-or-buy decision has become one of the most studied topics in the modern theory of the firm (Joskow, 2008; Klein, 2008). Building on Coase's original insight, the transaction cost approach emphasizes that vertical coordination can be an efficient means of protecting relationship-specific investments or mitigating other potential conflicts in the context of incomplete contracting (Klein *et al.*, 1978; Williamson, 1979) thereby ameliorating ex-ante disincentives to invest (Grossman and Hart, 1986). In addition, the efficient alignment between the governance mode and the attributes of the underlying transaction ameliorates ex-post haggling and maladaptation costs (Williamson, 1991).

However, the early models put forward by organization economists have primarily focused on just one of the attributes that distinguish markets from hierarchies. For example, Coase (1937) and Simon (1951) focus on the authority relationship between the asset owner and employees; Williamson (1979) and Hart and Moore (1990) emphasize lateral and vertical ownership of assets by the firm; while the principal–agent literature features monitoring and compensation systems to align the interests of agents and principals (e.g., Prendergast, 1999).

More recent versions of these theories have started to realize the multidimensional nature of governance. Williamson (1991, p. 271) suggests that 'each viable form of governance – market, hybrid, and hierarchy – is defined by a syndrome of attributes that bear a supporting relation to one another.' He advances the hypothesis that each generic form of governance is supported by a different form of contract law; and that there are crucial differences between markets, hybrids and hierarchies in how they adapt to changing circumstances and in the use of incentive and administrative control instruments.

Holmstrom and Milgrom (1994, p. 972) also hypothesize that markets and hierarchies are 'two alternative systems for managing incentives for the wide array of tasks for which a single worker may be responsible.' They further posit that firms deploy various instruments in these systems to manage coordination and motivation challenges. In their model, the purpose served by the incentive systems of alternative forms of governance is to minimize agency costs between principals and agents. Their system of attributes approach to the study of economic organization includes authority, asset ownership, incentives and job design as the key mechanisms of governance.

The mechanisms or instruments of governance

Conceptualizing governance modes as 'syndromes' or 'systems' of attributes begs the question – what are the main attributes? In other words, what are the instruments that should be considered in analysing alternative modes of governance? What are the organizational design decision variables that an organizational scholar should consider?

In reviewing various theoretical approaches in the literature on inter-organizational arrangements, Grandori and Soda (1995) identify and systematize a wide range of coordination and control mechanisms that are employed to sustain cooperation between firms and address the interdependence challenges laid out in the previous section. Their systematization of contributions from economics, sociology, social psychology and organization theory includes the following mechanisms: (i) communication, decision and negotiation mechanisms; (ii) social coordination and control; (iii) integration and linking-pin roles and units; (iv) common staff; (v) hierarchy and authority relations; (vi) planning and control systems; (vii) incentive systems; (viii) partner selection systems; (ix) information systems; and (x) public support and infrastructure. Their literature review suggests these ten mechanisms are used in inter-organizational relationships in various combinations and degrees. In addition, the substance of any relationship, in terms of the mix of mechanisms employed by partners, can vary substantially in degree of formalization. The extent to which an inter-organizational relationship is controlled and safeguarded by a formal contract is another important dimension of governance. Contractual arrangements may be used to regulate the horizontal association of cooperating firms and/or the vertical exchange between parties.

A more recent review of the literature on hybrid forms that are neither markets nor hierarchies is offered by Ménard (2004). His central proposition is that hybrid organizations form a specific class of governance structures combining contractual agreements and administrative entities with the purpose of coordinating partners' efforts to generate rents from mutual dependence while controlling for the risks of opportunism. The role of contracts in hybrid arrangements is crucial in coordinating partners' investments and decisions, and sharing rents. Contracts achieve these purposes by: (i) selecting partners; (ii) determining the duration of the relationship; (iii) specifying quantity and quality requirements; (iv) laying out procedures for regulating renegotiations when ex-post adaptation is required; and (v) specifying rules for distributing the expected gains from joint actions. Because contracts are unavoidably incomplete, the stability and continuity of hybrid arrangements require specific mechanisms to coordinate interdependent activities, organize transactions and solve disputes. According to Ménard (2004, p. 366), a core element in the architecture of hybrid organizations is the presence of private governments (or authorities) that 'pair the autonomy of partners with the transfer of subclasses of decisions to a distinct entity in charge of coordinating their action.' These authorities vary in degree of formalization and centralization of decision-making, ranging from trust to formal government.

This section builds on the foundations of Williamson (1991) and Holmstrom and Milgrom (1994) and complements their conceptualization of governance modes as systems of attributes with an additional set of attributes that were identified in the systematizations offered by Grandori and Soda (1995) and Ménard (2004). An enlarged and yet selective set of attributes to characterize distinct modes of governance is found in Table 2. This set of attributes is used to describe markets, hierarchies and a specific type of hybrid (the processing cooperative) in the next section.

The organizational architecture of markets, hierarchies and hybrids

The market

Consider the transaction between a farmer supplying a commodity to a downstream processor. In a pure market transaction, ownership is separated in the sense that the farmer owns farm assets and the processor owns processing assets (see 'market' column, Table 2). There is no authority relationship between the two parties and no central structure, common staff or administrative controls are deployed to govern the transaction. The price system provides high powered incentives for both parties to be efficient and adapt to changing market conditions. In a market with a large number of transacting agents, partner identity is not relevant and thus partner selection mechanisms are not needed. In such a 'thick' market there is no mutual dependency relation between exchange partners. Consequently, switching costs are negligible if the processor wants to buy raw material from another farmer or if the farmer wants to supply a rival processor.

Table 2. Markets, hierarchies and hybrids as systems of attributes.

Mechanisms / Instruments	MARKET	COOPERATIVE	HIERARCHY
1. Ownership (property rights)	Separated	Separated	Joint (Unified)
2. Authority (formal)	0	++	++
3. Incentive intensity	++	++	0
4. Administrative controls	0	++	++
5. Common staff (central structure)	0	++	++
6. Partner selection	0	0	++
7. Adaptation A	++	++	0
8. Adaptation C	0	+	++
9. Contract law	++	+	0
Degree of Formalization			
• Association (horizontal)	0	++	++
• Exchange (vertical)	0	+	++
Degree of Centralization	0	+	++

Note: ++ = strong; + = semi-strong; 0 = weak. This table builds on and extends Table 1 in Williamson (1991, p. 281).

Market transactions are supported by classical contract law, which relies on formal and rigid terms to solve disputes (Williamson, 1991). This inflexible contracting regime coupled with third party enforcement mechanisms is well suited when continuity is not relevant to exchange partners. In addition, adaptation to disturbances occurs in a decentralized fashion as exchange partners react and reposition to changing relative prices and other market signals. Because exchange partners respond independently to exogenous changes, without the need for any administrative or coordinating device, Williamson (1991) calls this 'adaptation A', where A denotes autonomous.

Hierarchy

As a result of market failures and transaction costs, a hierarchy might supersede the market. For example, the farmer might decide to vertically integrate downstream and acquire the processing assets. Joint or unified ownership results as now one party (in our example, the farmer) has residual claim and control rights over both farm and processing assets. The owner has now complete authority over decisions regarding farm and processing asset use and deployment. The 'visible hand' of the manager or central planner emerges. The interdependence (or mutual dependence) between upstream (i.e., farm) and downstream (i.e., processing) assets suggests that partner identity matters thus creating the need for partner selection mechanisms.

As the firm grows in scale and scope, common staff and administrative controls are needed to coordinate activities inside the vertically integrated firm. Administrative control mechanisms include planning, information-sharing, integration, monitoring and performance evaluation systems. In other words, coordinated adaptation (i.e., adaptation C) substitutes for autonomous adaptation to unanticipated disturbances. But as authority and centralized decision-making supersede the price system as coordinating devices, incentives inside the vertically integrated firm become low-powered and bureaucratic costs emerge. The implicit contract law of internal organization is known as 'forbearance' as courts following the business judgment rule refuse to hear disputes between internal divisions. In other words, the 'hierarchy is its own court of ultimate appeal' (Williamson, 1991, p. 274).

Hybrids and intermediate forms

In the transaction cost perspective, markets and hierarchies are defined as polar modes of governance, while hybrids display intermediate values in all governance attributes. In particular, the hybrid form is characterized by 'semi-strong incentives, an intermediate degree of administrative apparatus, displays semi-strong adaptations of both kinds, and works out of semi-legalistic contract law regime' (Williamson, 1991, p. 281).

But are all hybrid structures really intermediate forms that adopt intermediate values in all governance dimensions? In a recent article, Makadok and Coff (2009) take issue with this view. They observe that hierarchies increasingly use market-like instruments such as high-powered incentives, transfer-pricing schemes and decentralized decision-making. Also, some market transactions adopt hierarchy-like attributes including authority, administrative controls, and incentive systems less tied to short-term performance. This suggests that true hybrid forms are market-like in some dimensions and hierarchy-like in others. Makadok and Coff (2009) develop a taxonomy of hybrid governance forms based on three dimensions of governance – authority, ownership and incentives.

Chaddad (2009) also conceptualizes hybrids – focusing on farmer-owned cooperatives – as a distinct governance mode blending market-like attributes with hierarchy-like mechanisms. He shows that cooperatives may be usefully regarded as a true hybrid form exhibiting a particular mix of coordination and control mechanisms in various degrees. In other words, cooperatives have architectures of their own which are distinct from markets and hierarchies.

Processing (or marketing) cooperative

Because of some defensive or offensive economic motivation, farmers overcome initial collective action costs and decide to vertically integrate downstream to bypass the proprietary processor or add value to their farm commodity. In a processing (or marketing) cooperative, the farmer-members pool resources and risk capital to invest in downstream processing assets. The farm commodity is therefore stored, sorted, processed and/or marketed by the cooperative. Farm assets are owned independently by the individual farmers but the cooperative processing plant is jointly owned by the farmer-members. Note that this arrangement is not 'pure' vertical integration (or hierarchy) because farm assets and processing assets are not under a single entity with unified ownership. Put differently, the members remain economically independent as they do not merge their production activities in one large farm. Rather, the processing cooperative can be viewed as an inter-organizational relationship – in particular, an equity-based alliance (Table 1) – in the ownership dimension of governance. As farmer-members remain independent they are still subject to the high-powered incentives of the market.

Both the mechanism used by the cooperative to acquire risk capital from members and the amount of risk capital invested by members may vary across processing cooperatives. In general, traditional cooperatives rely on low amounts of upfront member-contributed equity capital and therefore depend on passive or quasi-passive internally generated capital. Another feature of most traditional processing cooperatives is the open membership policy; the adoption of partner (or member) selection mechanisms is not a common practice. Because the farmers now jointly own the processing assets, they have formal authority or residual control rights over how they are deployed.

In general, the farmer-members of a processing cooperative hire an agent (i.e., the general manager) to run the business, who in turn employs staff in a central structure (just like a hierarchy). In addition, administrative mechanisms are employed to coordinate activities in the central office but also to coordinate interdependencies among members and between members and the cooperative. The complex interdependencies present in most cooperatives suggest the need for a combination of several types of coordination mechanisms (Thompson, 1967; Lazzarini *et al.*, 2001).

This hybrid cooperative structure combines autonomous with coordinated adaptation. Autonomous adaptation occurs primarily at the farm level as independent farmers react to price and other market signals using local, specific knowledge. Coordinated adaptation is present both at the processing level and between the processing level and the farm level as a result of interdependencies that originate from resource-pooling and joint investment decisions in specific assets. A separate legal entity is established by the farmer-owners of the processing cooperative. They formalize their associational agreement with the cooperative incorporation statutes and bylaws. In addition to a formal horizontal agreement, the processing cooperative might introduce a formal contract to regulate vertical interdependencies with the members. The degree of formalization and the substance of this vertical agreement also vary widely across cooperatives.

Since specific investments of farmer-members lead to mutual dependence, the relationship requires continuity. The rigidity and inflexibility of classical contract law is ill-suited to this situation as it hinders adaptation to unforeseen contingencies. Consequently, partners in such a hybrid arrangement adopt a more flexible neoclassical contracting regime. The neoclassical contract in a hybrid arrangement works as a framework, i.e., 'an occasional guide in cases of doubt, and a norm of ultimate appeal when the relations cease in fact to work' (Williamson, 1991, p. 272).

In summary, the processing cooperative is a hybrid arrangement with some market-like attributes (high-powered incentives, no partner selection mechanisms, and autonomous adaptation), some hierarchy-like mechanisms (formal authority, administrative controls, and common staff in a central office) and intermediate levels in some governance dimensions such as ownership, contract law and coordinated adaptation.

Conclusions

This chapter adopted a micro-analytic approach to economic organization, which seeks to inform the comparative efficacy of alternative generic forms of governance – markets, hybrids and hierarchies. We are interested in a particular governance mode – inter-organizational relationships – and its efficiency relative to alternative forms of organization – in particular, markets and hierarchies (internal organization). But what kind of governance structure is an inter-organizational relationship and what makes it unique? This is a critical theoretical question that needs to be adequately addressed so that we can compare and contrast inter-organizational relationships with alternative governance modes. Ultimately organizations that economize on transaction costs, adapt effectively to changing market conditions and eliminate waste should survive and thrive.

This chapter first provides a typology of inter-organizational relationships as intermediate forms of governance. Subsequently it attempts to show that inter-organizational relationships actually blend market-like with hierarchy-like instruments and thus may be viewed as a 'true hybrid' mode. One obvious avenue for future research is to describe the diversity of inter-organizational relationship governance using the conceptual framework summarized in Table 2.

Having discussed the variability among hybrid inter-organizational arrangements, a logical follow-up research query is – what are the determinants of these forms? For example, what factors might explain the higher degrees of formalization and centralization in a joint venture compared to more informal relational agreements? According to Williamson (1991) and also Holmstrom and Milgrom (1994), each viable form of governance is defined by a 'syndrome' or 'system' of attributes exhibiting a complementary relation to one another. What is the logic behind different combinations of attributes? What purposes do they serve? Under what conditions will a particular 'syndrome of attributes' emerge and/or fit with organizational strategy and the business environment? Why do we observe different types of inter-organizational relationships (with different organizational architectures) co-existing simultaneously and often competing in agri-food systems? These are all relevant research questions that may be addressed from a transaction cost perspective.

References

Barney, J. and Hesterly, W. (1996) Organizational economics: understanding the relationship between organizations and economic analysis. In: Clegg, S., Hardy, C. and Nord, W. (eds) *Handbook of Organization Studies*. Sage Publications, London.

Chaddad, F. (2009) Both market and hierarchy: understanding the hybrid nature of cooperatives. Proceedings of the International Workshop Rural Cooperation in the 21st Century: Lessons from the Past, Pathways to the Future. Rehovot, Israel.

Coase, R. (1937) The nature of the firm. *Economica* 4, 386–405.

Coase, R. (1992) The institutional structure of production. *American Economic Review* 82(4), 713–719.

Cook, M. and Chaddad, F. (2000) Agroindustrialization of the global agrifood economy: bridging development economics and agribusiness research. *Agricultural Economics* 23(3), 207–218.

Cook, M., Klein, P. and Iliopoulos, C. (2008) Contracting and organization in food and agriculture. In: Brousscau, E. and Glachant, J. (eds) *New Institutional Economics: A Guidebook*. Cambridge University Press, Cambridge, pp. 292–304.

Doz, Y. and Hamel, G. (1998) *Alliance Advantage: The Art of Creating Value through Partnering*. Harvard Business School Press, Boston.

Dussauge, P. and Garrette, B. (1999) *Cooperative Strategy: Competing Successfully Through Strategic Alliances*. Wiley, New York.

Gomes-Casseres, B. (1994) Group versus group: how alliance networks compete. *Harvard Business Review* (July/August), 62–74.

Grandori, A. and Soda, G. (1995) Inter-firm networks: antecedents, mechanisms and forms. *Organization Studies* 16(2), 183–214.

Grossman, S. and Hart, O. (1986) The costs and benefits of ownership: a theory of vertical and lateral integration. *Journal of Political Economy* 94, 691–719.

Harbison, J. and Pekar, P. (1998) *Smart Alliances: A Practical Guide to Repeatable Success*. San Jossey-Bass Publishers, Francisco.

Hart, O. and Moore, J. (1990) Property rights and the nature of the firm. *Journal of Political Economy* 98, 1119–1158.

Hennart, J. (1988) A transaction cost theory of equity joint ventures. *Strategic Management Journal* 9, 361–374.

Holmstrom, B. and Milgrom, P. (1994) The firm as an incentive system. *American Economic Review* 84, 972–991.

Joskow, P. (2008) Vertical integration. In: Ménard, C. and Shirley, M. (eds) *Handbook of New Institutional Economics*. Springer, Berlin, pp. 319–348.

Klein, B., Crawford, R. and Alchian, A. (1978) Vertical integration, appropriable rents and the competitive contracting process. *Journal of Law and Economics* 21, 297–326.

Klein, P. (2008) The make-or-buy decision: lessons from empirical studies. In: Ménard, C. and Shirley, M. (eds) *Handbook of New Institutional Economics*. Springer, Berlin, pp. 435–464.

Lazzarini, S., Chaddad, F. and Cook, M. (2001) Integrating supply chain and network analysis: the study of netchains. *Journal on Chain and Network Science* 1(1), 7–22.

MacDonald, J. and Korb, P. (2008) Agricultural contracting update, 2005. *Economic Information Bulletin* 35. USDA Economic Research Service, Washington, DC.

Madhok, A. (2002) Reassessing the fundamentals and beyond: Ronald Coase, the transaction cost and resource-based theories of the firm and the institutional structure of production. *Strategic Management Journal* 23, 535–550.

Makadok, R. and Coff, R. (2009) Both market and hierarchy: an incentive-system theory of hybrid governance forms. *Academy of Management Review* 34(2), 297–319.

Ménard, C. (2004) The economics of hybrid organizations. *Journal of Institutional and Theoretical Economics* 160, 345–376.

Nohria, N. (1992) Introduction: is a network perspective a useful way to studying organizations? In: Nohria, N. and Eccles, R.G. (eds) *Networks and Organizations*. Harvard University Press, Boston, pp. 1–22.

Oliver, A. and Ebers, M. (1998) Networking network studies: An analysis of conceptual configurations in the study of inter-organizational relationships. *Organization Studies* 19, 549–583.

Oxley, J. and Silverman, B. (2008) Inter-firm alliances: a new institutional economics approach. In: Brousseau, E. and Glachant, J. (eds) *New Institutional Economics: A Guidebook*. Cambridge University Press, Cambridge, pp. 209–234.

Powell, W. (1990) Neither market nor hierarchy: network forms of organization. *Research in Organizational Behavior* 12, 295–336.

Prendergast, C. (1999) The provision of incentives in firms. *Journal of Economic Literature* 37, 7–63.

Simchi-Levi, D., Kaminski, P. and Simchi-Levi, E. (2000) *Designing and Managing the Supply Chain: Concepts, Strategies, and Case Studies*. Irwin McGraw-Hill, New York.

Simon, H. (1951) A formal theory of the employment relationship. *Econometrica* 19(3), 293–305.

Thompson, J. (1967) *Organizations in Action: Social Science Bases of Administrative Theory*. McGraw-Hill, New York.

Thorelli, H. (1986) Networks: between markets and hierarchies. *Strategic Management Journal* 7, 37–51.

Williamson, O. (1979) Transaction-cost economics: the governance of contractual relations. *Journal of Law and Economics* 22(2), 233–261.

Williamson, O. (1991) Comparative economic organization: the analysis of discrete structural alternatives. *Administrative Science Quarterly* 36, 269–296.

Williamson, O. (2000) The new institutional economics: taking stock, looking ahead. *Journal of Economic Literature* 38, 595–613.

Zylbersztajn, D. and Farina, E. (1999) Strictly coordinated food systems: exploring the limits of the Coasian firm. *International Food and Agribusiness Management Review* 2, 249–265.

Chapter 3

Behavioural Economics and the Theory of Social Structure: Relevance for Understanding Inter-organizational Relationships

Monika Hartmann, Julia Hoffmann and Johannes Simons

University of Bonn, Germany

Introduction

Coordination of economic action and its performance is one of the most relevant subjects of economics. According to neoclassical economics, markets can best coordinate the exchange between actors that are assumed to be rational, fully informed and only interested in their own utility.[1] The 'invisible hand' of the market is understood to allow for individual utility maximization with transactions involving no or negligible costs (Beckert, 2007; Rooks *et al.*, 2000).

Foregoing the unrealistic assumption of rational and fully informed actors, however, reveals that economic agents risk being exploited by their 'selfish' exchange partners. Transaction cost economics considers those problems. In a world characterized by bounded rationality and imperfect market transparency, individuals have incentives to protect themselves from the potentially opportunistic behaviour of their business partner by negotiating, drawing and enforcing contracts. Thus, transaction costs economics explains why markets are not always the most efficient governance structures. Depending on the characteristics of the product to be exchanged, hierarchy or hybrid forms such as franchises or contracts sometimes result in lower transaction costs than markets (Rooks *et al.*, 2000).

Neither neoclassical economics nor transaction cost theory considers that regulatory mechanisms exist that prevent actors from always and only following their own interests at the expense of others. Representatives of socio-economic theory argue that 'actors do not behave or decide as atoms outside a social contact' (Granovetter, 1985, p. 487) and underline the importance of social structures for analysing business relationships appropriately (Jackson and Wolinsky, 1996). Behavioural economics, in addition, provides a powerful empirical underpinning for the relevance of behaviour that is not purely self-

---

[1] In neoclassical economics, utility is assumed to be generated from the availability of goods, while social preferences are not taken into account.

interested, of people with considerable implications for many economic domains including the analysis of business relationships.

In this chapter, we analyse the extent to which social structure theory and behavioural economics provide explanations for the actual behaviour of economic agents and their motivation regarding business relationships. Within such a context issues such as trust, fairness-led reciprocity and reputation are relevant topics that will be addressed. The chapter is structured as follows. First, the concept of social structure will be presented focusing on its relevance for understanding the governance of inter-organizational relationships as well as the benefits linked to socially embedded inter-organizational relationships. This is followed by discussing insights from behavioural economics regarding the behaviour of economic agents in business relationships with a special focus on reciprocity and the interaction of reciprocity and reputation. The results of both concepts are synthesized and some conclusions for the future analysis of inter-organizational relationships are drawn.

The concept of social structure

According to the sociological perspective business relations are embedded[2] in social structures, which refer 'to the extent to which economic action is linked to or depends on action or institutions that are non-economic in content, goals or processes' (Granovetter, 2005, p. 35).[3] The focus on the atomized and purely self-interested, utility-maximizing individual is perceived as an 'undersocialized' (Granovetter, 1985) conception of business action. Embeddedness of market exchange 'reflects social conditions that help actors in addressing underlying problems of coordination' (Beckert, 2007, p. 14). In this respect, trust, reputation and reciprocity are considered as important issues to be examined (Uzzi, 1997; Powell, 1990; Sako, 1991). Trust is regarded as crucial to resolve problems of cooperation.

Trust can be defined as the 'mutual confidence that no party to an exchange will exploit another's vulnerabilities' (Barney and Hansen, 1994, p. 176). Vulnerabilities exist due to risk factors such as adverse selection, moral hazard or hold-up problems. According to Kollock (1994) and Fehr (2009), however, it is not belief but behaviour that proves the existence of trust. Thus, defining trust as an action that increases one's vulnerability to another person whose behaviour is not under one's control (Kollock, 1994) goes beyond the first, purely 'belief-based', definition in that it requires exposure to some level of risk for trust to develop. As risk provides opportunities for opportunism, abstaining from exploiting another person's vulnerability creates trust (Barney and Hansen, 1994; Kollock, 1994; Fehr, 2009). According to the sociological perspective, social relations are the most important driver for trust in business relationships.

---

[2] The concept of social embeddedness has its roots in the work of Karl Polanyi (1957), though the concept has undergone a great transformation. Since Granovetter's (1985) work on inter-firm networks the approach of adopting sociology concepts in economic theories became more acknowledged. Therefore, the idea of embeddedness of business relations also often traces back to Granovetter's work.

[3] Several forms of systematization of the embeddedness concept are discussed in the literature (Beckert, 2007; Rooks *et al.*, 2000; Halinen and Törnroos, 1998). For instance, Halinen and Törnroos (1998) distinguish between temporal, spatial, social, political, market and technological embeddedness. Though all of these elements have a different focus, social embeddedness is always perceived to be the most elemental part highly intervened with the other dimensions.

All businesses can be considered to be socially embedded since the individuals working in these firms have social ties beyond but also directly linked to their work. These individuals work as members in organizations and interact individually, or as a group of employees, with colleagues of their own firm as well as with members of other enterprises (Ahrne, 1994). Hence it is necessary to investigate the relevance of social embeddedness for understanding the governance as well as the economic outcome of inter-organizational relationships.

Governance of inter-organizational relations: the relevance of social structure

Depending on the characteristics of the exchange, social embeddedness is of varying relevance for business relationships. Assuming that no exchange hazards exist, there is no opportunity to exploit any vulnerability. In this case neither trust established by social ties nor the establishment of specific governance forms (e.g., contracts) is of relevance for safeguarding transactions. In fact, due to the lack of risk there is no test for trust and thus the characteristics of the market rather than those of the exchange partner lead to the occurrence of transactions (Kollock, 1994; Williamson, 1991; Granovetter, 1985).

In cases where significant vulnerabilities such as moral hazards exist, market transactions can nevertheless take place if appropriate market or contractual governance devices are implemented or trust is generated due to social ties. Market-based governance devices do not rely on third-party intervention but are self-enforcing. In a situation in which economic hostages are created intentionally (e.g., through symmetric investments in specialized assets), thus motivating the exchange partner to behave in a 'cooperative' way (Dyer and Singh, 1998) voluntary compliance results. Market-based government devices are sufficient if the gain from violation is assumed to be smaller than the expected future net benefits from adherence. Assuming selfish actors, the risk of being exploited is reduced or even eliminated, as is the necessity for trust to backup transactions.

Transaction cost economics assumes that in the case of strong vulnerabilities such as hold-up problems, opportunism can be best prevented by contractual safeguards (e.g., explicit contracts). Such safeguards cover issues of conduct and the respective penalties in case of non-compliance which are enforced by the State or a legitimate organization authority and thus rely on third-party enforcement (Dyer and Singh, 1998).[4] However, in a complex and highly unpredictable world, exchange partners will find it difficult and prohibitively expensive to foresee all possible contingencies, to write contracts that cover them and finally to enforce these contracts. This holds even more when outcomes are unobservable or difficult to verify by a third party (Hart, 1995; Guo and Jolly, 2008). As a consequence, formal contracts are in general not only costly, but also remain incomplete and thus give rise to ownership rights which refer to so-called residual rights that cannot be specified in advance as part of an enforceable contract (Grossman and Hart, 1986). If the cost of an explicit contract outweighs the potential benefits of the transaction, social embeddedness of inter-organizational relationships is a pre-condition for transactions to take place. Social ties are the most important driver for creating trust in business

---

[4] A second mechanism to safeguard against risk factors such as hold-up problems often proposed by transaction costs economics is vertical integration. However, in this chapter we are interested in discussing inter-organizational relationships so this governance form will not be further considered.

relationships. It can complement or substitute for unenforceable formal contracts, thereby reducing transaction costs (Granovetter, 1985).

According to the sociological view, trust will emerge due to social interactions among exchange partners with key contact people often being crucial to develop and maintain social structures (Bendapudi and Leone, 2002).[5] Over time, trustworthiness is backed by positive reciprocity,[6] which implies that in a relationship, knowledge, resources and/or other valued items are given *and* received (Portes and Sensenbrenner, 1993; Rooks *et al.*, 2000). Reputation will grow if promises made in the past are honoured, allowing trustworthiness to increase with the length and depth of a relationship (Uzzi, 1997; Powell, 1990; Sako, 1991; Beckert, 2007; Granovetter, 1985; Gulati, 1995; Dyer and Chu, 2000; Malecki and Tootle, 1996).

While positive reciprocity leads over time to a high level of trustworthiness of a business partner, negative reciprocity – punishing harmful acts – serves as an important safeguard. Various types of social sanctions can be imposed on business partners that behave in an opportunistic way. Substantial economic opportunity costs arise for an actor that for example gains the negative reputation of being a cheater. Even more, opportunistic behaviour can lead to high social costs for individuals or businesses that are deeply embedded in social networks if non-compliance leads to a withdrawal of respect and prestige or puts networks of relations at risk (Dyer and Chu, 2000; Barney and Hansen, 1994).

Trustworthy behaviour thus might only exist because actors aim at preventing the social or economic costs that could arise following opportunistic behaviour. Trustworthy behaviour, however, can also be motivated by internalized social norms.[7] The wish to prevent negative feelings such as failure or guilt rather than the desire to avoid externally imposed sanctions encourage this conduct (Barney and Hansen, 1994). Trustworthy businesses or individuals of this 'strong' type will not exploit their partner even if social ties are absent. In the context of business relationships, social embeddedness, however, still plays a considerable role. As for leading to a competitive advantage (see discussion below) it is essential that businesses are able to identify trustworthy exchange partners of this 'strong' type. Otherwise they need to rely on economic governance devices to prevent fraud. Developing social ties for example by past experiences with the exchange partner are crucial to reduce the costs of identifying these kind of trustworthy actors (Barney and Hansen, 1994).

The expositions above indicate that social structures influence the governance of inter-organizational relationships if vulnerabilities from adverse selection, moral hazard and/or hold-up exist. Other things being equal, social embeddedness will reduce the need for market and contractual governance devices as positive reciprocity strengthens trustworthiness leading eventually to the development of a positive reputation. Negative

---

[5] Employee turnover on the other hand can put inter-organizational relationships at risk (Bendapudi and Leone, 2002).

[6] The social norm of reciprocity is discussed in more detail in the section 'Relevance of 'homo reciprocity'', below. In general, it refers to the fact that people respond kindly to beneficial acts (positive reciprocity) while harmful behaviour induces retaliation (negative reciprocity).

[7] Social norms can be defined as: '(i) behavioural regularity that is (ii) based on a socially shared belief how one ought to behave which triggers (iii) the enforcement of the prescribed behaviour by social sanctions.' (Fehr and Gächter, 1998). 'Internalized' social norms imply that behaviour is not motivated by the sanction in the case malfeasance but by the personal belief regarding the value of the norm.

reciprocity, however, is as important, as it serves as an important safeguard since malfeasance can induce high economic and social costs. Finally, social structures facilitate the identification of trustworthy individuals or businesses, thus making economic governance devices redundant.

Benefits of social embeddedness for inter-organizational relationships

Several studies about the complex-product industry (e.g., the automobile sector) in Japan found that this country reached international competitiveness in different sectors through a system of highly embedded enterprises. It seems that a stronger group orientation than in Western societies exists, for cultural reasons (Hill, 1995), which makes the establishment of different types of mutual trust in inter-organizational relationships easier (Dyer and Chu, 2000; Dyer, 1996; Sako, 1991).

These, and other studies, indicate that social embeddedness can be a potential source of competitive advantage if social ties are different between competitors, and vulnerabilities from for example adverse selection, moral hazard and hold-up exist in the market (Barney and Hansen, 1994). This is for several reasons. First, companies with stable business relations have cost advantages, because of a reduced necessity to monitor, renegotiate and bargain as has been discussed above (Barney and Hansen, 1994; Hoang and Antoncic, 2003; Hill, 1995).[8] Second, as own experiences are considered the most reliable source of data, networks of business relationships generate the most secure information. A third positive effect of social embeddedness is the fact that exchange relations may be much deeper and richer than if they (exclusively) rely on contracts (Uzzi, 1997; Hoang and Antoncic, 2003) with the main rationale for this being summarized in the literature review of Dyer and Singh (1998).[9] Based on this review as well as more recent studies, three core reasons can be identified for the competitiveness-enhancing effect of social structures (e.g., Dyer and Singh, 1998; Hoang and Antoncic, 2003; Podolny and Page, 1998; Duschek, 2004):

(1) Inter-organizational relationship-specific assets. Organizations can realize routines and processes which are crucial sources of competitive advantage through the growing working experience with their exchange partner (Duschek, 2004; Dyer and Singh, 1998). Several studies show that the greater the willingness of organizations to make relationship-specific investments the greater is the potential to reap relational rents and thus competitive advantage for collaborating enterprises (see Dyer and Singh, 1998 and the references cited therein).

(2) Inter-organizational knowledge-sharing. Organizations can generate relational rents by developing patterns of inter-organizational interactions that allow the transfer, recombination and/or creation of specialized knowledge (Dyer and Singh, 1998; Duschek,

---

[8] Though social forms of governance imply no or fewer direct costs, they are linked to opportunity costs as relying on social forms of governance reduces the number of potential exchange partners.

[9] It should be mentioned that the authors are proponents of the relational view – the newest branch of strategic management theory that focuses on dyad and network routines and processes as important determinants of competitive advantage (Dyer and Singh, 1998). 'The relational view is based more or less implicitly on ideas of the social network perspective' (Duschek, 2004). Dyer and Singh for example talk about the 'network of relationships in which a firm is embedded'. Due to the fact that the discussion in the framework of the relational view in strategic management theory has a common ground with the social structure theory, they provide interesting insights regarding the benefits of social embedded inter-organizational relationships.

2004; Powell *et al.*, 1996; Valkokari and Helander, 2007). Inter-organizational knowledge-sharing routines rely on a high level of transparency and reciprocity. They can lead to the creation of tacit knowledge – a highly inimitable resource and thus an important source of competitive advantage. Firms are especially successful in utilizing such outside sources of knowledge if they have high absorptive capacity. This is the case if partners have been successful in building up an overlapping knowledge base as well as routines that allow for a high frequency and intensity of information sharing and 'socio-technical' interactions (Dyer and Singh, 1998). Thus, social embeddedness is not only a precondition for sharing tacit knowledge between organizations but the depth of the relationship, in addition, determines whether organizations can make use of such knowledge that leads to highly inimitable resources.

(3) Complementary endowments with resources and capabilities. Organizations can generate relational rents by combining distinctive strategic resources and capabilities owned by business partners that collectively are more valuable, rare and difficult to imitate than if used individually (Dyer and Singh, 1998; Duschek, 2004). As synergies can be captured Aristoteles' statement 'the whole is more than the sum of its parts' holds. The higher the organizational and cultural 'fit' between business partners, the more synergy-sensitive resources and capabilities exist, and the more successful partners are in identifying them, the more advantageous is the inter-organizational relationship for its members (Duschek, 2004; Dyer and Singh, 1998).

Social embeddedness of inter-organizational relationships based on trust, reliability and reciprocity are able to create the discussed intangible and mostly inimitable resources. Market and contractual governance devices (alone) could not totally stabilize uncertain expectations and thus the fear that vulnerabilities could be exploited would remain. As a consequence, for example the generation of tacit inter-organizational knowledge would not take place (Granovetter, 1985; Ullrich, 2004). Therefore, it can be concluded that social embeddedness can lead to social capital – in the form of intangible and mostly inimitable 'resources' (Nahapiet and Ghoshal, 1998) – which are a crucial source of competitive advantage (Sorama *et al.*, 2004).

Finally, it should be noted that the benefits of embedded business relationships often go far beyond purely monetary gains. Social embeddedness of transactions also implies intrinsic or process benefits in the form of personal relationships, grant and the pursuit of regard (Sage, 2003).

Behavioural economics

The model of rational optimization of traditional economic theory is also criticized by behavioural economics as an unrealistic view of human decision-making (Egidi, 2005). Behavioural economics is a combination of microeconomics and psychology. It is concerned with empirically testing neoclassical assumptions and explaining identified systematic deviations from those assumptions. Based on these empirical findings, behavioural economics can provide insights also for better understanding inter-organizational relationship. In this section, important deviations from the neoclassical model with relevance for inter-organizational relationship will be discussed. As in the expositions above, the section focuses on the impact of social preferences for reciprocity and of reputation as important incentives for contract enforcement.

Relevance of 'homo reciprocity'

Material pay-offs are powerful motivators for human behaviour. Nevertheless, people often deviate – in contrast to homo economicus – from exclusively self-interested behaviour in a strongly reciprocal way (Fehr and Gächter, 1998; Fehr *et al.*, 2002). Strong reciprocity means that in response to beneficial acts, people are much nicer (strong positive reciprocity) while harmful behaviour induces a more hostile response (strong negative reciprocity) than would be predicted assuming utility maximization.[10] Thus, strong reciprocity is fundamentally different from 'altruism' which is a form of unconditional kindness occurring independent of any prior benefits. It also differs from reciprocal behaviour of the 'cooperative'[11] or 'retaliatory' type in repeated interactions that only arise because people expect future gains from their behaviour (Fehr *et al.*, 2002; Fehr and Gächter, 2000). Regarding business relationships, reciprocity of the retaliatory type would be given if an organization punishes another organizations' malfeasance only if it will benefit from this act. Strong reciprocity, in contrast, implies that an organization would sanction opportunistic behaviour of a business partner, e.g., by terminating the contract, even if this implies a loss for the organization.

Numerous empirical studies provide clear evidence that a considerable share of people reward positive acts and punish hostile ones. This holds even if the interaction is among complete strangers with nobody able to observe and acknowledge individual choices, and despite the fact that the behaviour is (very) costly but neither linked to present nor future material gains (Fehr and Gächter, 1998, 2002). The driving motivation of strong reciprocity is 'fairness'.

A large body of trust or gift exchange games have demonstrated the existence of strong positive reciprocity. In a classical trust game, a proposer receives an amount of money x, and has the option to send anything between 0 and x forward to the responder. The experimenter multiplies the amount y sent to the responder for example by 3 ($3y$) who is free to return anything between 0 and $3y$. Strong reciprocal behaviour is in general beneficial for both players. However, the standard economic model assuming exclusively self-interested behaviour dictates that neither the proposer nor the responder makes a transfer. The evidence of hundreds of trials, however, reveals that many proposers have trust in the fairness of the responder and send money, and that many responders strongly reciprocate and give back some money. Moreover, often there is a positive correlation between the money sent and the money returned at the individual and aggregate levels in this one-shot kind of game (Fehr and Gächter, 1998, 2000).

Strong negative reciprocity has been documented in a large number of ultimatum bargaining games. In these games, two persons have to agree on the division of a fixed sum of money. The proposer can make exactly one proposal of how to divide the amount, which the responder can reject or accept. Rejection implies that both receive nothing. In the case of acceptance the proposal is implemented. Selfishness[12] predicts that the proposer would

---

[10] As mentioned above, utility is assumed to be generated from the availability of goods. For example, social preferences for fairness and the potential utility derived from fair behaviour are not considered.

[11] Fehr *et al.* refer to this form of cooperative behaviour also as 'reciprocal altruism'. However, actors behaving accordingly only help others, though it involves short-run costs, if they expect long-term net benefits (Fehr *et al.*, 2002).

[12] Selfishness is meant as exclusively considering the own material benefit. Utility derived from behaviour that is in accordance with social preferences is not considered.

offer an amount *x*, where *x* is the smallest money unit (e.g., 1 cent) and the responder would accept anything greater than 0. Again, there exists profound empirical evidence revealing that proposals of less than 30% of the available sum are rejected by a high probability though this rejection is costly for the responder (Fehr and Gächter, 1998). These results indicate that in decisions about personal gains, people not only consider their own benefits but also those of others. When they perceive that benefits may be distributed unfairly, people incur a psychological cost. If this is high enough they are willing to forgo the opportunity to increase their personal wealth if this prevents an increase of wealth of others (Frank, 2003). The findings are of relevance for business relationships as they provide an explanation why firms are willing to forego business opportunities that are considered to be unfair even if this implies material losses for themselves. Neuroscientific studies support the view that 'fair choices' are perfectly in accordance with preferences (Fehr *et al.*, 2009a).

According to the results of many empirical studies the share of people that show a concern for fairness and behave accordingly in a one-shot situation is between 40% and 66%. Furthermore, there is an emerging consensus that the tendency to punish hostile behaviour is stronger than the one to reward friendly actions (Fehr *et al.*, 2009a). However, those studies also reveal that 20% to 30% of subjects still behave in a selfish manner (Fehr and Gächter, 2000). As to inter-organizational relationships, it is interesting to analyse how the coexistence of strong reciprocal and selfish actors influence such relationships, the role the institutional environment plays, and how performance is affected.

The implications of social preferences on contract acceptance and enforcement, price rigidities as well as welfare consequences have been analysed in field studies but especially in numerous laboratory experiments. Most studies refer to the labour market and analyse employer–employee relationships in a 'gift-exchange' game framework. Not only intra-organizational relationships but also interactions among businesses are often complex, making it difficult to specify all dimensions of a contract. Thus, many of the results generated for intra-organizational interactions provide interesting insights also for complex inter-organizational relationships.

Fairness and strong reciprocity in spot interaction

In one-shot games, where reputation formation cannot play any role, and highly incomplete contracts with strong incentives for shirking exist, higher payment (e.g., wages) on average induce higher effort levels thus revealing concerns for fairness. A significant share of agents performs above the minimal effort level predicted by self-interest. Nevertheless, the achieved outcome is still far below the efficient level as the existence of selfish agents reduces trust on the side of the principals restraining many of them to offer payments that would be high enough to induce efficient effort levels from fair-minded agents (Fehr *et al.*, 2009a). The results also indicate that lowering payments are considered as unfair,[13] reducing agents' effort level. This effect is stronger in magnitude than the one following an increase in payments. Thus, the evidence from laboratory studies with students and non-student subjects as well as field experiments reveals fairness alone has a positive albeit small impact on performance (Fehr *et al.*, 2009a).

---

[13] The results of empirical studies also show that though low payments can be costly for the principal as they are in general 'punished' by the agents with lower effort levels this does not hold in all instances. If declining payments are not considered as unfair they will not induce adverse consequences (Fehr *et al.*, 2009a).

The efficiency-enhancing effects of contract enforcement due to strong reciprocity comes at a cost – a substantial price rigidity following demand and supply shocks (Fehr and Gächter, 1998, 2000; Fehr *et al.*, 2009b). Due to the lack of third-party enforcement, considerations of fairness and reciprocity are the major determinants of wage formation. Principals are reluctant to cut payments in a recession as they fear that this might inhibit agents' performance (Fehr and Gächter, 1998). In contrast, in the case of third-party enforcement, competition drives contract terms leading to much lower payments in situations of excess supply and higher outlays if there is a supply shortage (Fehr *et al.*, 2009b).

Several studies analysed whether the gap between the efficient and the actual effort level in a one-shot game can be reduced if not only the agent but the principal can reciprocate (punish or reward the agents' efforts ex-post), though at a cost (see Fehr *et al.*, 2009b and the studies cited there). Those experiments confirm the relevance of strong reciprocity, as despite the fact that rewarding or punishing is costly for principals, a high share of principals makes use of this opportunity (Fehr and Gächter, 2000). The opportunity to reciprocate on the side of the principal substantially contributes to the enforcement of contracts even in a one-shot game as it provides an incentive for selfish agents to provide non-minimal output thus increasing the monetary pay-off of the game. However, despite the fact that the gap between the efficient and the actual effort level provided is considerably reduced, efforts remain still far below the first best level (Fehr *et al.*, 2009b; Fehr and Gächter, 2000).

Other studies analysed how explicit material incentives incorporated in the contract (e.g., a fine to be paid in the case of verified shirking) are able to mitigate the enforcement problem (see Fehr and Gächter, 2000 and the studies cited there). It can be assumed that explicit incentives could discipline selfish agents. At the same time there is, however, the danger that explicit performance incentives may 'crowd out' reciprocity-based voluntary cooperation, thereby leading to counterproductive effects (Fehr and Gächter, 2000). Indeed the empirical evidence is mixed which indicates that 'providing explicit incentives is considerably more complicated than envisaged by standard principal–agent theory' (Fehr and Gächter, 2000, p. 15).[14]

Regarding inter-organizational relationships, the results indicate that in discrete one-off interactions among businesses that are characterized by exchange hazards, the existence of strong reciprocity is only able to slightly mitigate inefficiencies due to for example moral hazard.

The interdependencies between reputation and strong reciprocity in repeated interactions

In repeated interactions with a finite number of periods assuming again incomplete contracts and incentives for violation, we have an interaction of fairness and reputation effect. A series of papers has shown that reputation formation in relationships greatly strengthens the impact of fairness on performance bringing it closer to the optimal efficiency level. Firms' strategy of a contingent contract renewal rewarding high and

---

[14] A large body of literature deals with the inter-dependencies between 'extrinsic (the activity provides an indirect satisfaction, e.g., through monetary compensation)' and 'intrinsic incentives (immediate satisfaction, e.g., satisfaction of fairness concerns)' and the problem of 'crowding out' (e.g., Frey, 1997; Osterloh *et al.*, 2002).

satisfactory performance with contract renewal and high offers while sanctioning low effort with the termination of the relationship, serves as an effective incentive device. As reputation-building becomes profitable, even selfish agents have an incentive to mimic the behaviour of fair ones and thus gain a reputation of being fair. Only in the final period, where reputation does not matter can selfish agents be identified (Fehr *et al.*, 2009a; Brown *et al.*, 2004). An additional question is how the interaction between reciprocity and reputation influences price rigidity. Fehr *et al.* (2009b) come to the conclusion that compared to contracts that are only enforced by strong reciprocity, the provision of reputation incentives mitigates price stickiness with regard to supply and demand shocks. However, this in fact only seems to hold for entry-level payments. As is shown in Fehr *et al.* (2009a), changes in market conditions in the course of an ongoing relationship have little impact on principals' payment offer to incumbent agents. As the previous contract term become the reference for the actual offer, a cut in payments would be considered as unfair leading to a strong downward price rigidity in long-term relationships. This result confirms the view that people have not absolute but reference-dependent fairness preferences (Fehr *et al.*, 2009a).

The possibility, or lack of, third-party enforcement has a considerable impact on the length of a relationship. Experimental studies reveal that third-party enforceable contracts result in one-shot interaction not favouring incumbent agents. In contrast, in markets with incomplete contracts where principals face a moral hazard problem, long-term bilateral relationships dominate the market. As principals care about the identity of their business partner they consciously limit their options, tendering their offers exclusively to a particular agent with whom they form long-term relations. Obviously, as a consequence of incomplete contracts, relationship-specific reputational capital gains in relevance as a means to solve the enforcement problem, thereby fostering long-term relationships (Brown *et al.*, 2004, 762; Fehr *et al.*, 2009b).[15] In fact, even if a public reputation mechanism exists in the market, and thus all principals are able to observe all past payment and effort levels of all agents in the market, bilateral relations continue to play an important role, though to a somewhat lower degree (Fehr *et al.*, 2009b). Also the level of competition for agents (excess demand for agents) only reduces but does not eliminate the incidence of long-term relationships (Fehr *et al.*, 2009b). The results of experimental studies show that not only the aggregate gains from trade increase due to a long-term relationship but that both principals as well as agents earn more money the longer the relationship lasts (Brown *et al.*, 2004).

Most studies analysing the interaction of fairness preferences with repeated game incentives assume that principals have perfect information about trading partners' past effort. However, in reality random exogenous events often exist that influence agents' outputs. This parametric uncertainty makes it impossible for principals to use observed output level of agents as a precise indicator for their chosen effort levels. This ambiguity may reduce the possibility for agents to accumulate reputational capital and thus might also diminish the 'power of contingent renewal policy' (Fehr *et al.*, 2009b, p. 344). However, theory as well as a laboratory experiments indicate that despite the co-existence of behavioural and parametric uncertainty, reputation formation with high performance can be sustained (Fehr *et al.*, 2009b).

---

[15] While most of the studies discussed concentrate on the labour market some studies are also linked to the capital and product market. For example, Kollock (1994) shows that quality uncertainty in the product market in the absence of third-party enforcement contributes to the formation of long-term relationships.

The results indicate that in repeated inter-organizational interaction, the interaction between reciprocity and reputation building is able to considerably reduce the problems linked to exchange hazards. Actors favour long-term relationships as they prove to be beneficial for both sides.

Conclusions

The expositions in this chapter provide evidence that due to social relations and behavioural rules, economic behaviour can considerably deviate from the outcome assumed in neoclassical theory. The chapter shows that social embeddedness of inter-organizational relations allow transactions to take place, even in cases where considerable vulnerabilities exist that cannot (totally) be controlled for by explicit contracts, due to risk factors such as moral hazards. Social ties between organizations based on trust, reliability and reciprocity are able to create intangible mostly inimitable resources enhancing organizations' competitiveness. Fairness-driven reciprocity behaviour is a social norm with considerable implication for inter-organizational relationships as is shown in behavioural economics. It can explain why actors get involved or abstain from a business relationship as well as why they are willing to punish or reward their partner even if this implies personal losses. Numerous studies show that in markets with incomplete contracts, the interaction of fairness and reputation effects not only changes the outcome of a transaction, bringing it closer to the optimal efficiency level, but also influences the structure of relationships towards more long-term bilateral transactions.

The evidence we survey has powerful implications for understanding inter-organizational relationships. In this chapter, we combine insights regarding trust, reciprocity and reputation from social structure theory and behavioural economics and argue that they are important factors influencing the coordination and outcome of economic action. So far, there is still little empirical evidence from laboratory or field studies directly relating fairness concerns to inter-organizational relationships, as the studies in behavioural economics largely refer to the labour market. However, the structure of most laboratory experiments would not be expected to differ if inter- instead of intra-organizational relationships were analysed. This does not hold to the same extent for field studies, making these an important area for further research. Moreover, social structures and the relevance of social norms likely differ between sectors and countries. Comparing the relevance of for example fairness concerns across countries and sectors is an additional interesting area for future studies (see also Chapter 7).

References

Ahrne, G. (1994) *Social Organizations: Interaction inside, outside and between organizations*. Sage, London.

Barney, J. and Hansen, M. (1994) Trustworthiness as a source of competitive advantage. *Strategic Management* 15, 175–190.

Beckert, J. (2007) The Great Transformation of Embeddedness: Karl Polanyi and the New Economic Sociology. Max-Planck-Institut für Gesellschaftsforschung Discussion Paper (1).

Bendapudi, N. and Leone, R. (2002) Managing business-to-business customer relationships following key contact employee turnover in a vendor firm. *Journal of Marketing* 66, 83–101.

Brown, M., Falk, A. and Fehr, E. (2004) Relational contracts and the nature of market interactions. *Econometrica* 72(3), 747–780.

Duschek, S. (2004) Inter-firm resources and sustained competitive advantage. *Management Review* 15, 53–73.

Dyer, J. (1996) Does governance matter? Keiretsu alliances and asset specificity as sources of Japanese competitive advantage. *Organization Science* 7(6), 649–666.

Dyer, J. and Chu, W. (2000) The determinants of trust in supplier–automaker relationships in the U.S., Japan and Korea. *Journal of International Business Studies* 31(2), 259–285.

Dyer, J. and Singh, H. (1998) The relational view: co-operative strategy and sources of inter-organisational competitive advantage. *Academy of Management Review* 23(4), 660–679.

Egidi, M. (2005) From Bounded Rationality to Behavioral Economics. Experimental 0507002, Economics Working Paper Archive EconWPA.

Fehr, E. (2009) On the economics and biology of trust. *Journal of the European Economic Association* 7, 235–266.

Fehr, E. and Gächter, S. (1998) Reciprocity and economics: the economic implications of Homo Reciprocans. *European Economic Review* 42, 845–859.

Fehr, E. and Gächter, S. (2000) Fairness and Retaliation: The Economics of Reciprocity. CESifo Working Paper Series, No. 336. Munich.

Fehr, E., Fischbacher, U. and Gächter, S. (2002) Strong reciprocity, human cooperation and the enforcement of social norms. *Human Nature* 13, 1–25.

Fehr, E., Goette, L. and Zehnder, C. (2009a) A behavioural account of the labor market: the role of fairness concerns. *Annual Review of Economics* 1, 355–84.

Fehr, E., Brown, M. and Zehnder, C. (2009b) On reputation: a microfoundation of contract enforcement and price rigidity. *The Economic Journal* 119(536), 333–353.

Frank, R. (2003) *Microeconomics and Behavior*. McGraw-Hill, Boston.

Frey, B. (1997) *Not Just for the Money: An Economic Theory of Personal Motivation*. Edward Elgar, Cheltenham/Brookfield.

Granovetter, M. (1985) Economic action and social structure: the problem of embeddedness. *American Journal of Sociology* 91, 481–510.

Granovetter, M. (2005) The impact of social structure on economic outcomes. *Journal of Economic Perspectives* 19, 33–50.

Grossman, S. and Hart, O. (1986) The costs and benefits of ownership: a theory of vertical and lateral integration. *Journal of Political Economy* 94, 691–719.

Gulati, R. (1995) Does familiarity breed trust? The implications of repeated ties for contractual choice in alliances. *Academy of Management Journal* 38(1), 85–112.

Guo, H. and Jolly, R. (2008) Contractual arrangements and enforcement in transition agriculture: theory and evidence from China. *Food Policy* 33, 570–575.

Halinen, A. and Törnroos, J.-A. (1998) The role of embeddedness in the evolution of business networks. *Scandinavian Journal of Management* 14(3), 187–205.

Hart, O. (1995) *Firms Contracts and Financial Structure*. Clarendon Press, Oxford.

Hill, C. (1995) National institutional structures, transaction cost economizing, and competitive advantage: the case of Japan. *Organization Science* 6(1), 119–131.

Hoang, H. and Antoncic, B. (2003) Network-based research in entrepreneurship: a critical review. *Journal of Business Venturing* 18, 165–187.

Jackson, M. and Wolinsky, A. (1996) A strategic model of social and economic networks. *Journal of Economic Theory* 71, 44–74.

Kollock, P. (1994) The emergence of exchange structures: an experimental study of uncertainty. *The American Journal of Sociology* 100(2), 313–345.

Malecki, E. and Tootle, D. (1996) The role of networks in small firm competitiveness. *International Journal of Technology Management* 11, 43–58.

Nahapiet, J. and Ghoshal, S. (1998) Social capital, intellectual capital, and the organizational advantage. *The Academy of Management Review* 23(2), 242–266.

Osterloh, M., Frost, J. and Frey, B. (2002) The dynamics of motivation in new organisational forms. *International Journal of the Economics of Business* 9(1), 61–77.

Podolny, J. and Page, K. (1998) Network forms of organization. *Annual Review of Sociology* 24, 57–76.

Polanyi, K. (1957) *The Great Transformation*. Suhrkamp, Frankfurt am Main.

Portes, A. and Sensenbrenner, J. (1993) Embeddedness and immigration: notes on the social determinants of economic action. *American Journal of Sociology* 98, 1320–1350.

Powell, W. (1990) Neither market nor hierarchy: network forms of organization. *Research in Organizational Behaviour* 12, 295–336.

Powell, W., Koput, K. and Smith-Doerr, L. (1996) Interorganizational collaboration and the locus of innovation: networks of learning in biotechnology. *Administrative Science Quarterly* 41, 116–145.

Rooks, G., Raub, W., Selten, R. and Tazelaar, F. (2000) How inter-firm co-operation depends on social embeddedness: a vignette study. *Acta Sociologica* 43, 123–137.

Sage, C. (2003) Social embeddedness and relations of regard: alternative 'good food' networks in south-west Ireland. *Journal of Rural Studies* 19, 47–60.

Sako, M. (1991) The role of trust in Japanese buyer–supplier relationships. *Ricerche Economiche* 45(2–3), 449–474.

Sorama, K., Katajamäki, A. and Varamäki, E. (2004) Cooperation between SMEs: social capital and learning perspective. NCSB 2004 Conference, 13th Nordic Conference on Small Business Research.

Ullrich, C. (2004) *Die Dynamik von Coopetition*. DUV, Wiesbaden.

Uzzi, B. (1997) Social structure and competition in interfirm networks: the paradox of embeddedness. *Administrative Science Quarterly* 42(1), 35–67.

Valkokari, K. and Helander, N. (2007) Knowledge management in different types of strategic SME networks. *Management Research News* 30, 597–608.

Williamson, O. (1991) Comparative economic organization: the analysis of discrete structural alternatives. *Administrative Science Quarterly* 36(2), 269–296.

Chapter 4

Collaborative Advantage, Relational Risks and Sustainable Relationships: a Literature Review and Definition

Christian Fischer[1] and Nikolai Reynolds[2]

[1] Massey University, Auckland, New Zealand
[2] Synovate, Frankfurt, Germany

Introduction

During the last decades, the concept of vertical, producer-to-consumer 'chains' (supply/value/agri-food chains etc.) has been heavily promoted and applied in the food sector as a means of fostering agricultural development, and in particular of linking farmers to markets (see Chapter 1 of this volume and, among others, Webber, 2008; Will, 2008; GTZ, 2007; Jenkins *et al.*, 2007; Humphrey and Memedovic, 2006). However, somewhat less attention has been given to the question of how to actually enable organizations (farms, firms, research institutions, etc.) to build or to integrate successfully into global and local collaborative chains. Being part of a chain partnership (i.e., a group of independent organizations aligned in a non-hierarchical way which conduct business together and share knowledge) poses considerable management challenges. In particular, it requires all partners to develop and maintain close and sustainable relationships with each other.

Studies on inter- and intra-organizational relationships have been conducted in various academic disciplines, including sociology, psychology, law, economics, marketing, management, and combinations of these. Moreover, organizations can engage with each other horizontally as well as vertically (see Chapter 2). The analysis of the entire system of interactions is usually covered in the network literature. In this chapter we deal with vertical, buyer–seller relationships only. These relationships are sometimes also referred to as 'marketing channel' or 'chain' relationships. Strictly speaking, inter-organizational relationships are those between organizations, and not those between individuals from these organizations. In reality, it is possible that an inter-organizational relationship between two organizations is, for instance, highly competitive, while individuals from the two organizations can have rather collaborative inter-personal relationships. Given such complications, some authors prefer to use the more general concept of 'business relationships', defined as involving both inter-personal and inter-organizational interactions (Mouzas *et al.*, 2007).

In the following, the terms 'business relationships' and 'inter-organizational relationships' are used as synonyms. We define vertical business relationships in the agriculture-based food sector as 'agri-food chain relationships'. The focus is on professional interactions between organizations, which can have a profit-making objective (i.e., firms, farms) or not (e.g., public bodies, such as universities). In practice, we mean the professional relationships (which can have inter-personal elements) between individual leaders of the investigated organizations.

This chapter focuses on the literature from the marketing, management and applied economics fields and in particular the one that is of relevance for the agri-food sector (agricultural commodities as well as fast-moving consumer goods). We study agri-food chain relationships adopting a joint perspective of buyers and sellers, also referred to as dyadic perspective (see, e.g., Anderson and Narus, 1990). While the proposed theoretical framework is general (i.e., not sector-specific) we try to discuss it within the context of the agri-food sector which is characterized by the production and distribution of lowly priced, perishable goods of high purchasing frequency. Furthermore, the focus is on relationship sustainability – i.e., the characteristics which ensure that a relationship is long-lasting and rewarding for all involved parties.

The chapter aims at defining sustainable inter-organizational relationships (SIRs) in a way that they can be scientifically investigated. This means describing components that constitute SIRs and which can be measured so that a relationship can be quantitatively described and analysed. The common wisdom, and apparent advantage, behind this approach is that if an issue 'can be measured then it can be managed'. On the other hand, treating those invisible ties that connect organizations with scientific methods inevitably results in academic rigidity and some remoteness from the real world which can make it difficult for non-academics to see the practical relevance of the whole exercise.

We have structured this chapter as follows. The second section reviews the main advantages and risks of SIRs. Section three defines SIRs in a thorough way based on a review of the relevant literature. The last section concludes and discusses implications, in particular with regard to agri-food chains.

Collaborative advantages and relational risks

In successful buyer–seller partnerships, collaborative advantages, i.e., the competitive advantage derived from effective business relationships (Kanter, 1994), can be generated in different ways. In essence, effective business relationships can: (i) help to reduce environmental uncertainty (e.g., by securing a more stable or higher inflow of orders); (ii) contribute to obtain better access to crucial resources (e.g., raw materials, capital, specialized skills or knowledge); and/or (iii) result in higher business productivity (e.g., by enhancing loyalty among suppliers) (Dyer and Singh, 1998; Lank, 2006).

Alternative sources of competitive advantage, other than building and managing lasting inter-organizational relationships, exist however. There are operational options such as integration ('make') or procuring goods from markets ('buy') (Williamson, 1985). When favouring vertical collaboration it needs to be kept in mind that building long-term business relationships is not an end in itself. Rather they are a means to increase the profitability and thus competitiveness of the involved organizations. Without this added value, a relationship is an inferior transaction design and would not exist (Backhaus and Büschken, 1999). The economic advantage of long-term relationships can often only be achieved by relationship-specific (i.e., idiosyncratic) investments which in turn serve as a strong incentive for the involved business partners to closely collaborate at least until the idiosyncratic returns have recovered the incurred costs. Business partners are not in all cases willing or able to invest into long-term relationships. Rather they may opt for an 'arm's-length way of working' (Cox, 2004) with buyers or suppliers and thus prefer to 'buy'. However, if new complex products are to be developed where parts need to be specifically manufactured, or where credence attributes need to be preserved throughout an entire supply chain (e.g., in the case of food products), a 'collaborative way of working' (Cox, 2004) may be more appropriate. The sustainable inter-organizational relationship framework, discussed below, only applies to this latter situation.

The engagement into a collaborative relationship also bears relational risks which can lead to partnership failures. Empirical studies have until now paid little attention to this type of risk. According to Zsidisin (2003), there is still insufficient understanding of what risk means within a supply chain management context and given the growing importance and proliferation of partnerships, it appears that the relevance and significance of systematically assessing relational risk will grow in the future (Das and Teng, 2001). However, it is not possible to define a *typical* relationship risk (Hallikas *et al.*, 2002). The risks seem to vary a lot according to a number of factors but depend on informal rather than formal factors (Delerue, 2004).

Relational risk is generally defined as the probability and consequence of not having satisfactory cooperation (Das and Teng, 1996) or as the probability and consequence of opportunistic behaviour by the partner (Nooteboom *et al.*, 1997). Inter-organizational risks are dependent on the structure and coordination of the relationship which will fail if one of the actors takes advantage of the other's dependence in order to obtain benefits for himself. In other words, risks originate from opportunistic behaviour within a competitive environment and relate to the possibility of losing access to specific resources or assets. For instance, a buyer may have multiple competing suppliers where a supplier perceives a continuous risk of losing business to a competitor. Additionally, a business may perceive itself in a 'hostage' position if it lacks alternative options for relationships or specific resources and investments. Furthermore, relational risks comprise the concern that sensitive information and know-how (i.e., misuse of intellectual property) may be communicated and transferred to individuals or organizations outside of the dyadic relationship. Entering into partnership arrangements with larger partners within a supply chain can mean, on the one hand, having a competitive advantage through interchange of resources and know-how. On the other hand, there is the risk of becoming dependent on larger companies or even having to give up decision sovereignty.

Relational risks can be mitigated through safeguard mechanisms that can be either informal or formal. Informal mechanisms relate to trust-building activities such as fortifying transparency in transactions and communication within the relationship as well as joint conflict resolution activities when problems arise. Formal mechanisms attempt to address concerns and to limit opportunistic behaviour through contractual arrangements (see Chapter 2).

Theoretical foundations of relationship theory

The roots of relationship theory can be traced back to early Greek philosophy (Aristotle), and the concept of a relation or relationships between creatures both natural and social (animals, people, organizations, states, etc.) is fundamental in science (see Rychlak, 1984 for a discussion). Since relationships are generally defined as a relatively long-term association between two or more people, the concept is in particular of relevance in social science such as sociology and psychology. One can distinguish between inter-personal relationships and inter-group relationships but hybrid forms (e.g., an individual has a relationship with a company) are also possible. Inter-group relationships are thought to be more formal and handled on a more professional and rational (i.e., less emotional) basis than inter-personal relationships (Lambert and Knemeyer, 2006).

Different social science theories have dealt with relationships. Psychological attachment theory has argued that humans follow a fundamental drive to start and maintain social interactions with others and thus have a basic need for both stable and satisfying relationships (Baumeister and Leary, 1995). If either of these two components is missing, people will begin to feel anxious, lonely, depressed and unhappy. Social exchange theory defines relationships in the context of exchanged benefits. Relationships are judged in terms of obtained rewards relative to the potential benefits to be expected from alternative relationships (Blau, 1964). However, equity theory (Adams, 1965) has suggested that people care about more than just maximizing rewards and also want equity and fairness in their relationships. In particular this theory postulates that people strive for equity between the inputs which they contribute to a relationship (e.g., time, commitment, effort, etc.) and the outcomes (rewards/satisfaction) which they receive from it (such as friendship, love, commercial benefits, etc.). Newer developments have refined equity theory (Carrell and Dittrich, 1978) and extended it (Huseman *et al.*, 1987) but the fundamental proposition has remained unchanged.

The SIR construct which is discussed in the following is based in these theories. The focus is on inter-group (i.e., inter-organizational) relationships as discussed in the management, marketing and applied economics literature. Here, relationships are usually treated as constructs, in the sense of potentially multidimensional, complex concepts which usually cannot be directly observed. Finding a definition for such constructs requires identifying their components and to theoretically justify them (Law *et al.*, 1998).

Defining sustainable business relationships

A common definition of business relationships is: 'a series of market transactions and business-related interactions between a seller and a buyer which are not accidental. "Not accidental" means that there are reasons for both parties which make a planned linking of market transaction meaningful. It also means that there is an "internal connection" between the transactions' (Plinke, 1989, p. 307). From this definition, the special characteristics of business relationships can be derived (Kleinaltenkamp and Plinke, 1997): (i) a sequence of market transactions and other business-related interactions; (ii) an existing internal connection between the individual transactions; and (iii) investments by sellers and/or buyers in order to create and maintain the relationship.

Sustainability is a characteristic of a process or state that can be maintained, kept in existence or prolonged. Thus, SIRs can be seen as stable relationships of high quality which are responsive to changing business environments and which business partners continue as long as they are willing to do so. This does not mean that business relationships cannot be interrupted or terminated. The point is that SIRs end on good terms, so that they (potentially) may be continued at a later stage. In order to maintain business partners' willingness to sustain a relationship, or continue one after an interruption, the relationship must possess certain favourable components.

Relationship quality reflects the overall mean outcome of a sequence of commercial transactions/interactions which constitute a business-to-business (B2B) relationship. This static component describes the overall outcome level of a relationship. Relationship stability refers to the inter-temporal fluctuations characterizing these transactions/interactions. This dynamic component considers the nature of the transaction/interaction outcomes over time. Quality and stability are interrelated and are both necessary to make relationships sustainable.

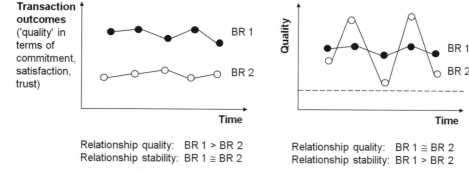

Relationship quality: BR 1 > BR 2   Relationship quality: BR 1 ≅ BR 2
Relationship stability: BR 1 ≅ BR 2   Relationship stability: BR 1 > BR 2

Fig. 1. Relationship quality and stability in business relationships (i.e., a sequence of commercial transactions/interactions).

Illustrating the concept graphically, Fig. 1 displays two business relationships (BR 1 and BR 2) as sequences of transactions/interactions. In addition to the temporal positioning of each transaction/interaction, their quality outcomes (in terms of satisfaction, commitment and trust) are depicted. The mean outcome across all transactions/interactions during a certain period reflects the overall quality of a relationship. Thus, in the left panel of Fig. 1, BR 1 is characterized by an overall higher level of relationship quality than BR 2. Relationship stability may be interpreted as the degree of inter-temporal quality fluctuations. The larger these fluctuations are, the lower the overall stability of a relationship. Thus, in the right panel of Fig. 1, BR 1 is characterized by a higher level of relationship stability than BR 2, although their overall quality level is about the same. Both relationship quality and stability determine the sustainability of a business relationship. Relationships where the overall quality is low may not be expected to last long. Relationships which are characterized by strongly fluctuating transaction quality may also not be sustainable. In particular, if there is something like a minimum expected quality level for each transaction/interaction (the dashed line in the right panel of Fig. 1), a relationship may break once a transaction/interaction has yielded a clearly unacceptable quality level.[1]

---

[1] A stable relationship does not mean it is 'static'. Transaction quality fluctuations in relationships can be considered as normal – but clearly among most partners these 'up and downs' must be within limits, in particular above a lower threshold.

Relationship quality

As a counterpoint to hard chain performance measures, such as return on investments or costs, the quality of relationships, while difficult to measure, can provide further insights into the well-functioning of corporate cooperation arrangements. Findings from a review of previous research by Roberts *et al.* (2003) suggest that satisfaction (gratification with another party), commitment (an engagement, voluntarily or by contract, towards another party) and trust (the belief in the fulfilment of obligations by another party) are important components of relationship quality, used in most studies. Relationship quality may therefore itself be seen as a three-dimensional construct, incorporating commitment, satisfaction and trust.

Commitment

Following psychological equity theory, relationships need continuous 'inputs', or 'investments' to sustain them. Different types of relationships may require different forms of inputs but commitment is important in most relations.

A general definition of commitment is the act of binding oneself (intellectually or emotionally) to a course of action and feeling dedicated or loyal to a longer-term endeavour (Morgan and Hunt, 1994). Commitment can take the form of an informal promise or of a formal pledge (e.g., a contract). In the context of inter-organizational relationships commitment can be defined as cooperative sentiments and existing affinity for the exchange partner with a preference for the continuation of the business relationship (Young and Denize, 1995). Another definition sees commitment as a tacit or expressed intention to support the persistence of the relationship between business partners (Wetzels *et al.*, 1998).

The importance of commitment in inter-organizational relationships results from the fact that highly committed chain actors 'stick' to the relationship and are less inclined to switch to other business partners (Barnes *et al.*, 2005). Commitment improves the sustainability of relationships since business partners are more likely to continue to work with their exchange partner.

Two general perspectives on commitment co-exist in the literature: (i) manifest (behavioural) commitment; and (ii) attitudinal commitment (Morgan and Hunt, 1994; Wetzels *et al.*, 1998). Attitudinal commitment comprises affective dimensions which influence the sense of unity. It can be understood as the inner psychological state of managers – i.e., their sensation of dedication and attachment to their business partners. Manifest commitment resembles a rational willingness to conduct business on the basis of verbal or written contracts and whether a relationship is commercially reasonable (Wetzels *et al.*, 1998). This means that a business partner may behave committed and continue to do business with the exchange partner as long as it is financially fruitful, however may not feel emotionally attached.

Satisfaction

Psychological equity theory stresses the importance of rewards in well-functioning relationships. Different types of relationships offer different 'outcomes' or benefits but whatever their exact nature, they need to translate into 'satisfaction' for the involved relationship parties.

Satisfaction is the feeling of contentment and gratification that arises when needs or desires have been fulfilled. Satisfaction also can be understood as a positive psychological state and response to the results of an evaluation process (Giese and Cote, 2000). In this evaluation process, individuals or businesses assess the degree to which their expectations have been met. Meeting or exceeding expectations is important for the sustainability of inter-enterprise relationships since it significantly influences the decision of the exchange partners to continue their business relationship (Selnes, 1998).

Two sub-dimensions of the evaluation of fulfilment of expectations exist: an affective social-emotional and a cognitive economic-rational one (Geyskens and Steenkamp, 2000; Ivens, 2004). The social-emotional component refers to how business partners emotionally perceive their expectations have been met by evaluating personal interactions and behaviour. Expectations refer here to factors such as equality, commitment of the seller, and non-opportunistic behaviour. Economic-rational satisfaction requires a specific level of knowledge about prices and products to assess if the economic outcomes meet one's own financial, economical expectations. Economical expectations can relate to product quality, price and service. Business relationships should meet or exceed both expectation dimensions to satisfy partners. Even if customers are economically satisfied with the performance of their suppliers, they may not feel that their social interactions are gratifying and may therefore switch their supplier (Bennett and Rundle-Thiele, 2004).

Trust

Generally speaking, trust is the inter-personal reliance gained from past experience which requires a previous engagement on a person's account, and recognition and acceptance that risk exists (Luhmann, 1988). That is, trust is a rational, experience-based concept, which is created, reinforced or decreased by bilateral, relational activities in a series of encounters.

In a business context, trust can be an important prerequisite for commercial exchange. When goods are not traded on spot markets trust in business partners is necessary as to whether they keep their promises (i.e., deliver the ordered goods in the agreed quality and quantity; or that payments are made as agreed). Trust has become increasingly important during the last decades given that commercial transactions nowadays take place in a global context. That is, business parties may not know each other personally and completely new trading infrastructures (e.g., e-commerce platforms) have emerged. Furthermore, products have become increasingly complex (e.g., the rising significance of 'credence' attributes for food products) implying increased information asymmetries between producers and consumers.

In business relationships trust is of a different nature to commitment and satisfaction which are rooted in psychological equity theory. Trust may be neither a relational 'input' nor 'outcome' but it has been characterized as a 'safeguard mechanism' (Dyer and Singh, 1998), serving as an efficient facilitator for the involved parties to receive what they expect from the relationship. The existence of trust between exchange parties may not strictly be necessary since other safeguard mechanisms such as contracts can be used.

Nevertheless, in collaborative inter-organizational relationships, trust is considered a crucial component (Beth *et al.*, 2006; Svensson, 2005), mostly because a lack of trust can have severe cost implications (see Chapter 3). If business partners can trust each other, contractual arrangements may be reduced or avoided, thereby implying lower costs (Chiles and McMackin, 1996). In particular, transaction cost economics suggests that trust can lower opportunistic behaviour and hence exchange and agency costs. Chen (2000) shows that trust is widely relied on in transactions involving relatively low monetary value and considerable resources are sometimes used in structuring contracts when the transactions involved have a relatively high monetary value.

Given these cost implications, trust is frequently defined as a willingness to take risks (Mayer *et al.*, 1995). Trust is warranted when the expected gain from placing oneself at risk by another is positive, and the decision to accept such a risk is taken to imply trust (Williamson, 1993). Trust in business relationships therefore requires accepting the danger of relatively small financial losses in order to avoid comparatively large costs which would arise from hedging against these losses. One way to hedge the risk of financial losses in business relationships can be the use of contracts. This implies that trust and contracts are substitutes. However, when the assets at stake are high (i.e., the potential financial losses large), business partners may prefer a contract even if, in principle, they trust each other. With contract costs then being small relative to the involved sums – and given these large sums, incentives to behave opportunistically are high – a 'better-safe-than-sorry' strategy may be the most rational option. In such a case, trust and contracts become complements. More specifically, since most contracts are incomplete, trust and contracts sometimes must co-exist, with the former being the more important the less complete the latter are.

Hence, trust in business relationships relates to the belief in the ability of a business partner to fulfil his/her business commitments and thus to be able to obtain the expected rewards. In order to assess this ability, business people often make inferences, based on previous experiences with a partner. Thus judgements are usually made about a business partner's honesty, integrity, sincerity, competence, reliability to keep promises, his/her concerns/considerations about other partners' interests, or his/her general responsibility.

Relationship stability

A direct assessment of the stability of a relationship requires acceptably accurate records of past transaction/interaction outcomes. In practice, such records rarely exist. Asking business people to recall complete transaction/interaction episodes would be unrealistic, in particular if relationships last for a long period of time, are characterized by high transaction/interaction frequency, and/or are handled by several staff simultaneously (as done in purchasing departments of large corporations). For these reasons, an indirect and more feasible way to assess relationship stability, and one which is conceptually similar to the one used for relationship quality, needs to be applied.

In the marketing literature, a comprehensive set of studies focus on relationship quality or chain performance issues. However, some aspects of inter-organizational relationships are commonly neglected, such as the degree of dependence between exchange partners, previous interaction episodes, and the susceptibility of relationships to conflicts – hence, stability issues. Other studies, found mainly in the management literature, often solely deal with the stability of relationships. From a systematic review of these studies typical components of relationship stability can be identified. Among the most commonly used are: mutual dependence (arising, for instance, through the existence of switching costs or asset specificity); conflict resolution capacity (the ability to endure and solve relationship problems); and positive collaboration history (the assumption that a relationship which has been successful in the past may be prolonged and continued).

Mutual dependence

Mutual dependence builds upon bi-directional perceptions of how individuals or organizations feel they rely on their exchange partner(s) to achieve a goal. Mutual dependence fosters collaborative attempts to create a win-win situation for the involved parties (Svensson, 2002). Therefore, inter-organizational relationships have been shown to move away from autocratic hierarchical structures to more cooperative partnerships characterized by mutual dependence between the exchange partners (Cox and Makin, 1994; Hu and Watkins, 1999).

The creation of mutual dependence can occur in multiple ways. Compatible technical standards, the adaptation of processes and know-how sharing can create mutual dependence between partners, as well as the existence of personal bonds or contracts (Hakansson and Snehota, 1995).

Mutual dependence involves switching costs for exchange partners when a relationship is terminated. Metge and Weiss (2008) differentiate switching costs into monetary and non-monetary (psychological) costs. Monetary costs arise when mutual financial dependence exists which was created through idiosyncratic investments which represent sunk costs (Burnham *et al.*, 2003). Non-monetary switching costs result from terminating personal relationships, and searching for new partners (Metge and Weiss, 2008). For instance, long-existing family businesses may have built personal bonds to their suppliers or buyers. Switching may then be perceived as an emotional barrier. Moreover, the necessity of gathering information and educational requirements for building a new relationship may represent a non-monetary barrier (Bhattacharya and Bolton, 2000).

Hence, monetary and non-monetary switching costs create powerful barriers towards disruption or termination of relationships even in critical relationship phases. In this way, higher degrees of mutual dependence are thought to increase the stability of business relationships.

Conflict resolution capacity

Conflicts can be understood as unsettled differences in opinions, leading to dispute between partners. Conflicts in business relationships arise from product or process problems, dependencies on mutual assets, non-conciliatory partner networks or unforeseen events in the natural or social environment (Peck, 2005). The intensity and frequency of these conflicts can vary depending on the nature of the relationship. More important is the effectiveness of how problems are handled – i.e., the degree of existing conflict resolution capacity (Anderson and Narus, 1990).

Five different methods of conflict resolution are discussed in the management literature (McKenna and Richardson, 1995): (i) competitive – dominating the partner for one's own needs; (ii) neglecting – avoiding the fact that a conflict exists; (iii) accommodation – appeasing the partner while maximizing one's own benefits; (iv) compromise – sharing what is available without finding a solution; and (v) collaborative behaviour – finding together a solution. Conflict resolution capacity also depends on national and business cultures, such as differences in dealing with uncertainty (Mello and Stank, 2005).

For the stability of inter-organizational relationships, collaborative behaviour is essential. If conflicts are not solved jointly or are neglected, the relationship may be terminated prematurely. Buyers and suppliers who recognize their mutual dependence can develop cooperative strategies to resolve conflicts more effectively than in relationships with competitive behaviour (Wong *et al.*, 1999). In this regard, Sachan *et al.* (2005) review agri-food chain relationship in India and identify powerful intermediaries as a source of conflict leading to high transaction costs. They conclude that the most sustainable relationship models are characterized by collaborative behaviour leading to lower supply chain costs for all stakeholders.

Positive collaboration history

Collaboration can be indirect or direct. Organizations may interact indirectly by sharing a common infrastructure (e.g., market), or collaborate directly through establishing contact with a potential partner and receiving a reaction. Here, the focus is on direct interactions.

Collaboration history comprises all positive and negative experiences made with the exchange partner and is used as a basis for deciding on future actions with the exchange partner. When companies transact for the first time, they commonly have no experience with an exchange partner and are limited in their evaluation possibilities (e.g., with regard to a partner's trustworthiness). While this may not be important for arm's-length transactions, such as in spot markets, adverse selection can cause severe hold-up problems at critical phases in longer-term relationships. Companies which have a positive collaboration history are characterized by commercially rewarding transactions, for all involved stakeholders, successful productive endeavours and critical phases which have been endured and successfully resolved. Thus, a positive collaboration history contributes to the stability of relationships by reducing the probability of partners switching to other buyers or suppliers (Bejou *et al.*, 1996; Anderson and Weitz, 1989).

Life-cycle models describe the typical development history of relationships. In theory, good relationships could continue endlessly as long as they are perceived as being rewarding and conflicts are resolved. However, more realistically, life-cycle models assume some kind of termination stage for a relationship. The models typically consist of three to five different stages, which include a 'birth' and a 'death' stage (Dwyer *et al.*, 1987). These models reflect an ideal and linear development of a relationship. In practice, relationships may not necessarily follow this strict progression.

The real life-cyle of a business relationship comprises routine and critical episodes (Storbacka *et al.*, 1994). Routine episodes are phases where behaviour and processes are mostly standardized and formalized. These episodes require only a low degree of involvement of partners (Storbacka *et al.*, 1994). In contrast, critical episodes are highly disruptive and difficult to handle. Organizations in stable, long-lasting relations usually can look back at a collaboration history in which critical episodes have been outlived and which is characterized by the existence of routine problem-solving processes.

The full SIR construct

In summary, sustainable inter-organizational relationships can be described as *high-quality* and *stable* relations which are responsive to changing environments and which business partners continue as long as the benefits derived from a relationship outweighs the costs of maintaining it. Relationship quality reflects the overall mean outcome of a sequence of commercial transactions/interactions which make up a B2B relationship. This static 'level' relationship component consists of inter-personal factors, such as commitment to, satisfaction with and trust in a business partner. Relationship stability considers the nature of the transaction/interaction outcomes over time, as reflected in the existence of mutual dependence, conflict resolution capacity and a positive collaboration history among business partners. Relationship quality and stability are interrelated and both are essential to make business relationships sustainable. Figure 2 depicts the components of the aggregate multidimensional SIR construct graphically.

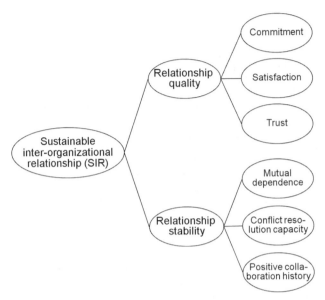

Fig. 2. The SIR construct.

Conclusion

This chapter has presented a theoretical discussion of sustainable inter-organizational relationships. Without an accepted definition of SIRs, as a basis for a valid, reliable and feasible measurement procedure, no thorough understanding of the relationship situation of either individual organizations or at the aggregate industry level can be obtained. Moreover, without such understanding, successful improvement of the status quo can hardly be possible. In general, a complete investigation will also need to take into account the characteristics of the involved relationship parties, and consider country-, commodity- and chain stage-specific particularities. Only then it would be feasible to derive meaningful recommendations for either improving business management practices or policy decisions for industry development.

The actual measurement of SIRs can be done in several ways. If large datasets obtained from standardized enterprise surveys are available, statistical methods such as structural equation modelling may be used. For the European agri-food sector, structural equation modelling involving the SIR construct has been applied successfully. Results from these empirical studies are reported in Chapter 7, Fischer *et al.* (2008, 2009) and Reynolds *et al.* (2009). In case a data-sparse (quantitative) case study approach is warranted for a particular analysis, the extent of the sustainability of the investigated business relationships may still be calculated and compared across study units, using an index method. This latter approach may also be used for larger datasets for benchmarking purposes, i.e., when SIR scores need to be compared between sub-groups such as different countries, agri-food chains or chain stages.

Further research may look at the issue of whether the subjective, self-reported measures commonly used to quantify the SIR construct could be replaced by more objective ones, such as automated measurements or those obtained from external observers.

References

Adams, J. (1965) Inequity in social exchange. In: Berkowitz, L. (ed) *Advances in Experimental Social Psychology*. Academic Press, New York, pp. 267–299.

Anderson, E. and Weitz, B. (1989) Determinants of continuity in conventional industrial channel dyads. *Marketing Science* 8(4), 310–323.

Anderson, J. and Narus, J. (1990) A model of distributor firm and manufacturer firm working partnerships. *Journal of Marketing* 54(1), 42–58.

Backhaus, K. and Büschken, J. (1999) The paradox of unsatisfying but stable relationships – a look at German car suppliers. *Journal of Business Research* 46(3), 245–257.

Barnes, B., Naudé, P. and Michell, P. (2005) Exploring commitment and dependency in dyadic relationships. *Journal of Business-to-Business Marketing* 12(3), 1–26.

Baumeister, R. and Leary, M. (1995) The need to belong: desire for interpersonal attachments as a fundamental human motivation. *Psychological Bulletin* 117, 497–529.

Bejou, D., Wray, B. and Ingram, T. (1996) Determinants of relationship quality: An artificial neural network analysis. *Journal of Business Research* 36(2), 137–143.

Bennett, R. and Rundle-Thiele, S. (2004) Customer satisfaction should not be the only goal. *Journal of Service Marketing* 18(7), 514–523.

Beth, S., Burt, D., Capacino, W., Gopal, C., Lee, H., Porter Lynch, R. and Morris, S. (2006) Supply chain challenges: building relationships. *Harvard Business Review on Supply Chain Management*. Harvard Business School Press, Boston, 65–86.

Bhattacharya, C. and Bolton, R. (2000) Relationship marketing in mass markets. In: Sheth, J. and Parvatiyar, A. (eds) *Handbook of Relationship Marketing*. Sage Publications, Canada.

Blau, P. (1964) *Exchange and power in social life*. Wiley, New York.

Burnham, T., Frels, J. and Mahajan, V. (2003) Consumer switching costs: a typology, antecendents, and consequences. *Journal of the Academy of Marketing Science* 31(2), 109–126.

Carrell, M. and Dittrich, J. (1978) Equity theory: the recent literature, methodological considerations, and new directions. *Academy of Management Review* 3(2), 202–210.

Chen, Y. (2000) Promises, trust, and contracts. *The Journal of Law, Economics & Organization* 16(1), 209–232.

Chiles, T. and McMackin, J. (1996) Integrating variable risk preferences, trust, and transaction cost economics. *Academy of Management Review* 21(1), 73–99.

Cox, A. (2004) The art of the possible: relationship management in power regimes and supply chains. *Supply Chain Management – An International Journal* 9(5), 346–356.

Cox, C. and Makin, P. (1994) Overcoming dependence with contingency contracting. *Leadership & Organization Development Journal* 15(5), 21–26.

Das, T. and Teng, B. (1996) Risk types and inter-firm alliance structures. *Journal of Management Studies* 33(6), 827–843.

Das, T. and Teng, B. (2001) Relational risk and its personal correlates in strategic alliances. *Journal of Business and Psychology* 15(3), 449–465.

Delerue, H. (2004) Relational risks perception in European biotechnology alliances: the effect of contextual factors. *European Management Journal* 22(5), 546–556.

Dwyer, F., Schurr, P. and Oh, S. (1987): Developing buyer–seller relationships. *Journal of Marketing* 5(1), 11–27.

Dyer, J. and Singh, H. (1998) The relational view: cooperative strategy and sources of interorganizational competitive advantage. *Academy of Management Review* 23(4), 660–679.

Fischer, C., Hartmann, M., Bavorova, M., Hockmann, H., Suvanto, H., Viitaharju, L., Leat, P., Revoredo-Giha, C., Henchion, M., McGee, C., Dybowski, G. and Kobuszynska, M. (2008) Business relationships and B2B communication in selected European agri-food chains – first empirical evidence. *International Food and Agribusiness Management Review* 11(2), 73–99.

Fischer, C., Hartmann, M., Reynolds, N., Leat, P., Revoredo-Giha, C., Henchion, M., Albisu, L. and Gracia, A. (2009) Factors influencing contractual choice and sustainable relationships in European agri-food supply chains. *European Review of Agricultural Economics* 36(4), 541–569.

Geyskens, I. and Steenkamp, J.-B. (2000) Economic and social satisfaction: measurement and relevance to marketing channel relationships. *Journal of Retailing* 76(1), 11–32.

Giese, J. and Cote, J. (2000) Defining consumer satisfaction. *Academy of Marketing Science Review* 4(2).

GTZ (2007) ValueLinks Manual – The Methodology of Value Chain Promotion. Division 45 Agriculture, Fisheries and Food. Deutsche Gesellschaft fuer technische Zusammenarbeit, Eschborn, Germany.

Hakansson, H. and Snehota, I. (1995) *Developing Relationships in Business Networks*. Routledge, London and New York.

Hallikas, J., Virolainen, V.-M. and Tuominen, M. (2002) Risk analysis and assessment in network environments: A dyadic case study. *International Journal of Production Economics* 78, 45–55.

Hu, X. and Watkins, D. (1999) The evolution of trade relationships between China and the EU since 1980s. *European Business Review* 99(3), 154–161.

Humphrey, J. and Memedovic, O. (2006) *Global Value Chains in the Agrifood Sector*. United Nations Industrial Development Organization, Vienna, Austria.

Huseman, R., Hatfield, J. and Miles, E. (1987) A new perspective on equity theory: the equity sensitivity construct. *Academy of Management Review* 12(2), 222–234.

Ivens, B. (2004) How relevant are different forms of relational behaviour? An empirical test based on Macneil's exchange framework. *Journal of Business & Industrial Marketing* 19(5), 300–309.

Jenkins, B., Akhalkatsi, A., Roberts, B. and Gardiner, A. (2007) *Business Linkages: Lessons, Opportunities, and Challenges*. IFC (International Finance Corporation), International Business Leaders Forum, and the Kennedy School of Government, Harvard University.

Kanter, R. (1994) Collaborative advantage: the art of alliances. *Harvard Business Review* July–August, 96–108.

Kleinaltenkamp, M. and Plinke, W. (1997) *Geschäftsbeziehungsmanagement*. Springer, Berlin.

Lambert, D. and Knemeyer, A. (2006) We're in this together. *Harvard Business Review on Supply Chain Management*. Harvard Business School Press, Boston: 1–22.

Lank, E. (2006) *Collaborative Advantage: How Organizations Win by Working Together*. Palgrave Macmillan, Hampshire, UK.

Law, K., Wong, C.-S. and Mobley, W. (1998) Towards a taxonomy of multidimensional constructs. *Academy of Management Review* 23(4), 741–755.

Luhmann, N. (1988) Familiarity, confidence, trust: problems and alternative. In: Gambetta, D. (ed) *Trust*. Basil Blackwell, New York, pp. 94–107.

Mayer, R., James, D., Schoormann, H. and David, F. (1995) An integrative model of organizational trust. *Academy of Management Review* 29(3), 709–734.

McKenna, S. and Richardson, J. (1995) Business values, management and conflict handling: issues in contemporary Singapore. *Journal of Management Development* 14(4), 56–70.

Mello, J. and Stank, T. (2005) Linking firm culture and orientation to supply chain success. *International Journal of Physical Distribution & Logistics Management* 35(8), 542–554.

Metge, J. and Weiss, P. (2008) Entry deterrence in markets with endogenous consumer switching costs. In: Shinnick, E. (ed) *New Public Finance and Market Issues*. INFER Research Perspectives, Vol. 3. LIT-Verlag, Berlin.

Morgan, R. and Hunt, S. (1994) The commitment–trust theory of relationship marketing. *Journal of Marketing* 58(3), 20–38.

Mouzas, S., Henneberg, S. and Naudé, P. (2007) Trust and reliance in business relationships. *European Journal of Marketing* 41(9/10), 1016–1032.

Nooteboom, B., Berger, H. and Noorderhaven, N. (1997) Effects of trust and governance on relational risk. *Academy of Management Journal* 40(2), 308–338.

Peck, H. (2005) Drivers of supply chain vulnerability: an integrated framework. *International Journal of Physical Distribution & Logistics Management* 35(4), 210–232.

Plinke, W. (1989) Die Geschäftsbeziehung als Investition. In: Specht, G., Silberer, G. and Engelhardt, W. (eds) *Marketing-Schnittstellen*. Schaeffer-Poeschel, Stuttgart, pp. 305–325.

Reynolds, N., Fischer, C. and Hartmann, M. (2009): Determinants of sustainable business relationships in selected German agri-food chains. *British Food Journal*, 111(8), 776–793.

Roberts, K., Varki, S. and Brodie, R. (2003) Measuring the quality of relationships in consumer services: an empirical study. *European Journal of Marketing* 37(1/2), 169–196.

Rychlak, J. (1984) Relationship theory: and historical development in psychology leading to a teleological image of humanity. *Journal of Social and Personal Relationships* 1, 363–386.

Sachan, A., Sahay, B. and Sharma, D. (2005) Developing Indian grain supply chain cost model: a system dynamics approach. *International Journal of Productivity and Performance Management* 54(3), 187–205.

Selnes, F. (1998) Antecedents and consequences of trust and satisfaction in buyer–seller relationships. *European Journal of Marketing* 32(3/4), 305–322.

Storbacka, K., Strandvik, T. and Grönroos, C. (1994) Managing customer relationships for profit: the dynamics of relationship quality. *International Journal of Service Industry Management* 5(5), 21–38.

Svensson, G. (2002) The measurement and evaluation of mutual dependence in specific dyadic business relationships. *Journal of Business & Industrial Marketing* 17(1), 56–74.

Svensson, G. (2005) Mutual and interactive trust in business dyads: condition and process. *European Business Review* 17(5), 411–427.

Webber, M. (2008) *Using Value Chain Approaches in Agribusiness and Agriculture in Sub-Saharan Africa – A Methodological Guide*. The World Bank, Washington DC, USA.

Wetzels, M., de Ruyter, K. and van Birgelen, M. (1998) Marketing service relationships: the role of commitment. *Journal of Business & Industrial Marketing* 13(4/5), 406–423.

Will, M. (2008) *Promoting Value Chains of Neglected and Underutilized Species for Pro-Poor Growth and Biodiversity Conservation – Guidelines and Good Practices*. Global Facilitation Unit for Underutilized Species, Rome, Italy.

Williamson, O. (1985) *The Economic Institutions of Capitalism – Firms, Markets, Relational Contracting*. Free Press, New York.

Williamson, O. (1993) Calculativeness, trust, and economic organization. *Journal of Law & Economics* 36, 453–486.

Wong, A., Tjosvold, D., Wong, W. and Liu, C. (1999) Relationships for quality improvement in the Hong Kong-China supply chain. *International Journal of Quality & Reliability Management* 16(1), 24–41.

Young, L. and Denize, S. (1995) A concept of commitment: alternative views of relational continuity in business service relationships. *Journal of Business & Industrial Marketing* 10(5), 22–37.

Zsidisin, G. (2003) A grounded theory of supply risk. *Journal of Purchasing & Supply Management* 9, 217–224.

Part II

Empirical Evidence on Trust and Sustainable Relationships in Agri-food Chains

Chapter 5

Trust and Relationships in Selected European Agri-food Chains

Philip Leat,[1] Maeve Henchion,[2] Luis Miguel Albisu[3] and Christian Fischer[4]

[1] Scottish Agricultural College (SAC), Edinburgh, United Kingdom
[2] Ashtown Food Research Centre, Teagasc, Dublin, Ireland
[3] Agri-food Research and Technology Center of Aragón (CITA), Zaragoza, Spain
[4] Massey University, Auckland, New Zealand

Introduction

Growing consumer expectations on the safety and quality of food, along with increasing regulatory requirements and intensifying competition, have encouraged many European agribusinesses to reorganize in integrated chains or networks. These structures imply increased mutual dependence and potentially add not only a new dimension to the overall performance of the chain, but also to the risk of business failure, since the performance of a whole chain/network might be jeopardized by a single chain partner.

A key issue for chain/network performance is that if business partners can have trust in each other, contractual arrangements may be reduced or avoided, thereby reducing business costs (Chiles and McMackin, 1996; Sodano, 2002) and securing competitive advantage. In particular, transaction cost theory has argued that trust has the important effect of lowering opportunistic behaviour and hence exchange and agency costs (Suh and Kwon, 2003). Chen (2000) argues that trust is widely relied on in transactions involving relatively low monetary value and considerable resources are sometimes used in structuring contracts when the transactions involved have a relatively high monetary value.

Trust, of some level, is a prerequisite for exchange. It is the inter-personal reliance gained from past experience and it differs from confidence because trust requires a previous engagement on a person's account, recognizing and accepting that risk exists (Luhmann, 1988). That is, trust is a relationship-based concept, which is created, reinforced, or decreased by bilateral, relational activities in a series of economic and social exchanges. Trust, therefore, is distinct from something involving goodwill and leads to higher levels of loyalty towards an exchange partner (Lindgreen, 2003).

Trust can be interpreted from several perspectives depending on the discipline of the researcher. Thus, personality psychologists traditionally regard trust as a characteristic trait within individuals; social psychologists may focus on expectation about the behaviour of others in transactions; and economists and sociologists frequently emphasize the creation of institutions and incentives to reduce uncertainty and increase trust (Sahay, 2003; Yee *et al.*, 2005). Consequently, it is to be expected that there is no unique definition of trust. Nevertheless, it is frequently defined in a general sense as a willingness to take risk (Johnson-George and Swap, 1982; Kee and Knox, 1970; Mayer *et al.*, 1995; Williamson, 1993). Trust is warranted when the expected gain from placing oneself at risk by another is positive, and the decision to accept such a risk is taken to imply trust (Williamson, 1993). Dapiran and Hogarth-Scott (2003) perceive it as a coordinating mechanism, based on shared norms and collaboration within an uncertain environment. Trust has also been conceptualized as involving the sub-dimensions of honesty, benevolence and competence (Anderson and Narus, 1990; Geyskens *et al.*, 1998).

In a similar manner, a variety of management and economics researchers have identified several different types of trust. Laeequddin *et al.* (2009) distinguish between: characteristics-based trust such as reliability, dependability, credibility, commitment, honesty, benevolence, fairness, goodwill, etc.; rational trust embracing the economics of relationships, the capabilities of partners and their technology; and institutional trust which derives from legal frameworks, commercial law, control systems, and agreements and contracts, etc. Ghosh and Fedorowicz (2008) focus on calculative trust (an ongoing assessment of the benefits and costs associated with a relationship); competence trust (the ability of chain member(s) to perform their required tasks); trust in integrity (a partner makes agreements in good faith, is truthful and fulfils promises); and trust in predictability (a trustee's actions are consistent and can be forecasted in a given situation). Similarly, Lindgreen (2003), drawing on Johnson and Grayson (2000), identifies four types of trust: generalized trust which is dictated by general shared norms of behaviour and enforced by social mechanisms; system trust which is written down in rules and is controlled by legislative and regulatory institutions; process-based trust arising from repeated interactions and their history, and is strengthened over time; and personality-based trust which is determined by the personality traits of the individuals concerned.

There are clearly similarities in these categorizations and there is some overlap in the ideas they embrace: characteristics-based trust resembles personality-based trust and trust in integrity, and may be strengthened by generalized trust; rational trust embraces competence trust and relates closely to calculative trust; institutional trust is similar to system trust when regulations are incorporated, and may help in furthering competence trust; trust in predictability can be considered as deriving from process-based trust.

This chapter explores the role of trust in selected agri-food chains in four EU countries. It is based on a qualitative assessment of expert interviews. After this introduction, the following section describes the methodology used. Thereafter, the main results are presented and discussed, before conclusions are drawn at the end.

Methodology

Secondary research, i.e., using existing literature and data, was undertaken to review and evaluate the four study countries in terms of social, cultural, economic and political factors (potentially) influencing business relationships and communication, with a focus on the food chain environment of a number of selected product sectors. The countries analysed were Germany, Ireland, Spain and the UK, selected on the basis of the economic relevance of their agri-food sectors and because of their diverse geographic locations. The selected agri-food chains were barley-to-beer, cattle-to-beef, pigs-to-pigmeat and pigmeat products, and cereals-to-bread. The product sectors were selected according to their importance to their respective national economies. Importance was determined on the basis of factors such as contribution to national agri-food output, export orientation and growth.

Primary data collection, involving expert interviews during summer/autumn 2005, was undertaken to complement the secondary research. Face-to-face interviews were the preferred data collection method, however the telephone was used where the preferred method was impracticable for logistical reasons, or not possible due to constraints on the part of the interviewee. A semi-structured interview guide was used to explore the chain environment, the nature of chain relationships and the influences on their development, the nature and importance of chain communication, the role of institutions in chain regulation and development, and features of chain performance. On average each interview lasted an hour.

Expert interviews were the preferred survey method because they allow issues to be explored in more detail with participants, who are encouraged to give the fullest possible answer. This interviewing technique may be more time consuming than other survey methods, samples are generally smaller and results are more difficult to be summarized and interpreted, but it offers some important advantages. It is a powerful tool for extracting information, in particular of a qualitative nature. With a semi-structured interview the interviewer is still free to explore, probe and ask questions deemed interesting to the researcher while permitting the interviewer to keep the interview within the parameters traced out by the aim of the study (Berry, 1999).

Interviewees were selected because of their ability to comment on the organization and functioning of their sectoral chains. In total 28 experts were interviewed. These were largely senior executives or directors of representative or trade associations, and some were senior personnel in significant enterprises in the respective countries.

Whilst the interviews produced a wide range of useful information, the focus here is on the role of trust within agri-food chains. The interview transcripts were evaluated by generating systematic assessments of the statements obtained, backed up by the secondary-research findings mentioned above.

Results

The results of two chains are reported for each country, with each product sector examined in two countries with the exception of cereals-to-bread, which is only examined in Spain, and pigs-to-pigmeat and pigmeat products, which is examined in three countries.

Germany – pigs-to-sausage chain[1]

Pig farmers in general exhibit a 'healthy distrust' attitude towards their business partners. Upstream, their suspicion towards feed suppliers results from their perception of being

[1] The key informants for this chain were: Michael Starp, Managing Head of Department for Livestock and Meat, German Farmers' Association (DBV); Thomas Vogelsang, Managing Director, German Association of Meat Product Manufacturers (BVdF); Angela Schillings-Schmitz, Meat Sector Specialist, European Retail Institute (EHI).

'wooed' by commercial feed agents. The main reason for the lack of trust regarding their downstream partners is the limited transparency regarding the grading of fattening pigs and the related invoicing. Processors, in general, fear that they are exposed to asymmetric information regarding the quality of fattening pigs, since farmer suppliers know more about how the pigs were fed.

With respect to distributors, processors' trust in them depends on their overall reputation. For retailers, the political environment regarding health risks and food safety has become more demanding as product liability regulation is enforced more strictly. Thus private label sellers generally aim to reduce liability risks by tending to prefer 'control' over 'trust' regarding their product manufacturers. As a result, suppliers are increasingly under pressure to comply with the quality specifications of retailers, which does not help to improve strained relationships. Overall, distrust seems to be distributed evenly across sausage chain actors.

Germany – barley-to-beer chain[2]

Malting barley farmers in general, rate the importance of personal relationships highly. In particular, personal relationships between farmers and barley traders have led to joint business activities. For instance, besides trading barley, farmers arrange the joint acquisition of costly machinery with traders (i.e., both sides engage in specific asset investments). Overall, farmers' trust in maltsters varies depending on the latter's size, past experience and existing relationship management initiatives. As brewers experience a decrease in profits, due to an increase in production costs and a reduction in real beer prices, trust is strained by financial worries.

While upstream 'healthy distrust' exists, brewers trust food retailers least. For instance, under-priced kegs sold by food retailers to catering enterprises can undermine existing beer delivery contracts between brewers and pubs. As beer wholesalers in general do not try to bypass the existing delivery contracts, processors trust them more. For distributors and brewers alike, beer has become an uncertain market as per capita consumption has decreased, and global and domestic players have started acquiring brewers and outlet channels in Germany. Therefore, currently, comparatively high levels of distrust exist in the whole chain. However, contrary to the pig-to-sausage chain, overall it appears that trust is more prevalent and personal relationships are more significant upstream.

Ireland – cattle-to-beef chain[3]

The business environment for Irish beef has changed dramatically in recent years with significant implications for farmer–processor relationships in particular. Historically, Irish beef producer–processor relationships were transaction-based, with price being the main transaction issue due to a commodity-market orientation arising from high levels of

---

[2] The key informants for this chain were: Herbert Siedler, Head of Section, Bavarian State Department of Agriculture and Forestry, Wuerzburg; Roland Demleitner, Managing Director, German Association of Middle-Sized Private Breweries; Günther Guder, Managing Director, German Federal Association of Beverage Wholesalers.

[3] The key informants for this chain were: Kevin Kinsella, Executive Secretary to Beef Committee, Irish Farmers' Association; Cormac Healy, Meat Industry Executive, Meat Industry Ireland; Pat Brady, Chief Executive, Irish Association of Craft Butchers; Michael Murphy, Manager of International Markets, Bord Bia.

Common Agricultural Policy (CAP) support. However, with CAP reform and the need to develop alternative outlets to intervention, there is evidence of closer relationships developing between producers and processors as a means of accessing specific commercial markets. These new relationships are based on verbal agreements, covering product specifications including production methods, e.g., feeding and management regimes, and price. Specific investments are generally involved in these new closer relationships and may include financial investments in housing and feeding equipment by a farmer, and investment in providing management and nutritional advice to the farmer on the part of the processor. The type of investment undertaken here is such that switching costs are not very high and thus relationships are not contractual in nature. Rather, they take the form of verbal agreements and depend on previous experience, involving mutually satisfying past exchanges. Favourable past experiences reduce uncertainty and perceived risk and facilitate the development of trust to a level sufficient to induce the parties to undertake these investments and assume the associated risk. Whilst unbalanced in terms of level of risk undertaken by the different partners, these specific investments create a degree of mutual dependence and trust, and allow the processor to develop sustainable markets abroad. Approximately 10% of Irish beef slaughterings are now based on such types of relationships. This figure is slowly increasing, however, there is still quite a high level of mistrust between farmers and processors in the sector, with occasional allegations by farmers of price collusion between processors.

Processor–retailer relationships involving multiple retailers are generally quite exclusive and are frequently called 'partnerships' by the trade press and the parties involved. The retail sector in Ireland and in the UK, Ireland's main export destination for beef, is highly concentrated, which means that such partnerships are generally directed by the multiple retailers who have considerable power, i.e., they act as the chain captain. Switching costs in such relationships tend to be high, as processors generally adapt to retailers' requirements and make specific investments to maintain and develop the relationships. Relationships involving independent retail outlets generally involve processors of much smaller scale than those dealing with multiple retailers. Such relationships are usually long-term in nature, with personal relations having a significant influence in supporting trust.

Ireland – pigs-to-pigmeat chain[4]

Producer–processor relationships are characterized by a lack of trust in the Irish pigmeat sector. A lack of price transparency is one of the reasons for this. Whilst pigs are sold to the slaughterhouse on a deadweight grading basis (with lean meat percentage within a specified weight range being the grading criteria), significant variation occurs in the net delivered price relative to the base price for pigs. Greater transparency could be achieved by increasing the base price and reducing the size of bonuses, as the latter can be altered by the slaughterhouse without any reference to market conditions. These relationships are also characterized by an absence of contracts. Transaction costs associated with enforcement are a significant barrier to contracts. Scarce supplies mean that processors cannot afford the

---

[4] The key informants for this chain were: James Brady, Executive Secretary to Pigmeat Commission, Irish Farmers' Association; Brendan Lynch, Head of Pig Advice & Research, Teagasc; Michael Barry, Meat Industry Executive, Irish Association of Pigmeat Processors.

negative publicity associated with enforcing contracts. Another reason is the high reliance of the industry on the domestic and UK markets (almost 90% of output is destined for these two markets) which have broadly similar requirements. Thus, unlike the beef industry, there is no need for contract production and associated specific investments to suit particular markets.

In relationships between processors and retailers/caterers there are moderate to fairly high degrees of trust, stable networks and regular and repeated streams of transactions. The large volume traded by the big retailers creates a mutual dependence, where the processor becomes dependent on a few large customers, whilst the retailer depends on its preferred supplier for safe, reliable and timely deliveries. None the less, it appears that the processor may be more dependent on the retailer than vice versa. This is because the processor will have invested in specific assets, including possibly transaction-specific equipment and dedicated assets (particularly in the case of the retailer's private label products), whilst the retailer will not have made reciprocal investments to the same extent.

Retailers regard their relationships with processors as close and quite open. Some retailers share information on end users, based on their sales data, with processors, especially in cases where processors supply products to the retailer's private label range. Information flow may also be high where value-added products are being produced, as dealing in value-added pork products typically requires the processor to conform to retailers' specific quality assurance schemes and to invest in retailer-specific processing and product development. The trust and commitment in the relationship can ensure that such transaction-specific investments are carried out, the effect of which is higher switching costs.

Spain – pigs-to-cured ham chain[5]

In the pigs-to-cured ham chain in Spain, reputation is an important source of trust, and has particularly become more relevant upstream, at the producer and processing level. A high percentage of pig farmers are integrated with feed suppliers and processors, who finalize their production activities, either for piglets or finished animals for slaughtering, under very precise and strict conditions. The contracts involved include penalty clauses if pig farmers break the stipulated conditions, so contracts replace trust among supply chain partners. Pig farmers rely on big companies, to the extent that they are able to continue enforcing their contractual arrangements and prices when market conditions are tight. Those pig farmers, not involved in vertical integration, trust their processors because there is a widely available information system in Spain, which adds a great deal of transparency to the market. The outbreak of livestock-related diseases, and their possible transfer to humans, has underlined the necessity for a reliable system to trace individual meat products back to their animal of origin. Many large processors are vertically integrated with pig producers to ensure traceability of the animals. Meat industry pressures for assurance on traceability and production techniques will help to promote confidence in the integrity and origin of their products.

Processors believe that their relationship with pig producers is more stable and trust driven than their relationship with the retail sector. The bargaining power of retailers affects

[5] The key informants for this chain were: José Oliván, General Manager, Llograsa, S.A.; Isidro Martín, General Manager, Turolense Ganadera, S.A.; José María Rubio, Meat Sector Manager, Sabeco, S.A.

the relationship with processors, who feel pressured to comply with retailer demands. The retailing sector is driven by economic factors. Retailers seek to offer a homogeneous product to their customers and therefore maintain stable relationships with a reduced group of suppliers. Retailers trust the integrity of this selected group of suppliers. New food safety and traceability measures have increased coordination along the chain, favouring a process of vertical integration and coordination. Also, the domestic market shows a clear upward trend in the demand for higher quality and higher priced cured hams. All actors in the chain seek to produce a marketable product complying with the stipulated safety and quality standards. This common goal helps to build commitment and trust between partners in the chain.

Spain – cereal-to-bread chain[6]

The cereal-to-bread supply chain in Spain is very fragmented at the producer level, which complicates communication and business relationships between farmers and processors. Spanish production of wheat is highly variable in terms of quantity, but also in terms of quality. Climate variability is one factor influencing the lack of homogeneity in the final product, however, another is the low level of certified seed use in Spain. Such variability affects the market negatively and creates a climate of insecurity for the milling industry. In the last few years there has been increased integration of farmers into cooperatives, which are also further integrated in second-tier cooperatives (groupings of affiliated cooperatives) as a way to assure product quality and quantity.

Processors tend to have stable suppliers, with trust-based exchanges being important. The lack of homogeneity of Spanish wheat production requires steady relationships between millers and a reduced number of producers to assure the quality of wheat produced. Interactions between big importing companies and millers are more distant; personal inter-relationships become less relevant and exchanges are bound by written contracts. Bakers seek to assure continuity of wheat flour supply. Since profit margins are very low all along the chain, personal relationships become important for bakers to establish verbal agreements with millers. The increasing size of the retail sector makes the relationship between bakers and retailers more impersonal, and trust shifts from being interpersonally constructed to more based on reputation.

UK – malting barley-to-beer chain[7]

In the UK barley-to-beer chain the business environment is characterized by considerable uncertainty. Grain prices can fluctuate markedly and the chain operates within a very competitive international trading environment. Excess malting capacity exists in Europe (80% utilization), and barley typically represents 60–66% of malt production costs (SAC, 2004). Businesses operate on very tight margins at all levels. In this environment, maltsters and brewers seek supply security, cost predictability and assured quality in their purchasing

---

[6] The key informants for this chain were: Rolando Pola, Cereal Sector Manager, Arento (cooperative); José Villamayor, General Manager, Harineras Villamayor, S.A.; Mario Moreno, General Manager, Panishop, S.A.

[7] The key informants for this chain were: Chris Barnes, Manager – Cereals Industry Forum, Home Grown Cereals Authority; Bill Dobson, Strategic Sourcing Manager, Coors Brewers; Pamela Bates, Senior Policy Adviser – Pubs & Leisure, British Beer and Pub Association; Elaine McCrimmon, Senior Policy Adviser – Brewing, British Beer and Pub Association.

activities. Consequently, they issue contracts to upstream suppliers to secure a significant proportion of their raw materials.

The chain involves close personal relationships at all stages. Although contracts are issued by grain merchants and cooperatives (to secure 30–60% of maltsters' requirements), trust in the competence and integrity of the grower is very important, because the transaction costs associated with contract enforcement are significant. Thus trust and the use of contracts are complementary phenomena. Generally speaking, farmers exhibit limited trust in the prices offered by maltsters. This is largely due to the sector's low profitability and limited transparency in the economics of the chain, or integration in its working. At the brewer–retailer end of the chain a degree of trust prevails, but contracts and supply agreements are used extensively to secure supplies, particularly of important brands.

UK – cattle-to-beef chain[8]

The UK beef chain also operates in a very competitive environment, in which there are significant imports (30% of supplies) and UK exports are very restricted. There is also marked excess slaughtering capacity. Whilst liveweight auction markets provide transparency to price setting, they cover only 23% of finished cattle marketings (MLC, 2005). The majority of cattle are sold direct to slaughterers on a deadweight basis (price depends on carcass weight and quality), whilst the retail market (70% of domestic market) is dominated by multiple retailers (>75% of sales). Large retailers seek competitive beef prices, appropriate quality for the segment concerned, supply continuity, efficient supply chain performance and innovation at reasonable cost.

There is considerable mistrust at the farmer–processor level, which centres on price. Farmers widely consider that multiple retailers are depressing their returns (e.g., capping prices by importing beef and taking disproportionately high margins). Such retailers believe, however, that many farmers do not understand the economics of the entire chain, and seek to drive chains in which all parties can benefit. Chains involving independent butchers generally have good personal relationships, considerable trust at all stages, and a reliance on spot market transactions, frequently with established suppliers. Multiple retailer chains are integrated through bilateral partnerships; retailers with slaughterer/processors, and slaughterer/processors with farmers through producer groups or clubs (Fearne, 1998). Such chains are highly directed and regulated by the retailers, which generates trust with respect to competence, but there are few contracts. Mutual benefit and necessity bind the partners. Contracts are most evident in the supply of caterers and food manufacturers, where supply continuity and cost predictability are important.

Discussion

Figure 1 summarizes the pattern of relationships and trust observed in the eight chains under review. Using 'Perceived trust levels' and 'Prevailing type of vertical chain

---

[8] The key informants for this chain were: Lisa Webb, Livestock Policy Manager, National Farmers' Union Scotland; John MacIntosh, Group Chairman and farmer, ANM Group Ltd; Alan McNaughton, Sales Director, McIntosh Donald (Grampian Country Food Group); Christopher Ling, Buying Manager (Beef and Lamb), Tesco; Duncan Sinclair, Economic Manager for Beef, Meat and Livestock Commission.

relationship' as axes, the relative positions of farmer–processor and processor–retailer interactions are depicted from the point of view of all considered stakeholders.

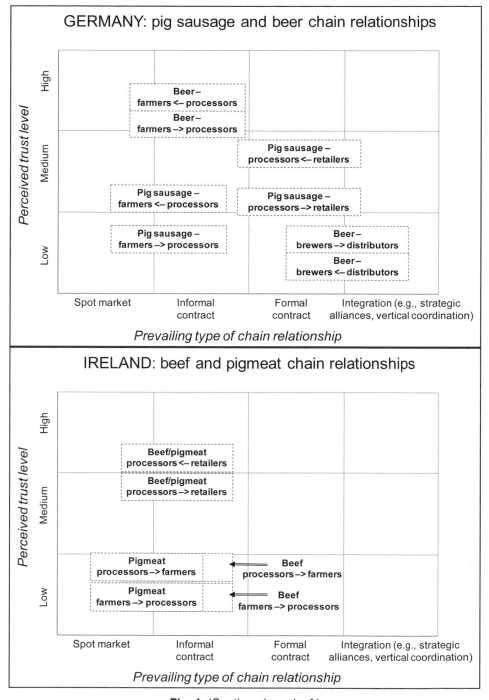

Fig. 1. (Continued overleaf.)

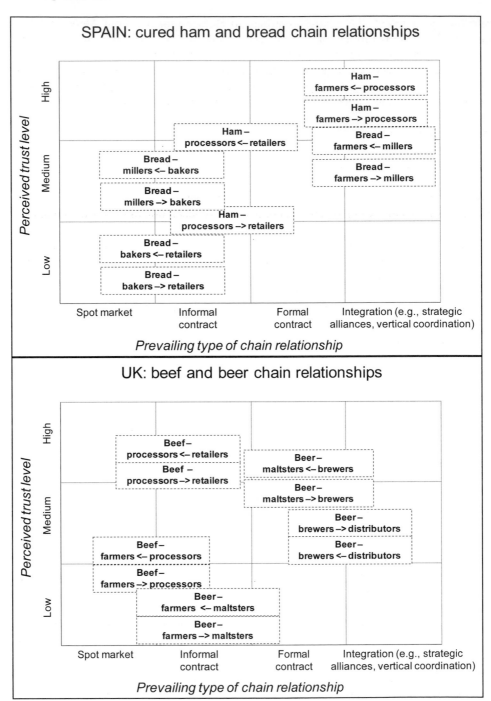

Fig. 1. Agri-food chain relationships and trust levels in eight analysed chains (direction of trust perception is indicated by arrow: e.g., 'farmers -> processors' indicates the trust of farmers in processors) actions are depicted from the point of view of all considered stakeholders.

The typology of vertical relationships types was adapted from Mohr and Nevin (1990).[9] We distinguish between spot market, informal (i.e., verbal) and formal (i.e., written) contracts and other, more integrated types of vertical relationships.

It should be recognized that these figures represent a generalization of the situations observed, and that there will be specific supply chains and relationships which fall outside the patterns presented. None the less, several general points are apparent.

Trust levels and types

Considerable mistrust is apparent at the farmer end of many of the observed chains, with numerous farmers having limited confidence in the fairness of their treatment by downstream customers (e.g., UK malting barley and beef, Irish beef and pigs, and German pigs). Several factors seem to play a major part in this generation of mistrust, including:

- price pressure in a very competitive international trading environment, which means that supply chain participants are competing for a share of the chain's margin which itself is under pressure (e.g., UK beef, Irish pigs and German beer);
- a lack of transparency in the quality achieved by agricultural products and the resultant price consequences, and of the financial pressures facing others in the chain (e.g., Irish and German pigs); and
- an imbalance in the scale and market power between farmers and many of the downstream supply chain participants. Small-scale production, combined with chain fragmentation, may create a feeling of insecurity amongst those concerned (e.g., Spanish wheat producers and millers, UK malting barley producers).

In the relationships between farmers and processors, generalized trust and characteristics-based trust are very important, but may be strengthened by various forms of system trust, for example, contracts such as those offered to malting barley growers in Britain, or the legislatively based traceability requirements imposed on the livestock industry. However, even in relationships where institutional or system trust is operating, the cost of enforcement or problem resolution may be such that characteristics-based trust, involving dependability and reliability, is still very important if not essential. As a result, for example, British grain merchants seek to offer contracts only to competent and reliable farmers, and German barley traders only engage in joint specific asset investment with farmers of high integrity and reliability. Similarly, having mutually satisfying past exchanges are important for Irish beef farmers engaging in specific asset investment with processors. This highlights the process-based aspect of trust, i.e., that it can be strengthened/weakened over time.

Further downstream the pattern of trust between chain participants is quite varied. System trust in the form of food safety and traceability regulation has driven considerable

---

[9] Other categorizations of vertical relationship types exist in the literature. Ménard (2004) argues that there may only be three main types (or 'governance structures'): markets, hybrids and hierarchies. Hybrids represent a collection of organizational forms, such as subcontracting, network of firms, franchising, collective trademarks, partnerships, cooperatives or alliances, which cannot be considered as unrelated supply–demand exchange mechanisms nor integrated corporate structures. Raynaud *et al.* (2005), building on work from Milgrom and Roberts (1992), classify 'spot market' exchanges and 'long-term relations' without written contracts as market governance structures; 'long-term relations with a qualified supplier', 'written contracts' and 'equity participation' as hybrid structures; and 'vertical integration' as firm governance structure.

chain integration in all countries, providing standards which set a minimum threshold for chain involvement. Major chain players, such as multiple retailers, have sought to add further supply conditions, narrowing their supply base and building closer relationships with relatively few suppliers. These developments might be regarded as assisting the development of trust in the competence and predictability of suppliers. Goodwill trust, where a party goes beyond contractual requirements to the mutual benefit of chain participants, is not generally apparent, although it may be constrained by inadequate communication on issues such as quality and price improvement, which also impairs calculative trust. Moreover, in very competitive chains, with margins under pressure, goodwill trust is likely to be suppressed by self-interest.

Trust and relationships

With respect to the nature of transaction relationships, it is evident that spot market transactions feature at all levels within the chains. For many farmers this is associated with maintaining their independence (this is particularly apparent in the UK and Germany). The findings suggest that while the development of trust based on personal relations can offer (transaction) cost advantages in business relationships, it may be complemented by formal contracts or other forms of integration as the involved stakes rise. The role of contracts in competitive environments may not be so crucial for those firms with extensive bargaining power, as they may be able to create partnerships which offer benefits such as a regular large volume outlet, and/or a share of a relatively good marketing margin. The exact level where the switch occurs from trust based on social relations to the use of formal contracts or institutional systems, depends on a variety of chain (business), cultural, social and personal characteristics.

Conclusions

More generally, our findings suggest that trust seems to be more pronounced in relationships involving SMEs, which are characterized by the existence of personal relationships between business partners. This is above all the case in the farm sector. However, as is clear for pig breeders, if a source of distrust exists, such as the lack of a clear definition of carcass quality and thus fair financial compensation, trust may not fully develop. Also, if the general economic situation is difficult, as is the case in many agri-food markets, the development of trust may be hampered, because all chain participants are struggling to command a share of a diminishing margin within the chain. Furthermore, if economic power is distributed unevenly in an agri-food chain, as is the case in Germany, Ireland, the UK and Spain where retailers dominate most chains, trust towards the more powerful may be limited. Finally, it is apparent that the dynamic nature of the agri-food market environment, and the generally low levels of information sharing within agri-food chains, acts to hinder the development of trust and relationships amongst participants.

Further research needs to use information on trust and relationships drawn from agribusinesses themselves. Industry experts are a valuable source of information. However, given that they are often obliged to offer 'official' and 'consensus' views, a richer picture (in terms of completeness and level of detail) may be obtained by surveying involved companies directly. Results from such an empirical in-depth analysis, based on a large survey dataset, are discussed in Chapter 7 of this volume.

Acknowledgements

This is an updated and revised version of an article which was previously published as: Fischer, C., Gonzalez, M., Henchion, M. and Leat, P. (2007) Trust and economic relationships in selected European agri-food chains. *Food Economics* 4(1), 40–49.

References

Anderson, J. and Narus, A. (1990) A model of distributor firm and manufacturer firm working partnerships. *Journal of Marketing* 54(1), 42–58.

Berry, R. (1999) Collecting data by in-depth interviewing. BERA 99 Conference Paper (The British Educational Research Association). Leeds, UK: Education-line. www.leeds.ac.uk/educ (accessed 10/05/2005).

Chen, Y. (2000) Promises, trust, and contracts. *The Journal of Law, Economics & Organization* 16(1), 209–232.

Chiles, T. and McMackin, J. (1996) Integrating variable risk preferences, trust, and transaction cost economics. *Academy of Management Review* 21(1), 73–99.

Dapiran, G. and Hogarth-Scott, S. (2003) Are co-operation and trust being confused with power? An analysis of food retailing in Australia and the UK. *International Journal of Retail & Distribution Management* 31(5), 256–267.

Fearne, A. (1998) The evolution of partnerships in the meat supply chain: insights from the British beef industry. *Supply Chain Management: An International Journal* 3(4), 214–231.

Geyskens, I., Steenkamp, J. and Kumar, N. (1998) Generalisations about trust in marketing channel relationships using meta analysis. *International Journal of Research in Marketing* 15(3), 223–248.

Ghosh, A. and Fedorowicz, J. (2008) The role of trust in supply chain governance. *Business Process Management Journal* 14(4), 453–470.

Johnson, D. and Grayson, K. (2000) Sources and dimensions of trust in service relationships. In: Swartz, T. and Iacobucci, D. (eds) *Handbook of Services Marketing and Management*. Sage, Thousand Oaks, California, pp. 357–370.

Johnson-George, C. and Swap, W. (1982) Measurement of specific interpersonal trust: construction and validation of a scale to assess trust. *Journal of Personality and Social Psychology* 43, 1306–1317.

Kee, H. and Knox, R. (1970) Conceptual and methodological considerations in the study of trust. *Journal of Conflict Resolution* 14, 357–366.

Laeequddin, M., Sardana, G., Sahay, B., Abdul Waheed, K. and Sahay, V. (2009) Supply chain partners' trust building process through risk evaluation: the perspectives of UAE packaged food industry. *Supply Chain Management: An International Journal* 14(4), 280–290.

Lindgreen, A. (2003) Trust as a valuable strategic variable in the food industry – different types of trust and their implementation. *British Food Journal* 105(6), 310–327.

Luhmann, N. (1988) Familiarity, confidence, trust: problems and alternatives. In: Gambetta, D. (ed) *Trust*. Basil Blackwell, New York, pp. 94–107.

Mayer, R., James, D., Schoorman, H. and David, F. (1995) An integrative model of organizational trust. *Academy of Management Review* 29(3), 709–734.

Ménard, C. (2004) The economics of hybrid forms. *Journal of Institutional and Theoretical Economics* 160, 345–376.

Milgrom, P. and Roberts, J. (1992) *Economics, Organization and Management*. Prentice Hall, Englewood Cliffs.

MLC (Meat and Livestock Commission) (2005) *A Pocketful of Meat Facts*. Milton Keynes.

Mohr, J. and Nevin, J. (1990) Communication strategies in marketing channels: a theoretical perspective. *Journal of Marketing* 54(4), 36–51.

Raynaud, E., Sauveè, L. and Valceschini, E. (2005) Alignment between quality, enforcement devices and governance structures in the agro-food vertical chains. *Journal of Management and Governance* 9(1), 47–77.

SAC (Scottish Agricultural College) (2004) Competitiveness of the UK Cereal Sector – Final Report to the Cereals Industry Forum.

Sahay, B. (2003) Understanding trust in supply chain relationships. *Industrial Management and Data Systems* 103(8), 553–563.

Sodano, V. (2002) Trust, economic performance and the food system: can trust lead up to unwanted results? In: *Paradoxes in Food Chains and Networks*. 5th International Conference on Chain and Network Management in Agribusiness and the Food Industry. Noorwijk, the Netherlands, 6–8 June, pp. 104–115.

Suh, T. and Kwon, I. (2003) The role of bilateral asset specificity and replaceability on trust in supply chain partners. *Proceedings in the American Marketing Association. Winter Conference.* http://business.slu.edu/centers/Consortium_Supply_Chain_Mgt/trust.pdf (accessed 14/07/05).

Williamson, O. (1993) Calculativeness, trust, and economic organization. *Journal of Law & Economics* 36, 453–486.

Yee, W., Yeung, R. and Morris, J. (2005) Food safety: building consumer trust in livestock farmers for potential purchase behaviour. *British Food Journal* 107(11), 841–854.

Chapter 6

A Review of the Trust Situation in Agri-food Chain Relationships in the Asia-Pacific with a Focus on the Philippines and Australia

Peter J. Batt

Curtin University of Technology, Perth, Australia

Introduction

In Kapatagan, a small village on the slopes of Mt Apo in southern Mindanao, some 1700 households depend upon the cultivation of temperate vegetable crops for their main source of income. Most farms are very small, ranging in size from 0.6 to 1.0 ha, with most being comprised of 1–5 small parcels of land, often at different altitudes. Growers either bring or consign the produce they have grown on their farms to the trading post. Here various middlemen facilitate the sale of produce to traders and wholesaler buyers who transport it down the mountain to various retail buyers, restaurants and other food service outlets. Generally, prices are determined on the day, but in some cases, wholesalers may pre-order vegetables, thereby making it the responsibility of the middlemen to source sufficient product.

As most growers have no direct contact with the market, they must rely upon the traders with whom they transact for market information. In a market where prices vary from day-to-day depending upon the quantity and quality of the produce available, there is some concern that traders may be in a position to take advantage of individual growers. Acting individually, smallholder growers can do little to influence the price, for they are widely dispersed and cultivate highly perishable crops. Marketing costs are high because of inefficient transport, inadequate cool storage capacity and significant variations in product form, variety and quality (Harris-White, 1995). Various informal credit arrangements may lock growers into relationships where the grower is, to varying degrees, more or less dependent upon one or more market intermediaries (Mendoza and Rosegrant, 1995). Smallholder growers can reduce the risk of exploitation by transacting with those traders who have a good reputation and/or with whom they have had a favourable transaction in the past.

In Perth, Western Australia, retailers have historically purchased the fresh fruit and vegetables they require from the central wholesale market. First established in 1926 under the Metropolitan Markets Act No. 55, the Perth Metropolitan Market Trust was appointed as the corporate body to manage the Perth Metropolitan Market (Caddy, 1978). The Trust owns and maintains all the buildings in the central market, whereupon it leases space to various market agents who dispose of the fresh produce consigned to the market by the many growers spread throughout the state. These market agents are licensed under the Metropolitan Market Trust Lease and Covenant to conduct the business of commission agents.

While keen rivalry and competition between the market agents have generally seen growers benefit from the market agent's endeavours to develop their business enterprises, in order to reduce costs, most have abandoned the auction as the principal price setting mechanism. With private negotiation becoming the primary means for establishing price, growers must rely to a much greater extent upon the integrity of their trading partners. Unlike the auction, where the price at which the produce is sold is public knowledge and, to some extent, where the identity of the buyer is revealed, the lack of transparency inherent in private negotiations has resulted in an underlying atmosphere of distrust between the growers and market agents. Growers feel there is inadequate disclosure of information about sales and, in particular, how the net price was calculated. Among other things, growers want to satisfy themselves that the market agents' margins are not excessive and that any other deductions are legitimate. Furthermore, many growers are dissatisfied by the reporting of a net price across a range of different qualities, sizes and varieties which obscure the market signals (Chamber of Fruit and Vegetable Industries, 1999). Consequently, distrust and disputation is common between the growers and downstream market intermediaries.

This chapter seeks to explore the various ways in which market intermediaries can build trust in their relationship with growers. From the market intermediaries' perspective, being able to attract and retain a greater share of the growers' business will have significant positive benefits on the profitability of their enterprise and their propensity to attract and retain downstream customers. From the growers' perspective, trust is expected to reduce much of the risk and uncertainty currently present in their transactions with market intermediaries.

Trust – a theoretical foundation

For any particular potential exchange, trust will be critical if two situational factors are present: (i) risk; and (ii) asymmetric information (Hawes *et al.*, 1989). Since most sales transactions present some degree of risk and uncertainty, trust acts as an information resource that reduces the perceived threat of information asymmetry and performance ambiguity.

Trust focuses on the belief or the expectation that the vulnerability arising from the acceptance of risk will not be used to advantage by an exchange partner (Lane, 2000; see also Chapter 3). These expectations may be calculative, value- or norm-based, or based on common cognitions.

Anderson and Narus (1990) define trust as the belief that an exchange partner will perform actions that will result in positive outcomes for the firm and will not take unexpected actions that may result in negative outcomes. Moorman *et al.* (1993) define trust as the willingness to rely on an exchange partner in whom one has confidence.

While both of these definitions view trust as a behavioural intention that reflects reliance on an exchange partner, both definitions, in part, capture quite different aspects of the construct. Moorman *et al.* (1993) describe trust as a belief, a sentiment or an expectation about an exchange partner that results from the partner's expertise, reliability and intentionality. This component of trust, which Ganesan (1994) describes as credibility, is based on the extent to which the buyer believes that the supplier has the necessary expertise to perform the activity effectively and reliably.

However, trust also relates to the focal firm's intention to rely on their exchange partner. Ganesan (1994) describes this component as benevolence, because it is based on the extent to which the focal firm believes that its partner has intentions and motives beneficial to it. A benevolent partner will subordinate immediate self-interest for the long-term benefit of both parties and will not take actions that may have a negative impact on the firm. Trust reduces the perception of risk associated with opportunistic behaviour, it increases the buyer's confidence that short-term inequities will be resolved over time and it reduces the transaction costs in an exchange relationship.

Trust is the critical determinant of many factors related to performance including: the more open exchange of relevant ideas and emotions; greater clarification of goals and problems; more extensive search for alternative courses of action; greater satisfaction with efforts; and, greater motivation to implement decisions (Achrol, 1997). When trust exists, buyers and suppliers believe that long-term idiosyncratic investments can be made with limited risk because both parties will refrain from using their power to renege on contracts or to use a change in circumstances to obtain profits in their own favour (Ganesan, 1994). Trust increases the partner's tolerance for each other's behaviour, facilitating the informal resolution of conflict, which in turn allows the partners to better adapt to the needs and capabilities of the counterpart firm (Hakansson and Sharma, 1996). Trust not only reduces the need for structural mechanisms of control, but firms learn to become more interdependent (Kumar, 1996). Once trust is established, firms learn that coordinated joint efforts lead to outcomes that exceed those that either firm could achieve if they were to act solely in their own best interest (Han *et al.*, 1993).

Trust-building behaviour

Satisfaction

According to the 'disconfirmation of expectations' model, satisfaction is the result of a comparison between an exchange partner's performance and the focal firm's expectations (Oliver, 1980). Whenever performance exceeds expectations, satisfaction will increase. Conversely, whenever performance falls below expectations, exchange partners will become dissatisfied.

Satisfaction has been defined as a positive affective state resulting from an appraisal of all aspects of the firm's working relationship with another (Frazier *et al.*, 1989). Geyskens *et al.* (1999) propose that a satisfied exchange partner considers the relationship to be a success when it is satisfied with the effectiveness and productivity of the relationship and the resulting positive financial outcomes. Exchange partners that are highly satisfied with the economic rewards that flow from their relationship generally perceive their exchange partner as being more trustworthy.

However, satisfaction with past outcomes also indicates equity in the exchange (Anderson and Narus, 1990). Equity generally refers to the fairness or rightness of something in comparison to others (Halstead, 1999). Equitable outcomes provide confidence that neither party has been taken advantage of in the relationship and that both parties are concerned about their mutual welfare (Ganesan, 1994).

While satisfaction with the exchange also affects an exchange partner's morale and their incentive to participate in collaborative activities (Geyskens *et al.*, 1999), conflict is one of the few constructs which have a direct negative effect on satisfaction (Frazier *et al.*, 1989). Firms that are able to lower the overall level of conflict in their relationship experience greater satisfaction (Anderson and Narus, 1990). Satisfactory conflict resolution increases mutual trust and reinforces each member's commitment to the relationship (Thorelli, 1986).

Satisfaction also increases when non-coercive sources of power are employed. When a channel partner frequently pressures or coerces the focal firm into taking some action that it would not have otherwise taken or forces it to forgo some positive outcome, the focal firm will experience tension and frustration because its decision autonomy is constrained (Frazier, 1983). It is therefore hypothesized that there will be a positive association between a grower's satisfaction with the exchange and the trust the grower places in their preferred market intermediary.

Communication and information exchange

Communication has been described as the glue that holds together a channel of distribution (Mohr and Nevin, 1990). Communication in marketing channels serves as the process by which persuasive information is transmitted (Frazier and Summers, 1984), participative decision-making is fostered, programmes are coordinated (Anderson and Narus, 1990), power is exercised (Gaski, 1984) and commitment and loyalty are encouraged (Anderson and Weitz, 1992). Communication enables information to be exchanged that may reduce certain types of risk perceived by either one of the partners to the transaction (McQuiston, 1989). The more information a partner has and feels they can obtain, the more likely they are to trust their exchange partner (Moore, 1999).

However, trust-building is a dynamic process. In the early stages of a relationship, commitments are usually less extensive and there will be little need for trust and information (Wilson, 1995). However, as the relationship matures, there will be a positive association between trust and information. Meaningful communication in a relationship is therefore a necessary antecedent of trust (Anderson and Narus, 1990). Consequently, it is hypothesized that there will be a positive association between the extent to which the preferred market intermediary communicates with the grower and the grower's trust in the preferred market intermediary.

Personal relationships

Inter-personal trust in business-to-business relationships is rarely offered spontaneously: rather, it results from an extended period of experience with an exchange partner (Lane, 2000). During this time, knowledge about the exchange partner is accumulated, either through direct contact, or indirectly through reliable third parties (see Chapter 3). Inter-personal trust between individuals is based on familiarity, developed either from previous interactions or derived from the membership of similar social groups. Zucker (1986) describes how characteristics-based trust rests on social similarities that assume cultural congruence because both parties belong to the same social group or community. They may share a common religion, ethnic status, or family background. Fukuyama (1995) describes how trust evolves in relationships where common values and norms, often based on kinship, familiarity and common interests and backgrounds predominate. In Ghana, Lyon (2000) describes how personal friendships provide the supplier with greater confidence that the exchange partner will follow through because of moral obligations to reciprocate. It is therefore hypothesized that there will be a positive association between the personal relationship that the grower has with their preferred market intermediary and the grower's trust in their preferred market intermediary.

Reputation

When neither party to the exchange has any hands-on experience of the other's performance, reputation may furnish the partner with a belief system that enables it to assuage anxieties, resolve ambiguities and lend predictability towards situations. Herbig and Milewicz (1995) view reputation as a customer's estimation of the consistency over time of an attribute, based on an evaluation of the exchange partners willingness and ability to perform an activity repeatedly in a similar fashion. Derived primarily from personal experience, perceptions of past performance may also be drawn from the various signalling behaviours an exchange partner undertakes to both develop and maintain its reputation (Fombrun and Shanley, 1990).

Reputation creates expectations, not only about the key attributes of an exchange partner, but about how that exchange partner will behave in the future. In the transitional economies, Fafchamps (1996) describes reputation as a collective coordination and information-sharing device which ensures contracts are complied with. In its simplest form, it suggests that individuals will choose not to interact with those firms who are known not to comply with their contractual obligations. Reputation is a form of social collateral that can guarantee contract performance without prior acquaintance. Concern for one's reputation may be sufficient to ensure compliance and to enable firms to offer credit or take orders without knowing each other personally. Therefore, it is hypothesized that there will be a positive association between the preferred market intermediary's reputation and the grower's trust in their preferred market intermediary.

The making of relationship-specific investments

If a firm wishes to improve its relationship with another to achieve future benefits, then in all probability, the firm will need to commit various resources to the relationship, whether expressed in terms of managerial or sales force time, product or service development, process, financial or administrative adaptations (Ford *et al.*, 1996). Any resource committed above and beyond that required to execute the current exchange transaction can be regarded as an investment (Campbell and Wilson, 1996).

An investment is the process in which resources are committed in order to create, build or acquire resources that may be used in the future (Easton and Araujo, 1994). Through interacting with other firms and committing resources to specific relationships, firms have the opportunity to use relationships as a resource for the creation of other resources, product adaptations and innovations, process improvements, or to provide access to third parties (Hakansson and Snehota, 1995).

Williamson (1985) suggests that investments stabilize relationships by altering the firm's incentive structure. By making relationship-specific investments, the firm creates an incentive to maintain the relationship. Engaging in opportunistic behaviour and thereby risking the dissolution of the relationship is contrary to the self-interest of a channel member, for if the opportunism is detected and the relationship terminated, the investment may not have generated adequate returns (Lohtia and Krapfel, 1994).

Furthermore, the making of such relationship-specific investments may also provide a powerful signal to the other party. Observing the other party's pledges causes the channel member to be more confident in the other party's commitment to the relationship, because the other party will sustain considerable economic loss if the relationship is terminated (Anderson and Weitz, 1992). Relationship-specific investments offer tangible evidence that an exchange partner can be believed and that it cares for the relationship (Ganesan, 1994). In this regard, the making of relationship-specific investments provides strong signals of a channel partner's trustworthiness.

However, the making of relationship-specific investments and subsequent locking-in of channel partners is not sufficient in itself to develop a long-term relationship. Long-term relationships bound only by dependence and investments may signify forced collaboration rather than cooperation (Ganesan, 1994). It is therefore hypothesized that there will be a positive association between the willingness of the preferred market intermediary to make relationship-specific investments and the grower's trust in the preferred market intermediary.

Power and dependence

Dependence and power are integral components of channel relationships. Dependence refers to a firm's need to maintain a channel relationship in order to achieve desired goals (Frazier *et al.*, 1989). When the outcomes obtained from the relationship are important or highly valued, the focal firm becomes more dependent (Heide and John, 1988). The same is true when the magnitude of the exchange itself is higher (Lohtia and Krapfel, 1994). The higher the percentage of sales and profits that are contributed by handling an exchange partner's product line and the greater the expectations of sales and profits in the future, the greater the focal firm's dependence (Frazier *et al.*, 1989).

Dependence is also increased when the outcomes from the relationship are comparatively better than the outcomes available from alternative relationships. Firms dealing with the best supplier are more dependent because the outcomes associated from dealing with that supplier are better than those available from poor-performing suppliers (Heide and John, 1988).

When fewer alternative sources of exchange are available to the focal firm, or when replacing or substituting a current exchange partner is difficult because there are fewer potential alternatives, dependence increases (Heide and John, 1988; Frazier *et al.*, 1989).

However, it is the firm's perception of its dependence relative to its partner that is of most interest in channel relationships (Anderson and Narus, 1990). Relative dependence determines the extent to which a firm will have influence over or be influenced by its partner, for with increasing dependence comes greater vulnerability. Asymmetric dependence in an exchange relationship may make one firm more susceptible to the power and influence of another. The more powerful partner may be in a position to create more favourable terms of trade for itself (Heide and John, 1988; Frazier *et al.*, 1989).

Power resides in the ability of one party to make another do what s/he would not have otherwise done (Gaski, 1984). According to French and Raven (1959), power is derived from the more dependent firm's perception of the dominant firm's ability to mediate rewards, mediate punishment, its legitimate right to prescribe behaviour, some specific knowledge or expertise and the extent to which the more dependent firm identifies with the dominant firm. More recently, Johnson *et al.* (1993) have classified power as either mediated power (reward, coercion and legal legitimate power) or non-mediated power (expertise, referent, information and traditional legitimate power). This dichotomy reflects whether the source does or does not control the reinforcements that guide the target firm's behaviour (Brown *et al.*, 1995).

While the power to coordinate is the prerogative of the dominant firm (Achrol, 1997) the subsequent use of that power will impact on the exchange partner's perception of relationalism (Brown *et al.*, 1995). Channel leaders often use various reward and coercive powers and legitimate authority to cajole and coerce cooperation among channel members. However, it is the use of non-mediated power that inevitably builds social bonds and close relationships. It is therefore hypothesized that there will be a negative association between the extent to which the preferred market intermediary uses coercive power and the grower's trust in the preferred market intermediary.

Forbearance

Achieving trust in any relationship requires a deliberate strategy of forbearance with a view towards future pay-offs and accumulated evidence of non-reneging behaviour (Parkhe, 1993). Opportunism refers to the incomplete or distorted disclosure of information, especially calculated efforts to mislead, distort, disguise, obfuscate or otherwise confuse (Williamson, 1985). The incentive to engage in opportunistic behaviour arises because one party finds it advantageous to maximize their own gains at the expense of the relationship (Gundlach *et al.*, 1995). However, if either party in the relationship chooses to behave opportunistically and that opportunism is detected, it will provoke retaliatory behaviour. When trust is betrayed, the aggrieved partner may react with great emotion and trust will be lost. With trust and confidence in the relationship undermined, the aggrieved party will seek to withdraw or to limit their commitment to the relationship. Not unexpectedly, it is hypothesized that there will a negative association between the extent to which the preferred market intermediary seeks to take advantage of the grower and the grower's trust in their preferred market intermediary.

An empirical evaluation of trust in fresh produce supply chains

In order to identify how the various trust-building behaviours impact on trust, a comprehensive questionnaire was developed. In the Philippines, the survey instrument was translated by academic staff at the University of the Philippines in Mindanao and administered through personal face-to-face interviews with vegetable farmers in Kapatagan. A total of 75 smallholder growers were interviewed.

In Western Australia, all 1260 fresh fruit and vegetable growers dealing with market agents in the Perth Metropolitan Market were asked to complete a mail questionnaire. Although a self-addressed reply paid envelope was included to facilitate responses, a total of 198 responses were received – a response rate of 15.7%.

Growers were first asked to respond to a number of open-ended questions about the nature of their farming enterprise and the means by which they disposed of their crops. Growers were then asked to respond to a number of questions about the nature of their relationship with their preferred trading partner.

In the Philippines, growers responded to the various item measures on a scale of 1 to 6 where 1 was 'I disagree a lot' and 6 was 'I agree a lot'. However, in Western Australia, a seven-point scale was employed.

Trust was assessed by five item measures derived from Moorman et al. (1992) and Doney and Cannon (1997). Principal component analysis (with Kaiser normalization and varimax rotation) produced a robust single factor solution that collectively explained 88% of the variance for the construct in the Philippines and 74% in Western Australia (Table 1).

Measures for the 19 trust-building behaviours were developed from the literature reported by Anderson and Narus (1990), Anderson and Weitz (1992), Doney and Cannon (1997), Frazier (1983), Frazier et al. (1989), Heide and John (1988), Ganesan (1994) and Gundlach et al. (1995). These were then regressed against trust as the dependent variable using stepwise regression.

Stepwise regression provides a convenient means of selecting from a large number of predictor variables, a smaller subset of variables that account for most of the variation in a single dependent variable (Malhotra et al., 2006). In this procedure, the predictor variables enter or are removed from the regression equation one at a time without appreciably increasing the residual sum of squares.

Table 1. The trust construct in the Philippines and Western Australia.

	Philippines	Australia
Factor loadings		
• My preferred trading partner always considers my best interests	0.953	0.900
• I have confidence in my preferred trading partner	0.946	0.886
• My preferred trading partner always keeps his promises	0.943	0.872
• I believe in the information provided by my preferred trading partner	0.927	0.860
• I trust my preferred buyer	0.911	0.770
Percentage of variance explained	87.67	73.76
Cronbach's alpha	0.965	0.910

While stepwise procedures do not result in regression equations that are optimal, in the sense of producing the largest R^2 for a given number of predictors, the technique is valuable when multi-collinearity is present. In stepwise regression, the order in which the predictors enter or are removed from the regression equation is used to infer their relative importance (Malhotra et al., 2006). Nevertheless, to quantify the severity of multi-collinearity, a variance inflation factor (VIF) analysis was applied.

In the Philippines, trust between the grower and their preferred downstream market intermediary was facilitated by three trust-building behaviours: a willingness of the market intermediary to advise the grower of supply problems; a close personal friendship and a perception that the buyer had treated the grower fairly and equitably (Table 2).

Table 2. Step-wise regression analysis in the Philippines.

Dependent variable: trust	Beta	Sig	VIF
Constant		0.963	
My buyer often advises me of supply problems	0.530	0.000	2.751
My buyer and I have a close personal friendship	0.360	0.000	2.575
My buyer treats me fairly and equitably	0.155	0.008	1.515
Adj R^2 = 0.875; SE = 0.431; F = 138.42; Sig = 0.000			

In a market where prices are determined primarily by supply and demand, where the demand itself is relatively stable, but supply is very volatile, depending upon the seasonality of production and the incidence of typhoons (in Northern Luzon), prior knowledge of the prevailing market prices may enable individual growers, as their crop approaches maturity, to harvest a larger proportion of their crop and to take advantage of the higher prices. Anecdotal evidence suggests that smallholder subsistence farmers in the Philippines often plant in the expectation that unseasonal or unexpected events elsewhere will reduce the supply of fresh vegetables thereby resulting in a significant increase in price.

There is abundant evidence in the literature to support the value of close personal friendships and their subsequent impact on trust. Fafchamps (1996) describes how in a transitional economy, when individuals feel uncertain about the reliability of a customer, the individual will express an overwhelming desire to conduct business with people they already know. Personal relationships are important because of the lack of codified and diffused public information, a general distrust towards strangers, the presence of reciprocity as a norm and the development of trust prior to and commensurate with economic exchange (Bjorkman and Koch, 1995). Personal relationships arise from a common background where the exchange partners went to school together, studied together, came from the same town/area, or worked together. Fafchamps (1996) describes how those traders who have established good relationships with customers and suppliers make higher margins, not because they abuse their market power, but because the relationships they have made make them more efficient. Even so, the ability of traders to manipulate the exchange process depends upon their control over supply, access to information, capital and credit, transport, knowledge of marketing procedures and the organization of the market (Lyon, 2000).

However, Luhmann (1979) argues that trust involves a learning process that is only complete when the person to be trusted has had the opportunity to betray trust. Not unexpectedly, should a trader choose to behave opportunistically and to take advantage of the grower by incorrectly reporting prices and withholding money that should rightfully be paid to the grower, the grower's trust in their preferred trading partner will very rapidly decline. For this reason, personal relationships are not only an antecedent to trust, but they are continually assessed and reassessed with each subsequent exchange transaction.

In a similar vein, in the absence of complete information, the grower's perception as having been treated fairly and equitably will have a significant and positive impact on trust. When economic outcomes are high, channel members may attribute a great deal of credit to their exchange partner and thus the channel member's attraction to and trust in their exchange partner will increase (Geyskens *et al.*, 1998).

While reputation was found to be a consequence of a good long-term relationship in the Philippines, in Western Australia, a good reputation was the key antecedent in building trust between growers and the preferred market agent (Table 3).

Table 3. Step-wise regression analysis in Western Australia.

Dependent variable: trust	Beta	Sig	VIF
Constant		0.011	
My agent has a good reputation	0.417	0.000	2.372
I feel I am adequately rewarded by my agent	0.250	0.000	4.200
My agent treats me fairly and equitably	0.180	0.011	5.131
My agent keeps me well informed on technical matters	0.155	0.000	1.539
Having a long-term relationship with my agent reduces the uncertainty of price	0.084	0.012	1.153
My agent has all the power in our relationship	−0.066	0.050	1.172
My agent often acts opportunistically	−0.080	0.015	1.121
Adj R^2 = 0.861; SE = 0.545; F = 131.22; Sig = 0.000			

Such is not to be unexpected, given that Moorman *et al.* (1992) regard reputation as an indicator of reliability. Furthermore, Fombrun and Shanley (1990) view reputation not only as a signal of an exchange partner's ability to deliver valued outcomes, but also as an important cue about how the focal firm's offer quality compares with competing firms.

In Western Australia, growers were more likely to trust their preferred market agents when they believed they had been adequately rewarded for their efforts and treated fairly and equitably. Since the majority of growers transact with more than one market agent and since most growers re-evaluate their relationship with their preferred market agent after each transaction, the grower's satisfaction with the exchange will be contingent upon their preferred market agent providing returns that are comparable to those achieved from competing market agents. Where there is some difference in the net price received between two or more market agents, the grower can be expected to allocate a greater proportion of their crop to that agent from whom they received the highest price. While it is unlikely that small differences in the price received will cause the farmer to abandon the relationship, over time, where the prices received are consistently lower than those offered by other traders, farmers may begin to feel that their preferred trading partner is no longer representing their best interests.

Since prices are largely determined by supply and demand, there will always be some uncertainty as to what price the grower will ultimately receive for their produce. Market agents can attempt to reduce some of the price uncertainty and thereby generate greater trust by making various relationship-specific investments, whereupon they advise growers in advance of what quantities and quality of produce are required by customers. Market agents can also advise growers of what varieties are in greatest demand and assist them on-farm to improve the quality of the produce they offer for sale.

There is also some evidence that the relationship itself can potentially reduce some of the uncertainty associated with price. Although fresh produce is often regarded as a largely undifferentiated commodity, there are significant differences in the quality and maturity of the produce between individual growers. Where growers can maintain a sufficient supply, market agents are able to identify retailers who may not only prefer a particular grower's produce but are also willing to pay a price premium to attain it. For most retailers, fresh produce is regarded as the key determinant in the consumer's choice of store because it provides an attractive, fresh and colourful display and is a symbol of the pervading quality standards throughout the store (Retail Business, 1997). While shopper's accord great importance to the quality, price, range and availability of fresh produce (Hughes and Merton, 1996), fresh produce also generates some of the highest profit margins of any product category in store. However, while fresh produce is highly profitable, the products are highly perishable and very sensitive to mishandling and damage at all levels of the supply chain (White, 2000). Simpson *et al.* (2002) describe how suppliers have the potential to positively or negatively affect customer inventories, product quality, product cost and delivery times. As prudent supplier selection leads to improved performance, price is becoming less important in the decision to purchase.

Conversely, should a market agent choose to behave opportunistically and to take advantage of the grower by incorrectly reporting prices and withholding money that should rightfully be paid to the grower, the grower's trust in their preferred trading partner will very rapidly decline. Should such opportunism be detected, growers will also certainly terminate their relationship. Since growers also communicate among themselves, those market agents who have been found to engage in opportunistic trading will find it more difficult to prevent other growers from switching.

While power is the ability of one firm to influence the decisions and/or the behaviour of another (El-Ansary and Stern, 1972), Ogbonna and Wilkinson (1998) argue that the power to control is not only dependent on the possession of power, but also the extent to which other exchange partners have countervailing market power. The presence of countervailing power forces exchange partners to differentiate between the possession of power and its use.

Power is ever present whether actors choose to use it or not (Hingley, 2005). While the power to coordinate is the prerogative of the dominant firm (Achrol, 1997) the frequent use of that power will impact negatively upon the exchange partner's perception of trust. For this reason, there is a significant negative relationship between trust and the extent to which the preferred market agent uses its reward power, its coercive power and its legitimate authority to cajole and coerce cooperation among growers.

Hardy *et al.* (2000) describe power as a functional equivalent of trust. However, trust and power may accrue to the dominant partner only, for when the weaker party has no alternative, it can be trusted to behave in a particular way. Where coordination is achieved through the use of legitimate power, the weaker partner may concede and undertake what is required of it. In other circumstances, trust will decline when the weaker party perceives that it has no alternative or has been coerced into the exchange. However, in most instances, strong moral and social pressures will persuade the dominant exchange partner not to exercise coercive influence strategies. Invariably, in the fresh produce industry, power is imbalanced and favours the downstream buyers (Hingley, 2005). However, as this power provides the means for achieving coordination and cooperation in the supply chain, growers will generally accommodate the power imbalance providing that they receive a reasonable proportion of the relationship value.

Concluding comments

Trust is the critical determinant of a good relationship. In the context of the fresh produce industry, where prices are largely determined by supply and demand, there will always be some uncertainty as to what price a grower will receive for the produce they have grown. Traders and market intermediaries can reduce some of this uncertainty and thereby generate greater trust by advising farmers in advance of the quantities and quality of the produce required by downstream customers. Sharing sensitive market information will not only improve transparency in the exchange, but also signal the traders' desire to cooperate.

However, within the literature, there is considerable debate in who people trust. Plank *et al.* (1999) contest that trust is comprised of three individual components: salesperson trust, product trust and company trust. Similarly, Anderson and Narus (1990) and Doney and Cannon (1997) find it necessary to differentiate between trust in an individual and trust in an organization. In the Philippines, it is abundantly clear that smallholder vegetable growers primarily trust the individual as evidenced by the importance placed on personal friendships.

For the market intermediaries, Swan *et al.* (1985) indicate how competence, customer orientation, honesty, dependability and likeability facilitate the development of trust between sales representatives and their customers. Moorman *et al.* (1993) argue that the inter-personal factors that most affect trust include perceived expertise, sincerity, integrity, tactfulness, timeliness and confidentiality. Crosby *et al.* (1990) contend that mutual disclosure, a cooperative rather than a competitive intention and the style and intensity of the communication between individuals is critical in establishing and maintaining inter-personal relationships.

References

Achrol, R. (1997) Changes in the theory of inter-organizational relations in marketing: Toward a network paradigm. *Journal of Academy of Marketing Science* 25(1), 56–71.

Anderson, E. and Weitz, B. (1992) The use of pledges to build and sustain commitment in distribution channels. *Journal of Marketing Research* 29, 18–34.

Anderson, J. and Narus, J. (1990) A model of distributor firm and manufacturing firm working relationships. *Journal of Marketing* 54(1), 42–58.

Bjorkman, I. and Koch, S. (1995) Social relationships and business networks: the case of Western companies in China. *International Business Review* 4(4), 519–535.

Brown, J., Lusch, R. and Nicholson, C. (1995) Power and relationship commitment: their impact on marketing channel member performance. *Journal of Retailing* 71(4), 363–392.

Caddy, J. (1978) *Perth Markets*. Perth Metropolitan Market Trust.

Campbell, A. and Wilson, D. (1996) Managed networks: creating strategic advantage. In: Iacobucci, D. (ed) *Networks in Marketing*. Sage, Thousand Oaks, California, pp. 125–143.

Chamber of Fruit and Vegetable Industries (1999) *Submission to the Review of the Perth Market Act (1926). Part 1. Transparency.*

Crosby, L., Evans, K. and Cowles, D. (1990) Relationship quality in services selling: an interpersonal influence perspective. *Journal of Marketing* 54(July), 68–81.

Doney, P. and Cannon, J. (1997) An examination of the nature of trust in buyer–seller relationships. *Journal of Marketing* 61(April), 35–51.

Easton, G. and Araujo, L. (1994) Market exchange, social structures and time. *European Journal of Marketing* 28(3), 72–84.

El-Ansary, A. and Stern, L. (1972) Power measurement in distribution channels. *Journal of Marketing Research* 9, 47–52.

Fafchamps, M. (1996) The enforcement of commercial contracts in Ghana. *World Development* 24(3), 427–448.

Fombrun, C. and Shanley, M. (1990) What's in a name? Reputation building and corporate strategy. *Academy of Management Journal* 33(2), 233–258.

Ford, D., McDowell, R. and Tomkins, C. (1996) Relationship strategy, investments and decision making. In: Iacobucci, D. (ed) *Networks in Marketing*. Sage, Thousand Oaks, California, pp. 144–176.

Frazier, G. (1983) Inter-organizational exchange behaviour in marketing channels: a broadened perspective. *Journal of Marketing* 47(Fall), 68–78.

Frazier, G. and Summers, J. (1984) Interfirm influence strategies and their application within distribution channels. *Journal of Marketing* 48(Summer), 43–55.

Frazier, G., Gill, J. and Kale, S. (1989) Dealer dependence levels and reciprocal actions in a channel of distribution in a developing country. *Journal of Marketing* 53(January), 50–69.

French, R. and Raven, B. (1959) The bases of social power. In: Cartwright, D. (ed) *Studies in Social Power*. University of Michigan Press, Ann Arbor, MI.

Fukuyama, F. (1995) *Trust: The Social Virtues and the Creation of Prosperity*. Hamish Hamilton, London.

Ganesan, S. (1994) Determinants of long-term orientation in buyer–seller relationships. *Journal of Marketing* 58(April), 1–19.

Gaski, J. (1984) The theory of power and conflict in channels of distribution. *Journal of Marketing* 48(Summer), 9–29.

Geyskens, I., Steenkamp, J. and Kumar, N. (1998) Generalisations about trust in marketing channel relationships using meta-analysis. *International Journal of Research in Marketing* 15(3), 303–318.

Geyskens, I., Steenkamp, J. and Kumar, N. (1999) A meta-analysis of satisfaction in marketing channel relationships. *Journal of Marketing Research* 36(May), 223–238.

Gundlach, G., Achrol, R. and Mentzer, J. (1995) The structure of commitment in exchange. *Journal of Marketing* 59(January), 78–92.

Hakansson, H. and Sharma, D. (1996) Strategic alliances in a network perspective. In: Iacobucci, D. (ed) *Networks in Marketing*. Sage, Thousand Oaks, California, pp. 108–124.

Hakansson, H. and Snehota, I. (1995) *Developing Relationships in Business Networks*. International Thomson Business Press, London.

Halstead, D. (1999) The use of comparison standards in customer satisfaction research and management: a review and proposed typology. *Journal of Marketing Theory and Practice* 7(3), 13–26.

Han, S., Wilson, D. and Dant, S. (1993) Buyer–supplier relationships today. *Industrial Marketing Management* 22, 331–338.

Hardy, C., Phillips, N. and Lawrence, T. (2000) Distinguishing trust and power in interorganizational relations: forms and facades of trust. In: Lane, C. and Bachman, R. (eds) *Trust Within and Between Organizations*. Oxford University Press, Oxford.

Harris-White, B. (1995) Efficiency and complexity: distributive margins and the profits of market enterprise. In: Scott, G. (ed) *Prices, Products and People. Analysing Agricultural Markets in Developing Countries*. Lynne Rienner, Boulder, Colorado, pp. 301–324.

Hawes, J., Mast, K. and Swan, J. (1989) Trust earning perceptions of sellers and buyers. *Journal of Personal Selling and Sales Management* 9(Spring), 1–8.

Heide, J. and John, G. (1988) The role of dependence balancing in safeguarding transaction-specific assets in conventional channels. *Journal of Marketing* 52(January), 20–35.

Herbig, P. and Milewicz, J. (1995) To be or not to be...credible that is: A model of reputation and credibility among competing firms. *Marketing Intelligence and Planning* 13(6), 24–33.

Hingley, M. (2005) Power to all our friends? Living with imbalance in supplier-retailer relationships. *Industrial Marketing Management* 34, 848–858.

Hughes, D. and Merton, I. (1996) Partnership in produce: the J. Sainsbury approach to managing the fresh produce supply chain. *Supply Chain Management* 1(2), 4–6.

Johnson, J., Sakano, T., Cote, J. and Onzo, N. (1993) The exercise of interfirm power and its repercussions in US–Japanese channel relationships, *Journal of Marketing* 57(April), 1–10.

Kumar, N. (1996) The power of trust in manufacturer-retailer relationships. *Harvard Business Review*, November-December, 92–106.

Lane, C. (2000) Introduction: theories and issues in the study of trust. In: Lane, C. and Bachmann, R. (eds) *Trust Within and Between Organizations*. Oxford University Press, Oxford, pp. 1–30.

Lohtia, R. and Krapfel, R. (1994) The impact of transaction-specific investments on buyer–seller relationships, *Journal of Business and Industrial Marketing* 9(1), 6–16.

Luhmann, N. (1979) *Trust and Power*. Wiley, Chichester.

Lyon, F. (2000) Trust, networks and norms: the creation of social capital in agricultural economies in Ghana. *World Development* 28(4), 663–681.

Malhotra, N., Hall, J., Shaw, M. and Oppenheim, P. (2006) *Marketing Research: An Applied Orientation*. 3rd edition. Pearson Prentice Hall, French Forest, NSW.

McQuiston, D. (1989) Novelty, complexity and importance as causal determinants of industrial buyer behaviour. *Journal of Marketing* 53(2), 66–79.

Mendoza, M. and Rosegrant, M. (1995) Pricing conduct of spatially differentiated markets. In: Scott, G. (ed) *Prices, Products and People. Analysing Agricultural Markets in Developing Countries.* Lynne Rienner, Boulder, Colorado, pp. 343–360.

Mohr, J. and Nevin, J. (1990) Communication strategies in marketing channels: a theoretical perspective. *Journal of Marketing* 54(October), 36–51.

Moore, M. (1999) Truth, trust and market transactions: What do we know? *Journal of Development Studies* 36(1), 74–88.

Moorman, C., Zaltman, G. and Deshpande, R. (1992) Relationships between providers and users of market research: the dynamics of trust within and between organizations. *Journal of Marketing Research* 29(3), 314–328.

Moorman, C., Deshpande, R. and Zaltman, G. (1993) Factors affecting trust in market research relationships. *Journal of Marketing* 57(January), 81–101.

Ogbonna, E. and Wilkinson, B. (1998) Power relations in the UK grocery supply chain. Developments in the 1990's. *Journal of Retailing and Consumer Services* 5(2), 77–86.

Oliver, R. (1980) A cognitive model of the antecedents and consequences of satisfaction decisions. *Journal of Marketing Research* 17(November), 460–69.

Parkhe, A. (1993) Strategic alliance structuring: a game theoretic and transaction cost examination of interfirm cooperation. *Academy of Management Journal* 36(4), 794–829.

Plank, R., Reid, D. and Pullins, E. (1999) Perceived trust in business-to-business sales: a new measure. *Journal of Personal Selling and Sales Management* 14(3), 61–71.

Retail Business (1997) Corporate intelligence, market survey, fresh fruit and vegetables, Part 1. *Retail Business* 469, March.

Simpson, P., Siguaw, J. and White, S. (2002) Measuring the performance of suppliers: an analysis of evaluation processes. *Journal of Supply Chain Management* 38(Winter), 29–41.

Swan, J., Trawick, I. and Silva, D. (1985) How industrial salespeople gain customer trust. *Industrial Marketing Management* 14, 203–211.

Thorelli, H. (1986) Networks: between markets and hierarchies. *Strategic Management Journal* 7(1), 37–51.

White, H. (2000) Buyer–seller relationships in the UK fresh produce industry. *British Food Journal* 102(1), 6–17.

Williamson, O. (1985) *The Economic Institutions of Capitalism.* The Free Press, New York.

Wilson, D. (1995) An integrated model of buyer–seller relationships. *Journal of the Academy of Marketing Science* 23(4), 335–345.

Zucker, L. (1986) Production of trust: institutional sources of economic structure, 1840–1920. *Research in Organizational Behaviour* 8, 53–111.

Chapter 7

Determinants of Sustainable Agri-food Chain Relationships in Europe

Christian Fischer,[1] Monika Hartmann,[2] Nikolai Reynolds,[3] Philip Leat,[4] César Revoredo-Giha,[4] Maeve Henchion,[5] Azucena Gracia[6] and Luis Miguel Albisu[6]

[1] Massey University, Auckland, New Zealand
[2] University of Bonn, Germany
[3] Synovate, Frankfurt, Germany
[4] Scottish Agricultural College (SAC), Aberdeen and Edinburgh, United Kingdom
[5] Ashtown Food Research Centre, Teagasc, Dublin, Ireland
[6] Agri-Food Research and Technology Center of Aragón (CITA), Zaragoza, Spain

Introduction

Several research efforts have already been undertaken to gain a better understanding of business relationships and thereby enable more effective management (e.g., Lee *et al.*, 2001; Prahalad and Ramaswamy, 2004). In the agri-food sector, for instance, Schulze *et al.* (2007a) investigated business relationships in the German pork sector and revealed a relatively low level of vertical coordination within this chain. The same authors highlighted the crucial roles of communication and personal relationships in the German pork and dairy chains (Schulze *et al.*, 2007b). The analysis of Lindgreen *et al.* (2004) indicated for the Netherlands that power imbalance and information asymmetry posed obstacles for the development of trust between chain actors.

While studies exist for individual countries (e.g., Leat and Revoredo-Giha, 2008; Reynolds *et al.*, 2009), extensive cross-country research on the determinants of chain relationships in the European agri-food sector has been scarce. A thorough understanding of the current status of chain relations across the EU agri-food system and their key drivers would be useful in order to further increase the efficiency, effectiveness and competitiveness of supply chains. However, such an investigation needs to take into account country, commodity and chain stage-specific aspects. Implementation of these findings can help agri-food chain members to successfully improve their commercial interactions and relationships, and to deal with critical policy issues, such as enhancing food quality (including safety).

In this chapter we present empirical results related to the main factors that determine sustainable inter-organizational relationships in agri-food chains. A thorough definition of the Relationship Sustainability construct is described in Chapter 4 of this book. The theoretical underpinnings justifying the determinants of sustainable relationships used in the empirical analysis of this chapter are discussed in Part I of this volume, and in Fischer *et al.* (2009). This chapter's structure is as follows. After providing a brief overview regarding the methodologies used for data collection and model estimation (second section), the empirical results are presented (third section). In the fourth section the findings are discussed and conclusions are drawn.

Data and methodology

The sustainability of agri-food chain relationships is examined for six different EU countries (Germany, UK, Spain, Ireland, Finland and Poland), two different commodities (meat and cereals) and two chain stages (upstream: farmer–processor, and downstream: processor–retailer). While the results of a pooled data analysis, combining all countries into a single dataset, has already been reported in Fischer *et al.* (2009), this chapter describes disaggregated findings.

Data

Based on pilot study findings involving expert interviews (see Chapter 5), a company survey was developed. The questionnaire was pre-tested separately in each participating country. Where feasible, personal interviews were conducted (mostly with farmers) or respondents were interviewed by telephone. In addition, questionnaires were sent by mail (followed-up by telephone calls and/or a subsequent reminder mailing).

Most of the obtained samples were drawn from existing sampling frames, i.e., all businesses whose contact details were available were approached.[1] The data were collected from November 2006 to April 2007. In total, the surveys yielded 1442 usable responses (see Table 1 for the sample profile).

The representativeness of the obtained sample was assessed using two criteria for which complete target population information was available across the different countries: first, geographic distribution of farm/company location and second, farm/company size. Despite some differences across the countries,[2] overall the collected responses reflect existing population variation with regard to these two indicators.

As to data quality, in total 86% of survey respondents claimed to be in upper management positions or (part-) owners of the surveyed businesses. Non-response bias (Armstrong and Overton, 1977) was assessed by comparing early survey responses with later ones, using multivariate analysis of variance on key demographic characteristics (company/farm size, geographic location and business activity). However, no significant differences were found.

---

[1] In the Finnish case a sample of contact details was randomly selected from the total population of relevant businesses.

[2] The Spanish sample was collected in the Aragón region. In the UK sample more than 80% are from Scotland. Despite these regional biases we refer to these samples as Spanish and UK ones.

Methodology

The relevance of different factors potentially affecting the sustainability of agri-food chain relationships were analysed by estimating a structural equation model (SEM).[3] SEM in its most general form consists of a set of linear equations that simultaneously test two or more relationships among directly observable and/or un-measurable latent variables (Bollen, 1989).

Table 1. Number of collected survey responses by country, agri-food chain and activity.

Country and agri-food chain	Economic activity				Total
	Farmers	Processors	Retailers	N.s.[†] or others	
Spain	206	79	51		336
pig meat	102	35	25		162
cereals	104	44	26		174
Poland	209	35	91		335
pig meat	100	17	48		165
beef	109	18	43		170
UK	229	12	6	7	254
beef	171	5	6	2	184
cereals	58	7		4	69
n.s.[†] or mixed				1	1
Finland	156	43	24	3	226
pig meat	75	16	8	2	101
cereals	81	27	16		124
n.s.[†] or mixed				1	1
Ireland	120	14	17		151
pig meat	49	7	6		62
beef	71	7	11		89
Germany	42	88	9	1	140
pig meat	13	26	3		42
cereals	24	60	6	1	91
n.s.[†] or mixed	5	2			7
Total	**962**	**271**	**198**	**11**	**1442**

[†]N.s. = not specified (i.e., respondent did not reveal affiliation).

All SEMs discussed in this chapter were estimated as multi-group models, keeping the model specification constant for all groups. The selected generic model performed best on theoretical and statistical grounds for the overall (pooled) dataset. One of the drawbacks of this approach is, however, that a generic model may not necessarily be the best specification for each group (country, commodity chain, chain stage). Optimized country-specific SEMs have been estimated, and their findings are discussed, in FOODCOMM (2008). Here, a generic SEM is most appropriate because the aim of the analyses in this chapter is to compare the relevance of central determinants of Relationship Sustainability between countries, commodities and chain stages.

The generic model specification is depicted in Fig. 1. Based on the theoretical work in Part I of this volume, and a qualitative study (Chapter 5), Relationship Sustainability is assumed to be a function of Effective Communication, the Existence of Personal Bonds, the

---

[3] The AMOS software package (version 6.0) was used with unbiased covariances as the input matrix. Given the existence of missing values in the dataset maximum likelihood estimation was conducted.

Impact of Key People Leaving, and Equal Power Distribution among chain partners. For a detailed discussion of the underlying hypotheses, and the data for these variables, see Fischer *et al.* (2009).

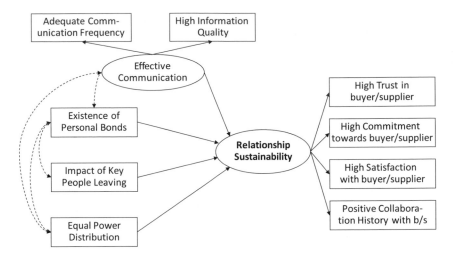

Fig. 1. Determinants of Relationship Sustainability – the generic model.

Two constructs (multi-item, latent variables) were used in the specified model. All other variables were measured as single items. The Relationship Sustainability construct was specified as a one-level, four-item latent variable.[4] For the generic model, construct reliability (Anderson and Gerbing, 1988) was assessed by Cronbach's alpha. Construct validity was assessed using principal component analysis (PCA) on the four items. Group-specific factor loadings and significance levels for these items are reported in the results section below. For the Effective Communication construct, two items were used – Adequate Communication Frequency and High Information Quality. Given only two items, PCA is not meaningful for this construct. Finally, common method bias was assessed, i.e., whether significant measurement error due to multiple-item questions limited the validity of our results, despite great care undertaken in questionnaire design and item wording to minimize this problem. Following Podsakoff *et al.* (2003), the 'single-common-method-factor approach' was used to test whether the generic SEM results would significantly change when explicitly controlling for the effects of an unmeasured latent methods factor. This was not the case. Hence, taken together these statistical tests suggest that there is no major problem with either the reliability or validity of the two constructs.

---

[4] We also tested a two-level, six-item Relationship Sustainability construct, made up from two latent factors (Relationship Quality and Relationship Stability, both measured by three different items, as discussed in Chapter 4). However, the final overall model fit was less satisfactory, thus we decided to estimate the structural equation model based on the one-level, four-item Relationship Sustainability construct.

Empirical results

Relationship Sustainability index scores

Relationship Sustainability (RS) index scores are reported for the meat and cereal chains for all countries and separately for two chain stages. These scores relate to respondents' relationships with their most important suppliers or buyers only.

   In the meat chain (Table 2), overall RS scores are high, and higher in the downstream relationship (5.9) than between farmers and processors (5.5). The differences between downstream and upstream relationships (across all countries) are statistically significant at the 99% confidence level (using a univariate ANOVA test). Downstream relationships seem to be better than upstream ones in all countries but Germany and the UK. In the upstream relationship, Finland has the lowest RS score and Germany has the highest. In the downstream relationship, Poland scores highest and Germany and the UK lowest.

Table 2. Sustainability levels[†] of B2B relationships in selected EU meat (beef, pig) chains.

| Country | Chain stage | | | | | |
| | Farmer–processor | | | Processor–retailer | | |
	Mean	Std dev	(n)	Mean	Std dev	(n)
Germany[††]	5.8	0.8	(23)	5.2	1.6	(10)
UK[††]	5.7	1.0	(139)	5.2	1.2	(6)
Spain[††]	5.5	0.9	(116)	5.7	0.7	(46)
Poland[††]	5.6	0.8	(208)	6.1	0.6	(99)
Ireland[††]	5.3	1.2	(113)	6.0	0.6	(27)
Finland[††]	5.1	1.0	(71)	5.3	0.6	(9)
Total[††]	**5.5**	**1.0**	**(670)**	**5.9**	**0.7**	**(197)**

[†]Mean values of index score calculated on the basis of four individual components (items), each one measured on a rating scale (1 = very poor, ..., 7 = very good). The items are 'Our trust in this supplier/buyer', 'Our commitment towards this buyer/supplier', 'Our satisfaction with this buyer/supplier' and 'Our collaboration with this buyer/supplier in the past'. The index scores are unweighted averages of the four obtained item scores. [††]Significant difference at the 99% confidence level. Only cases with valid observations on all four items were considered.

   In the cereal chain (Table 3), overall RS scores are also relatively high, and again generally higher in the downstream relationship. As in the meat chain, these differences are statistically significant, but only at the 95% confidence level. No significant differences can be detected when comparing the RS scores between the meat and cereal chains. This holds for the aggregated data over both chain stages as well as for each of the individual two chain stages, implying that RS only differs across chain stages but not between the two analysed commodities. As to the individual countries in the cereal chain, relationships seem to be more sustainable between processors and retailers than between farmers and processors, again with the exception of Germany. In the upstream relationship, the highest (lowest) RS scores are revealed for the UK (Finland). In the downstream case, relationships seem to be best in the UK[5] and worst in Germany. However, as in the meat chain, the differences are relatively small and may not carry real practical implications. It is also important to take into consideration that comparative results from rating scales across countries have limitations because respondents might have consistent cultural biases to judge the proposed statements.

---

[5] The UK situation is only reflected by one observation and thus this finding is not reliable.

Table 3. Sustainability levels[t] of B2B relationships in selected EU cereals (wheat, barley, rye) chains.

Country	Chain stage					
	Farmer–processor			Processor–retailer		
	Mean	Std dev	(*n*)	Mean	Std dev	(*n*)
Germany[tt]	5.8	0.8	(59)	5.6	0.9	(28)
UK	6.0	0.8	(60)	7.0	–	(1)
Spain[tt]	5.5	0.9	(117)	5.7	0.7	(50)
Finland[tt]	5.2	0.9	(82)	6.0	0.7	(38)
Total[tt]	**5.6**	**0.9**	**(319)**	**5.8**	**0.8**	**(117)**

[t]Mean values of index score calculated on the basis of four individual components (items), each one measured on a rating scale (1 = very poor, ..., 7 = very good). The items are 'Our trust in this supplier/buyer', 'Our commitment towards this buyer/supplier', 'Our satisfaction with this buyer/supplier' and 'Our collaboration with this buyer/supplier in the past'. The index scores are unweighted averages of the four obtained item scores. [tt]Significant difference at the 95% confidence level. Only cases with valid observations on all four items were considered.

In summary, RS differs mainly across chain stages with downstream relationships being generally perceived as more sustainable than upstream ones. In the case of the meat chain this holds for all countries except for the UK and Germany. In the case of the cereal chain the only exception is Germany. No significant differences in RS levels could be detected between the analysed agri-food commodities. Finally, the results indicate that respondents evaluate their most important supplier/buyer relationship as comparatively sustainable. This holds for all investigated countries, commodities and chain stages.

Structural equation modelling results

The factors which potentially affect RS are analysed by estimating the generic SEM separately, first for the different countries, second for the meat and cereal chain (across all countries) and third for the two chain stages (farmer–processor and processor–retailer).

Results for individual countries

The country-specific SEM (Table 4) fits the collected data quite well, with most goodness-of-fit measures exceeding the recommended acceptance levels (CMIN/DF = 1.336; NFI = 0.970; RMSEA = 0.015).[6] The country-specific results reveal interesting differences between the analysed countries, which can be attributed, at least partly, to national dissimilarities.[7]

---

[6] For an assessment of how well a specified SEM fits the analysed data, the following criteria are commonly used (Shook *et al.*, 2004): (i) the chi-square index; (ii) the normed fit index (NFI); and (iii) the root mean square error of approximation (RMSEA). The chi-square index tests the hypothesis whether an unconstrained specified model fits the covariance/correlation matrix as well as the given data. It should not be significant for a good model fit. If it is, there is a significant deviation of the model from the data. The chi-square fit test (CMIN/DF) adjusts the chi-square index for the degrees of freedom. Values as large as five are accepted as adequate fit, but more conservative thresholds are two or three. The NFI varies from zero to one, with one representing perfect fit. By convention, NFI values below 0.90 indicate a need to re-specify the model. The RMSEA incorporates a discrepancy function criterion (comparing observed and predicted covariance matrices) and a parsimony criterion. Most recommendations demand the RMSEA to be less than or equal to 0.05 (0.08) for a good (adequate) model fit.

[7] Differences between the country-specific SEM results can only be assumed to reflect fundamental national dissimilarities as long as the underlying sample data are representative for the analysed countries. The collected country samples are comparatively small and therefore the identified differences may reflect sample-specific particularities as well as fundamental national dissimilarities.

Table 4. Country-specific SEM estimation results – standardized parameters† and significance levels.

	Germany (n = 140)	UK (n = 254)	Ireland (n = 151)	Finland (n = 226)	Poland (n = 335)	Spain (n = 336)
Structural model						
Effective Communication –> Relationship Sustainability	0.29***	0.55***	0.32***	0.37***	0.66***	0.48**
Existence of Personal Bonds –> Relationship Sustainability	0.38***	0.15**	0.36***	0.30***	0.21***	
Impact of Key People Leaving –> Relationship Sustainability	–0.11*		–0.18**	–0.11**		
Equal Power Distribution –> Relationship Sustainability	0.42***	0.18**		0.32***		0.24***
Effective Communication <–> Existence of Personal Bonds	0.33***	0.49**	0.46***	0.36***	0.30***	0.35***
Effective Communication <–> Equal Power Distribution	0.26**	0.42***	0.37***	0.65***	0.22***	0.25***
Existence of Personal Bonds <–> Impact of Key People Leaving		0.27***		0.31***	0.60***	0.30***
Existence of Personal Bonds <–> Equal Power Distribution	0.24**	0.43***	0.45***	0.31***	0.12**	0.31***
R² Relationship Sustainability	**0.625**	**0.543**	**0.414**	**0.608**	**0.551**	**0.392**
Measurement models						
Communication Frequency <– Effective Communication	1.0+	0.86+	0.81+	0.87+	0.73+	0.89+
Information Quality <– Effective Communication	0.84***	0.87***	0.97***	0.85***	0.86***	0.88***
Trust <– Relationship Sustainability	0.86+	0.89+	0.84+	0.89+	0.81+	0.85+
Commitment <– Relationship Sustainability	0.75***	0.81***	0.77***	0.65***	0.69***	0.79***
Satisfaction <– Relationship Sustainability	0.92***	0.93***	0.89***	0.91***	0.86***	0.89***
Collaboration History <– Relationship Sustainability	0.92***	0.84***	0.83***	0.67***	0.78***	0.70***
R² Communication Frequency	1.04	0.733	0.662	0.752	0.532	0.787
R² Information Quality	0.700	0.759	0.934	0.728	0.733	0.771
R² Trust	0.745	0.792	0.712	0.783	0.653	0.719
R² Commitment	0.559	0.660	0.593	0.428	0.474	0.623
R² Satisfaction	0.848	0.871	0.796	0.829	0.730	0.796
R² Collaboration History	0.841	0.697	0.686	0.470	0.610	0.495
Overall fit						
CMIN/DF			1.336			
p			0.006			
NFI			0.970			
RSMEA			0.015			

† In the structural model, –> are regression weights and <–> are correlation coefficients; in the measurement model <– are factor loadings, –> are regression weights and <–> are correlation coefficients. R² are squared multiple correlations in the structural model and communalities in the measurement models. *** (**, *) means statistically significantly different from zero at the 99% (95%, 90%) confidence level. Only significant parameters are reported. + Parameter was constrained to 1 before estimation, thus no significance levels are available.

In Germany, the most important RS determinant is Equal Power Distribution between suppliers and buyers (with a standardized regression weight of 0.42), followed by the Existence of Personal Bonds (0.38) and Effective Communication (0.29). The Impact of Key People Leaving is not significant (at the 95% confidence level). Together, the three determinants explain about 63% of the variance of the RS score. The three significant determinants are positively and significantly correlated with each other. There is no significant correlation between the Existence of Personal Bonds and the Impact of Key People Leaving. The measurement models perform well, although there is one estimate (for Communication Frequency) which is not permissible (i.e., larger than 1). This might be due to the small sample size. The most important RS components are Satisfaction and Positive Collaboration History.

In the UK, Effective Communication is by far the most important determinant (0.55) for a sustainable chain relationship, followed by Equal Power Distribution (0.18) and the Existence of Personal Bonds (0.15). The Impact of Key People Leaving is not significant. All three determinants are positively and highly significantly correlated with each other and together explain about 54% of the variance of the RS score. The measurement models perform well with Satisfaction and Trust being the most important components of the RS construct, and Communication Frequency and Information Quality being equally important in the Effective Communication construct.

In Ireland, RS is determined by the Existence of Personal Bonds (0.36), Effective Communication (0.32), and the Impact of Key People Leaving (–0.18). However, some other factors must be important in the Irish context because the three identified determinants, taken together, only explain about 41% of the variance of the RS score. Interestingly, despite the importance of the Impact of Key People Leaving, the variable is not significantly correlated with Personal Bonds. This implies that key people in the Irish context are not necessarily those who build and look after personal relationships. All other determinants are positively and significantly correlated with each other. The measurement models perform generally well, with Satisfaction and Trust being the most important components in the RS construct. In the Effective Communication construct, Information Quality is considerably more important than Communication Frequency.

In Finland, the estimations show that three variables are about equally important for explaining RS: Effective Communication (0.37), Equal Power Distribution (0.32), and Personal Bonds (0.30). In addition, there is a significant negative effect of the Key People Leaving variable (–0.11). As to the correlations, it is striking that there is a high correlation (0.65) between Effective Communication and Equal Power Distribution. This suggests that communication is considerably better between business partners who share bargaining power. All other determinants are positively and significantly correlated with each other. Together they explain about 61% of the variance of the RS score. In the measurement models, Satisfaction and Trust are the most important components of the RS construct, while Communication Frequency and Information Quality are about similarly important in the Effective Communication construct.

In Poland, Effective Communication (0.66) is the most important determinant of RS, followed by the Existence of Personal Bonds (0.21). Both variables taken together explain about 55% of the variance in the RS score. There seems to be no effect of Key People Leaving on RS. However, this variable is highly (0.60) positively correlated with the Personal Bonds variable, suggesting that key people in the Polish context are those who are responsible for developing and maintaining relationships (and personal bonds) with business partners. The effect of Key People Leaving is then mainly reflected in the Personal Bonds variable. Interestingly, Equal Power Distribution does not seem to be of major importance in the Polish context, since it does not have a significant impact on RS nor is there a considerable correlation of this variable with Communication Quality (0.22) or the Existence of Personal Bonds (0.12). In the measurement models, Satisfaction and Trust are the most important components in the RS construct, while Communication Frequency and Information Quality are about equally important in the Effective Communication construct.

In Spain, there are also only two variables which have a significant impact on RS: Effective Communication (0.48) and Equal Power Distribution (0.24) between business partners. However, taken together, these two variables only explain about 39% of the variance in the RS score. Thus, similar to Ireland, some other, non-included factors must be important in the Spanish context. As in most of the other discussed countries, the four considered determinants are positively and significantly correlated with each other, suggesting that, for example, Effective Communication is positively related to the Existence of Personal Bonds as well as to Equal Power Distribution between business partners. Also similar to the other countries, in the measurement models, Satisfaction and Trust are the two most important components in the RS construct, while Communication Frequency and Information Quality are about equally important in the Effective Communication construct.

Overall, the country-specific estimations suggest that Effective Communication is a particularly important determinant for RS in Poland and the UK. Equal Power Distribution between business partners is mostly an issue in Germany and, to a smaller extent, in Finland, Spain and the UK. The Existence of Personal Bonds is of importance for RS in all countries except for Spain. However, the Impact of Key People Leaving is only significant in Ireland and Finland.

Results for individual commodities

The commodity-specific estimates (Table 5) reflect best the correlation structure in the collected dataset among all estimated SEMs. All model fit statistics exceed the recommended levels (CMIN/DF = 1.221; NFI = 0.990; RSEMA = 0.012). Moreover, there is an insignificant difference in correlation structures, underlining the good model fit.

As the results of the two estimated commodity SEMs are fairly similar they are discussed together. In both commodity chains, Effective Communication (0.55 in the meat chain and 0.46 in the cereals chains) is the most important RS determinant, followed by the Existence of Personal Bonds (0.26/0.23) and an Equal Power Distribution between buyers and suppliers (0.12/0.20). In the meat chain, there is also a significant negative Impact of Key People Leaving, but it is small (−0.07). Taken together, these determinants explain about 51% (meat chain) and 43% (cereal chain) of the RS score's variance. All four determinants are positively and significantly correlated with each other and the measurement models are also fairly similar with Satisfaction and Trust being the most important components in the RS construct, and Communication Frequency and Information Quality being about equally important in the Effective Communication construct.

Hence, overall, the main difference between the two commodity chains seems to be that key people are more crucial to achieve sustainable relationships in the meat chain than they are in the cereal chain.

Table 5. Commodity-specific SEM estimation results – standardized parameters† and significance levels.

		Meat (n = 967)	Cereals (n = 463)
Structural model	Effective Communication –> Relationship Sustainability	0.55***	0.46***
	Existence of Personal Bonds –> Relationship Sustainability	0.26***	0.23***
	Impact of Key People Leaving –> Relationship Sustainability	–0.07**	
	Equal Power Distribution –> Relationship Sustainability	0.12***	0.20***
	Effective Communication <–> Existence of Personal Bonds	0.29***	0.34***
	Effective Communication <–> Equal Power Distribution	0.36***	0.19***
	Existence of Personal Bonds <–> Impact of Key People Leaving	0.39***	0.36***
	Existence of Personal Bonds <–> Equal Power Distribution	0.17***	0.25***
	R^2 Relationship Sustainability	**0.512**	**0.428**
Measurement models	Communication Frequency <– Effective Communication	0.85+	0.91+
	Information Quality <– Effective Communication	0.89***	0.85***
	Trust <– Relationship Sustainability	0.86+	0.87+
	Commitment <– Relationship Sustainability	0.75***	0.75***
	Satisfaction <– Relationship Sustainability	0.89***	0.91***
	Collaboration History <– Relationship Sustainability	0.77***	0.78***
	R^2 Communication Frequency	0.727	0.833
	R^2 Information Quality	0.796	0.730
	R^2 Trust	0.737	0.757
	R^2 Commitment	0.566	0.555
	R^2 Satisfaction	0.785	0.829
	R^2 Collaboration History	0.586	0.610
Overall fit	CMIN/DF	1.221	
	p	0.149	
	NFI	0.990	
	RSMEA	0.012	

† In the structural model, –> are regression weights and <–> are correlation coefficients; in the measurement model <– are factor loadings, –> are regression weights and <–> are correlation coefficients. R^2 are squared multiple correlations in the structural model and communalities in the measurement models. *** (**, *) means statistically significantly different from zero at the 99% (95%, 90%) confidence level. Only significant parameters are reported. + Parameter was constrained to 1 before estimation, thus no significance levels are available.

Results for individual chain stages

The chain stage-specific estimates (Table 6) also yield a model fit which is highly satisfactory. With a CMIN/DF of 1.349, a NFI of 0.989 and a RSMEA of 0.016, and an insignificant deviation in correlations structures (at the 95% confidence level), all model fit measures exceed recommendations.

In the farmer–processor stage, Effective Communication (0.49) is the most important RS determinant, followed by the Existence of Personal Bonds (0.29), Equal Power Distribution between business partners (0.15), and the negative Impact of Key People Leaving (–0.07). Taken together, these four determinants explain about 48% of the variance in the RS score. These determinants are positively and significantly correlated with each other. The measurement models reflect the findings of the country- and commodity-specific models, with Satisfaction and Trust being the most important components in the RS construct, and Communication Frequency and Information Quality being equally important in the Effective Communication construct.

At the processor–retailer level, there are only two significant determinants of RS: Effective Communication (0.62) and Equal Power Distribution (0.14). Taken together, these two determinants explain about 50% of the variance in the RS score. Nevertheless, all four considered determinants are positively and significantly correlated with each other, indicating that in particular Effective Communication is improved by the Existence of Personal Bonds and an Equal Power Distribution between business partners. The obtained estimates for the measurement models are similar to the ones at the farmer–processor stage.

Overall, the Existence of Personal Bonds (and the Impact of Key People Leaving) do not seem to be important at the processor–retailer chain stage but Effective Communication is crucial. Equal Power Distribution is equally of relevance at both chain stages. Taking into consideration the findings from the Relationship Sustainability scores (Tables 2 and 3) it appears that at the processor–retailer chain stage more sustainable relationships can be achieved as between farmers and processors, and that fewer factors are responsible for achieving this higher outcome. Thus it seems that as organizations become larger, inter-organizational relationships become more dependent on good communication practices.

Conclusions

In this chapter, the sustainability of relationships for primarily small- and medium-sized enterprises in several European agri-food chains was analysed in two steps. First, sustainability levels were calculated using an index. Due to statistical reasons, the index incorporates four out of six theoretically justified components (see Chapter 4) and considers trust, satisfaction, commitment and a positive collaboration history between buyers and suppliers. Second, the relevance of important RS determinants was identified and compared between countries, commodity chains and chain stages, using structural equation modelling.

Our results indicate that the sustainability of agri-food chain relationships with their most important suppliers/buyers is generally high. Major differences in RS index scores exist only between the two chain stages. In general, processor–retailer relationships are more sustainable than farmer–processor ones. This is true for all countries except for Germany in both commodity chains, and for the UK in the meat chain. As to Germany, it is

Table 6. Chain stage-specific SEM estimation results – standardized parameters† and significance levels.

		Farmer–processor (n = 1,086)	Processor–retailer (n = 344)
Structural model	Effective Communication –> Relationship Sustainability	0.49***	0.62***
	Existence of Personal Bonds –> Relationship Sustainability	0.29***	0.10*
	Impact of Key People Leaving –> Relationship Sustainability	−0.07**	
	Equal Power Distribution –> Relationship Sustainability	0.15***	0.14**
	Effective Communication <–> Existence of Personal Bonds	0.29***	0.29***
	Effective Communication <–> Equal Power Distribution	0.29***	0.23***
	Existence of Personal Bonds <–> Impact of Key People Leaving	0.38***	0.38***
	Existence of Personal Bonds <–> Equal Power Distribution	0.19***	0.22***
	R^2 Relationship Sustainability	**0.475**	**0.499**
Measurement models	Communication Frequency <– Effective Communication	0.86+	0.90+
	Information Quality <– Effective Communication	0.87***	0.91***
	Trust <– Relationship Sustainability	0.87+	0.80+
	Commitment <– Relationship Sustainability	0.74***	0.72***
	Satisfaction <– Relationship Sustainability	0.89***	0.87***
	Collaboration History <– Relationship Sustainability	0.77***	0.78***
	R^2 Communication Frequency	0.742	0.801
	R^2 Information Quality	0.761	0.826
	R^2 Trust	0.755	0.647
	R^2 Commitment	0.551	0.519
	R^2 Satisfaction	0.799	0.752
	R^2 Collaboration History	0.589	0.603
Overall fit	CMIN/DF	1.349	
	p	0.061	
	NFI	0.989	
	RSMEA	0.016	

† In the structural model, –> are regression weights and <–> are correlation coefficients; in the measurement model <– are factor loadings, –> are regression weights and <–> are correlation coefficients. R^2 are squared multiple correlations in the structural model and communalities in the measurement models. *** (**, *) means statistically significantly different from zero at the 99% (95%, 90%) confidence level. Only significant parameters are reported. + Parameter was constrained to 1 before estimation, thus no significance levels are available.

well-known that the food retail situation is highly competitive, marked by high price competition, a predominance of low-service discounters and a high relevance of retail brands. In fact, the situation is so competitive that even Walmart, the world biggest retail chain, had to abandon the German market in 2006. This special situation helps to explain the lower RS scores at the downstream level.

Regarding the relative importance of the considered RS determinants, the SEM estimations identified Effective Communication as being crucial in most of the estimated models. It was the most important determinant in Poland, the UK and Spain, in both analysed chains as well as at both chain stages. This finding confirms the results from other studies which found Effective Communication as the most important factor in achieving successful inter-organizational cooperation (Bleeke and Ernst, 1993; Mohr *et al.*, 1996).

The Existence of Personal Bonds component was found to be the second-most important determinant for sustainable business relationships in most of the analysed agri-food chains and thus supports the discussion on social structure theory in Chapter 3 of this book. In Ireland, it is even the most crucial factor and it is also positively correlated with Effective Communication. This suggests that the two effects reinforce each other, indicating that for example the existence of personal bonds improves the effectiveness of communication and vice versa.

The effect of key people leaving an organization is from a theoretical viewpoint related to the importance of personal bonds and indeed our findings show that both factors are often positively and significantly correlated with each other. This indicates that key people are in general those who develop personal bonds with business partners. The Impact of Key People Leaving on RS has been consistently estimated as being negative, but it is not always significant and generally low in magnitude. Only in the models for Ireland and Finland, the farmer–processor chain stage and in the meat chain does the effect prove to be significant.

Equal distribution of power between business partners was found to be the third-most important determinant for RS in both commodity chains and at the farmer–processor stage. It is of particular relevance in Germany and the second-most important one in Finland, the UK and Spain, but it seems to have no impact on RS in Ireland and Poland. Market power asymmetries between businesses, often due to differences in the scale of the companies, can create a feeling of insecurity and vulnerability among smaller partners. This is frequently linked to limited confidence in the fairness of treatment by the more powerful business partners. These results confirm the insights from behavioural economics discussed in Chapter 3 as people not only care about their own material pay-off but also about fairness. Thus, power asymmetries can lower trust and commitment and can be detrimental to the sustainability of a relationship. None the less, it is acknowledged that where fair treatment occurs, the effect of unequal power distribution may be reduced.

Overall, the results of our analysis comply with the findings of Schulze *et al.* (2007a, b) for Germany, and Lindgreen *et al.* (2004) for the Netherlands (as outlined in the introduction). Moreover, they represent the first cross-country empirical results in this particular research area.

From a managerial viewpoint, our results offer a number of useful insights. They confirm the postulates from social structure theory that business relationships have a personal dimension that is potentially significant for its sustainability. The development of personal bonds between owners/managers/staff of organizations needs to be acknowledged by managers as it can help to improve inter-organizational relationships by strengthening mutual trust and commitment. Trust is often an important pre-condition for the willingness to exchange information, especially if it is confidential. Thus, there is also a close link between the existence of personal bonds and the effectiveness of communication, which is confirmed by our results. Personal bonds can be particularly helpful when conflicts arise within a business relationship. Given the relevance of personal bonds, an organization should be fully prepared when staff who have dealt with customers/suppliers leave, or change positions within a company. We found that especially for Finland and Ireland, the departure of key staff can cause a disruption in agri-food chain customer or supplier relationships. The problem of key staff leaving might be mitigated by measures such as advance notification of customers/suppliers and securing transition periods which allow for the 'handing over' of a relationship to new staff (Bendapudi and Leone, 2002). In addition, organizations could try to create 'structural' bonds, by offering to business partners value-added benefits that are not readily available from competitors. Accordingly, structural bonds lead to the creation of multiple links between organizations and thus reduce the potential impact of key contact employee turnover (Bendapudi and Leone, 2002). As an alternative, organizations can reduce the reliance on key people by a number of measures such as pursuing a staff rotation strategy or having teams rather than single persons dealing with customers/suppliers. However, such practices may hamper the development of vital personal bonds (Bendapudi and Leone, 2002).

The results also indicate that larger companies can make their business relationships more sustainable by achieving win–win situations and a fair distribution of rewards. In addition, improving communication and transparency and/or developing personal bonds with business partners can help to offset a potential negative impact of unequal power distribution. Where large business partners have a superior understanding of market requirements, conditions and developments, sharing such information with smaller suppliers may lead to benefits for all concerned. Moreover, enhancing technological and human communication capacities, leading to higher information quality and more adequate transmission frequencies, could help to build and maintain stronger agri-food chains.

From a policy perspective, the results highlight that policies that focus on improving effective business-to-business communication are likely to have a positive impact on the stability of chain relationships across all member states, commodities and chain stages. In addition, competition policies that prevent market power exploitation or measures that allow SMEs to build up countervailing market power could improve the sustainability of business relationships (see Chapter 16). However, differences between countries, chains and chain stages in the relative importance of the various determinants analysed indicate that a fine-tuning of policy interventions is required in many instances to be most effective. Given the identified significant differences of RS between chain stages, improvements should be targeted at the farmer–processor stage more than the processor–retailer stage. This will be more challenging due to lower levels of concentration, greater geographic dispersion and lower levels of information and communication technology adoption in the farming sector.

Further research should consider the specific situation for each country and chain in order to better understand why some determinants are more important than others. The concentration degree at different chain levels, the existence and level of policy intervention, etc. are some factors that could affect the implementation of specific actions at regional and national levels. Future empirical investigations in this area may fruitfully consider other agri-food chains. Our analysis has explored agricultural commodities (cereals and red meat) that are mostly characterized by discrete production periods. For more continuously produced products (or those with short production cycles) such as dairy products or certain vegetables (tomatoes, cucumbers etc.) commercial interactions between farmers/growers and processors/marketers are more frequent and regular, and the nature of their business relationships may therefore be different. Large-scale longitudinal studies on the evolution of vertical inter-organizational relationships in the agri-food sector may also be valuable. They can help to better understand how relationship sustainability develops over time.

Acknowledgements

This publication derives from the research project on 'Key factors influencing economic relationships and communication in European food chains' (FOODCOMM, SSPE-CT-2005-006458) which was funded by the European Commission as part of the Sixth Framework Programme. The collaborating laboratories are: Institute for Food and Research Economics, University of Bonn, Germany (coordinator); Land Economy and Environment Research Group, Scottish Agricultural College (SAC), Edinburgh and Aberdeen, UK; Institute for Agricultural Development in Central and Eastern Europe (IAMO), Germany; The Ashtown Food Research Centre (AFRC), Teagasc, Dublin, Ireland; Ruralia Institute, University of Helsinki, Finland; Institute of Agricultural and Food Economics (IAFE), Poland; and Unit of Agri-food Economics and Natural Resources, Agri-food Research and Technology Center of Aragón (CITA), Spain.

References

Anderson, H. and Gerbing, D. (1988) Structural equation modeling in practice: a review and recommended two-step approach. *Psychological Bulletin* 103(3), 411–423.

Armstrong, J. and Overton, T. (1977) Estimating nonresponse bias in mail surveys. *Journal of Marketing Research* 14(Aug), 396–402.

Bendapudi, N. and Leone, P. (2002) Managing business-to-business customer relationships following key contact employee turnover in a vendor firm. *Journal of Marketing* 66, 83–101.

Bleeke, J. and Ernst, D. (1993) *Collaborating to Compete*. John Wiley & Sons, New York.

Bollen, K. (1989) *Structural Equations with Latent Variables*. John Wiley & Sons, New York.

Fischer, C., Hartmann, M., Reynolds, R., Leat, P., Revoredo-Giha, C., Henchion, M., Albisu, L. and Gracia, A. (2009) Factors influencing contractual choice and sustainable relationships in selected agri-food supply chains. *European Review of Agricultural Economics* 36(4), 541–569.

FOODCOMM (2008) Analysis of survey data and identification of issues for country-specific research, Report No 4, Key Factors Influencing Economic Relationships and Communication in European Food Chains (FOODCOMM), European Commission: The Sixth Framework Programme 2002–2006, unpublished.

Leat, P. and Revoredo-Giha, C. (2008) Building collaborative agri-food supply chains: the challenge of relationship development in the Scottish red meat chain. *British Food Journal* 110(4/5), 395–411.

Lee, D., Pae, J. and Wong, Y. (2001) A model of close business relationships in China (guanxi). *European Journal of Marketing* 35(1/2), 51–69.

Lindgreen, A., Trienekens, J. and Vellinga, K. (2004) Contemporary marketing practice: a case study of the dutch pork supply chain. In: Bremmers, H., Omta, S., Trienekens, J. and Wubben, E. (eds) *Dynamics in Chains and Networks: Proceedings of the 5th International Conference on Chain and Network Management in Agribusiness and the Food Industry*. Wageningen, the Netherlands, pp. 273–279.

Mohr, J., Fisher, R. and Nevin, J. (1996) Collaborative communication in interfirm relationships: moderating effects of integration and control. *Journal of Marketing* 60(6), 103–115.

Podsakoff, P., McKenzie, S., Lee, J.-Y. and Podsakoff, N. (2003) Common method biases in behavioral research: a critical review of the literature and recommended remedies. *Journal of Applied Psychology* 88(5), 879–903.

Prahalad, C. and Ramaswamy, V. (2004) Co-creating unique value with customers. *Strategy & Leadership* 32(3), 4–9.

Reynolds, N., Fischer, C. and Hartmann, M. (2009) Determinants of sustainable business relationships in selected German agri-food chains, *British Food Journal* 111(8), 776–793.

Schulze, B., Spiller, A. and Theuvsen, L. (2007a) A broader view on vertical coordination: lessons from German pork production. *Journal on Chain and Network Science* 7(1), 35–53.

Schulze, B., Wocken, C. and Spiller, A. (2007b) Relationship quality in agri-food chains: Supplier management in the German pork and dairy sector. *Journal on Chain and Network Science* 6(1), 55–68.

Shook, C., Ketchen, D., Hult, G. and Kacmar, K. (2004) An assessment of the use of structural equation modeling in strategic management research. *Strategic Management Journal* 25(4), 397–404.

Chapter 8

Enhancing the Integration of Agri-food Chains: Challenges for UK Malting Barley

César Revoredo-Giha and Philip Leat

Scottish Agricultural College (SAC), Edinburgh and Aberdeen, United Kingdom

Introduction

Despite their importance for describing how an industry is organized, issues related to relationships amongst firms (e.g., how they are integrated and how they coordinate their plans) have been largely absent from the traditional theory of the firm in economics. Instead, according to Richardson (1972), the traditional theory has portrayed firms within the economy 'as islands of planned coordination in a sea of market relations' (p. 895).

Whilst the aforementioned representation allows market structures to be depicted where the number of firms goes from one (monopoly) to infinity (perfect competition), with the price system as the organizing mechanism, it leaves aside the fact that several markets or industries can be described as a 'dense network of cooperation and affiliation by which firms are inter-related' (Richardson, 1972, p. 883) and where prices are only one possible way of coordination amongst them. Furthermore, the problem with the traditional view is that not only does it not describe the different relationships between firms, commonly found in the market, but also as a consequence, it is not capable of analysing them and formulating recommendations to improve their efficiency and effectiveness.[1]

As pointed out by Hobbs and Young (2000), agri-food markets in many countries are moving away from traditional spot markets (where the description of the traditional firm theory fits well), towards closer vertical integration arrangements. However, this process takes a variety of forms and involves a diverse number of partners, as spot markets and vertical integration can be viewed as the two extremes of a continuum of inter-firm arrangements. Thus, in some cases such as the well researched case of poultry in the USA, it consists of a fully vertically integrated arrangement; while in others, it may take the form of partnerships, strategic alliances, etc. (for examples of collaborative arrangements in the UK see Hughes, 1994).

---

[1] Within economics a reaction to the traditional theory of the firm can be found in Coase's seminal work on the nature of the firm (Coase, 1937) and which has evolved to become the 'new institutional economics'. An overview of the approach to supply chain management can be found in Hobbs (1996).

In this context, this chapter seeks to explore the theoretical and practical issues that may affect the relationships within an agri-food supply chain, and therefore the chain's degree of cohesion and coordination. The main motivation behind this topic is that improved supply chain coordination and cooperation amongst the different segments of a supply chain can improve its efficiency and effectiveness, and therefore, its competitiveness and long-term sustainability.

This chapter, which focuses on the UK barley-to-beer and whisky supply chain, draws on information collected as part of the EU-funded project FOODCOMM, and is based on two complementary analyses. First, using postal survey data from businesses operating at various stages along the supply chain, a structural equation model is formulated to determine those factors that affect the sustainability of relationships in the chain (and therefore its degree of integration). The second analysis, which helped to improve our understanding of the model's findings, consists of an in-depth case study based on an important malting barley-to-beer supply chain in eastern England.

The chapter starts with a brief theoretical discussion of the factors that may affect business relationships within supply chains. This is followed by the empirical section, which comprises three parts: first, a brief overview of the malting barley supply chain in the UK so as to set the context of the analysis; second, the results of a structural equation model (SEM) of sustainable relationships; and third, a case study analysis based on the aforementioned English supply chain. Finally, the main conclusions from the analysis are presented.

Coordination of supply chains: findings from the literature

A starting point for the factors affecting the coordination of supply chains can be found in the 'new institutional economics' and one of its major components: the presence of transaction costs in the use of market instruments.[2]

According to Hobbs (1996) 'transaction costs, and their reduction, lie at the heart of the interest in supply chain management' (p. 26). In this sense, proactive moves to enhance the management of supply chains are fundamentally concerned with improving their efficiency to gain competitive advantage. Thus, on the one hand, examples of factors reducing transaction costs can be found in cooperation, teamwork and the rapid interchange of data among companies. On the other hand, adversarial relationships along the supply chain, for instance, may increase transaction costs.

Whilst the literature on 'new institutional economics' has focused more on understanding the reasons behind the existence of different types of firm arrangements using economic analysis (see for instance Milgrom and Roberts (1992) for an overview), it has not dwelt on factors affecting the business relationships within supply chains. In this respect, as pointed out by Hobbs (1996), 'supply chain management offers many insights into how industries are organized and into efficiency gains which can be made under different organizational structures, pointing out that this is a multidisciplinary concept,

---

[2] Transaction costs can be divided into three main categories: *information costs* (i.e., costs faced by firms and individuals in the search for information about products, prices, inputs and buyers and sellers); *negotiation costs* (i.e., costs that arise from the physical act of the transactions such as negotiating and writing a contract or paying for the services of an intermediary to the transaction); and *monitoring or enforcement costs* (i.e., costs that arise after the terms of the contract have been negotiated, and may involve controlling quality of the products to ensure that the terms of the contract are satisfied or the costs of legally enforcing the terms of a contract) (Hobbs, 1996).

drawing on aspects of marketing, economics, logistics, organizational behaviour, etc.' (p. 15). Therefore, the supply chain management literature is an appropriate source to search for factors affecting business relationships, and consequently, the degree of integration within supply chains.

Figure 1, taken from Leat and Revoredo-Giha (2008), is an effort to summarize the interaction of factors which influence the development of good, sustainable supply chain relationships and performance, based on issues in the literature.

As shown in the figure, the supply chain relationships within which decision-making is integrated, invariably involve the development of inter-organizational relationships. Such relationships, if they are to be sustainable, should be stable and mutually beneficial amongst the partners. Furthermore, in recent times it has become widely recognized that the pro-active management of such relationships can present a critical source of competitive advantage (e.g., Dyer and Singh, 1998; Sahay, 2003; Power, 2005).

The different elements in Fig. 1 can be categorized as follows: characteristics of the supply chain members including awareness, trust, commitment and satisfaction; devices to facilitate relationships and performance, such as communication, planning and reward distribution policies; and the interventions of the chain leader or focal enterprise.

At the outset it should be recognized that supply chain relationships take place within a social, cultural, political, economic and technological environment, which in Fig. 1 is represented by the dotted and dashed line that frames the diagram. In the wider scope of economic activity – be it production, exchange or consumption – such activity is regarded as 'embedded' in patterns of social organization, relationships and cultural characteristics (Granovetter, 1985).

A fundamental pre-requisite of good marketing performance is that of awareness of the customer, and their needs. Harmsen *et al.* (2000) note that market orientation involves a focus on, and responsiveness to, customers and competitors, as part of an external orientation. Within the context of supply chains and their performance, this awareness should be extended to embrace the needs of other chain participants as well. Such awareness invariably involves information sharing (Brown, 1984; Peterson *et al.*, 2000).

Assessing the quality of inter-organizational relationships has been the focus of many recent studies. Roberts *et al.* (2003) reviewed several of them, which along with other studies have illustrated the importance of 'soft' factors – as opposed to 'hard' economic or financial measures of performance – as indicators of relationship quality (Lagace *et al.*, 1991; Moorman *et al.*, 1992; Wray *et al.*, 1994; Bejou *et al.*, 1996; Hennig-Thurau and Klee, 1997; Boles *et al.*, 1997; Dorsch *et al.*, 1998; Rosen and Suprenant, 1998; Lang and Colgate, 2003; Bennet and Barkensjo, 2005). Collectively these studies show the importance of: satisfaction (cognitive and affective evaluation based on the personal experience across all episodes within a relationship, Storbacka *et al.*, 1994); commitment (an enduring desire to maintain a valued relationship, Moorman *et al.*, 1992) and trust (a willingness to rely on an exchange partner in whom one has confidence, Lewin and Johnston, 1997).

Moving away from the attributes of supply chain participants to the mechanisms which can further supply chain relationships and performance, communication has emerged as an important factor in achieving successful inter-organizational cooperation (e.g., Bleeke and Ernst, 1993; Mohr *et al.*, 1996; Tuten and Urban, 2001). Since communication allows chain participants to learn about and react to changes in the requirements and expectations of other chain participants, superior chain performance, enabled by modern information technologies, is of prime importance to the continued development of inter-organizational relationships. Enhanced transparency, through an information-sharing mechanism linking supply chain partners, is one of the most critical drivers of supply chain success (Min and Zhou, 2002). Increasingly, communication of comparative performance information, which enables benchmarking, can also play a role in furthering enterprise and chain performance.

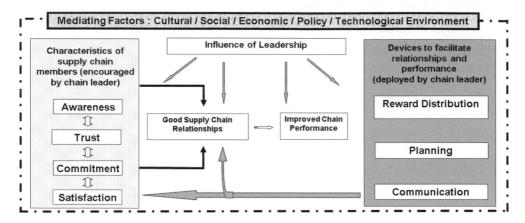

Fig. 1. Conceptual framework: relationships in an effective supply chain (based on Leat and Revoredo-Giha, 2008).

The concept of sharing rewards and penalties within the chain is a mechanism for driving chain efficiency and unity (Peterson *et al.*, 2000). This might be regarded as particularly important within agri-food chains where the overall supply chain margin is under pressure. O'Keeffe (1998), in presenting lessons from supply chain partnerships in Australian agribusiness, identifies the importance of rewards being shared equitably for partnership success. Similarly, Fearne (1998), in looking specifically at supply chain partnerships in the British beef industry, stresses the importance of the premise that all will benefit and all will be winners.

Peterson *et al.* (2000) stress that whole chain planning is necessary for whole chain success and that all chain members should be involved in the planning process if a chain's potential is to be realized. Similarly, Fearne (1998) in his British beef sector research concludes that partners need to share a common vision of how to work together and to meet their volume and quality requirements. The practical details of such planning activities in the meat industry are reported by Sadler and Hines (2002), who conclude that 'it is necessary to work with all partners in a number of supply chains to complete the design and practical steps required to enable the whole supply chain to plan its operations and logistics in one process' (p. 238).

The value of leadership to successful supply chain relationships, which appears in Fig. 1 as facilitating the interaction of all the elements in the supply chain, has been highlighted in a number of studies related to marketing channels (Brown, 1984). Its importance is summarized (Peterson *et al.*, 2000) in the statement that 'leaderless chains lack vision, direction and unity and are characterized by a high failure rate. The leader's role is to provide the focus and coordination and to ensure that all participants know, and are committed to, the customer's objectives' (p. 10). Furthermore, the quality of leadership within supply chain firms is an important driver of development and improvement, as this helps to shape the culture of the firm as well as managing the perceptions held by staff of 'us and them' in their alliances (Kidd *et al.*, 2003).

Empirical results

Overview of the UK malting barley-to-beer chain

The purpose of this section is to provide a brief overview of the UK malting barley-to-beer chain at the time of the study (2006) and the relationships amongst its different constituent agents.

Malting barley is grown mainly down the (drier) east side of Great Britain, with the distilling industry being a major buyer of malt in Scotland and brewers being the main customers in England. Of the total usage of barley in 2006, 1.7 million t were classified as used in domestic brewing and distilling, with the remainder of the crop going for feed (about 3.5 million t), seed, other uses and export (between 0.7 to 1.0 million t). Specialist growers in the main production areas produce high-quality malting barley, although new varieties have widened the geographical area where malting barley can be successfully grown. However, the recent CAP reforms and price increases of other cereals may well lead to a reduction in malting barley production in marginal areas and those more distant from customer outlets.

Malting barley is normally purchased from farmers through cooperatives and merchants. Such purchases are most frequently made using contracts, which may set 'relative' prices (relative to other grain prices) along with conditions relating to quality and service. These contracts tend to be issued by merchants on behalf of maltsters. They provide maltsters with some predictability regarding prices, quality and service attributes.

While many specialist growers have good relationships with their merchants, a significant proportion fails to meet quality or service requirements. Moreover, the small size of the malting barley premium (usually in years when grain prices are high) can disrupt the spot market, with farmers reluctant to release grain.

Table 1 presents the demand–supply balance of malt in Great Britain. There are three types of maltsters: brewer-maltsters (own and operate maltings for their own brewing needs, they represent 11% of the total production), distiller-maltsters (own and operate maltings for their own distilling needs, 12% of the total production) and sales maltsters (make malt to customers' specification, for the brewing, distilling and food industries, 78% of the total production).

The malting sector is generally operating with relatively old plants and is achieving low margins, although some exceptions do exist. There is a degree of integration in the upstream part of the chain: some maltsters own, or have a stake in, their merchant suppliers. However, there is little vertical integration through ownership between maltsters and brewers. Only two significant brewers own maltings (subsequently falling to one).

The contracts issued by a merchant on behalf of a maltster are normally matched by those of brewers for malt purchase. Relationships between maltsters and brewers are well established and close associations have developed. However, UK barley suffers from high drying and storage costs, and price remains a key factor in maltster–brewer relationships.

The beer-brewing sector is generally achieving modest returns. Differentiation within particular segments of the market (e.g., premium lager) is limited, so brand promotion and efficiency in production and cost control are very important for business performance.

Table 1. Great Britain – malt production, usage and trade, 2003–2005 (thousand t).

	2003	2004	2005
Malt production:	1608	1577	1451
Brewer-maltsters	164	163	129
Distiller-maltsters	179	166	163
Sales maltsters	1265	1248	1159
Malt imports	11	17	4
Total availability	**1619**	**1594**	**1455**
Malt requirements:	1184	1183	1190
For brewing	661	666	637
For distilling	458	448	481
Other purposes	65	69	72
Malt exports	381	403	288
Total usage	**1565**	**1586**	**1478**

Source: Maltsters' Association of Great Britain (MAGB).

The biggest differences in brewers' cost bases occur between the large national brewers and small local and regional breweries. The largest brewers have achieved most of the readily available efficiencies in production and the biggest future improvement in performance appears to be in packaging. The business relationships between brewers and major pub chains tend to be based on supply agreements (contracts) of 3–5 years. Brewers also act as wholesalers, selling their own beers and beers and spirits of other producers. Brewers frequently have supply agreements with retailers. There is a long-term trend away from beer consumption in pubs and clubs (the On-Trade) towards consumption through the Off-Trade (e.g., supermarkets and off-licences). Beer retailing is also under pressure from personal imports, which may account for 8% of UK beer consumption.

Overall, the barley-to-beer chain produces good-quality malting barley and malt but has been under economic pressure, which stems largely from international competition and challenging domestic demand conditions. Moreover, weaknesses exist in the supply of barley from farms, and in the generally old plants and low margins of the malting sector. Most improvement can probably be achieved through 'vendor-assured grain', which should facilitate cost savings further down the chain. However, adoption of this approach will require some of the benefits to be passed back to farmers in enhanced prices.

The greatest area of weakness in the barley-to-beer chain is generally at the interface between barley producers and their customers. Farmers rely on personal contact with the staff of their purchaser (merchant or maltster). Improved two-way communications on issues such as farm production costs, quality and performance standards, market conditions, customer requirements, etc., are widely regarded as important for relationship development and improved chain performance.

Modelling of relationships for the UK malting barley-to-beer supply chain

In order to study the determinants of sustainable relationships in the UK malting barley-to-beer supply chain, a postal survey was undertaken during 2006/07. The survey resulted in 69 stakeholders' responses (58 farmers, 7 processors and 4 middlemen).

Before proceeding to the structural equation modelling (SEM) analysis some results from the survey are worth noting. According to the survey, the most important relationships in the supply chain were 'formal contracts' and 'repeated transactions with the same partner'. The main reason for the use of formal contracts was that they give farmers security of demand, and to processors, security of supply and cost predictability.

The respondents' main relationship (i.e., the relationship that explained more than 50% of the respondents' turnover) was considered to be 'commercially rewarding' and also based on a 'strong personal relationship(s)' component. Furthermore, it was mentioned that the main relationship has a positive effect on different performance aspects of the firm such as 'profitability' (70% of the respondents mentioned a positive effect), 'product or process quality' (60%) and 'turnover' (44%).

As regards communication, the most common means used were 'telephone' and 'e-mail'. Respondents indicated that they were quite satisfied with the communication features (e.g., frequency, quality, relevance) that they operated with. Moreover, they asserted that communication had a positive effect on 'profitability' (77% of respondents indicated this to be the case) and 'product quality' (60%).

The determinants of sustainability in business relationships in the malting barley-to-beer chain were studied using the information collected in the survey and SEM. The model consisted of one structural equation for explanation of the latent dependent variable Relationship Sustainability. The full model is presented in Table 2.

The construct Relationship Sustainability was constructed from two other latent variables: Relationship Quality and Relationship Stability (see Chapters 4 and 7 for a more detailed description of this construct). On the one hand, Relationship Quality was built using indicators from the survey related to Trust in the Buyer/Seller, Commitment Towards the Buyer/Seller and Satisfaction as Regards the Relationship with Buyer/Seller. On the other hand, Relationship Stability was constructed on the basis of the Collaboration History of the Partners and the Ability of the Relationship to Endure Conflict.

Several variables were included in the model to explain Relationship Sustainability in order to test a number of hypotheses. These variables were: Communication Quality, the existence of strong Personal Bonds, both partners have Equal Power in the relationship, Age of Relationship, Age of Business, the company operates in a market under Strong Competition, the relationship with the partner is Commercially Rewarding, the company tries to avoid uncertainty whenever possible (Risk Aversion), the firm has important roots in the local economy (Local Embeddedness), and the percentage use of repeated transactions with the same partner (Use of Repeated Transactions).

Two variables entering into the structural model were constructed as latent variables. The variable Communication Quality was constructed from two indicators from the survey: Information Quality and Communication Frequency. The factor Local Embeddedness was constructed using four indicators: whether the products of the firm are part of a local brand (Local Products), whether the firm's suppliers were from the local area (Local Suppliers), whether their buyer was from the local area (Local Buyer) and whether the firm participates in the local community (Other Local Ties).

The overall results indicate a resonable fit to the data in terms of: the minimum discrepancy divided by its degrees of freedom (i.e., CMIN/DF), which was equal to 1.5; the normed fit index (i.e., NFI) equal to 0.634; and the square error of approximation (i.e., RMSEA) equal to 0.086. Despite the model seeming acceptable in statistical terms,

Table 2. SEM estimation results – standardized parameters[†] and significance level.

Dependent construct: Relationship Sustainability	Barley-to-beer chain (*n* = 69)	
	Parameters	Significance
Structural model		
Communication Quality	0.777	***
Personal Bonds	0.063	0.43
Equal Power	−0.033	0.73
Age of Relationship	−0.112	0.17
Age of Business	0.041	0.61
Competition	−0.103	0.20
Commercially Rewarding	0.358	***
Risk Aversion	−0.147	*
Local Embeddedness	−0.073	0.47
Use of Repeated Transactions	0.002	0.16
Measurement models for latent variables		
• Relationship Sustainability		
Relationship Quality	1.000	+
Relationship Stability	1.000	***
• Relationship Quality		
Trust	0.863	+
Commitment	0.745	***
Satisfaction	0.925	***
• Relationship Stability		
History of Collaboration	0.909	+
Endurance of Conflict	0.781	***
• Communication Quality		
Communication Frequency	0.819	***
Information Quality	0.766	+
• Local Embeddedness		
Local Products	0.667	***
Local Suppliers	0.671	***
Local Buyers	0.479	***
Other Local Ties	0.372	+
R^2 Relationship Stability	0.802	
R^2 Communication Frequency	0.641	
R^2 Information Quality	0.613	
R^2 Endurance of Conflict	0.609	
R^2 History of Collaboration	0.826	
R^2 Satisfaction	0.856	
R^2 Commitment	0.555	
R^2 Trust	0.744	
R^2 Local Products	0.139	
R^2 Local Suppliers	0.450	
R^2 Local Buyers	0.445	
R^2 Other Local Ties	0.230	
Overall model fit indicators		
CMIN/DF	1.505	***
NFI	0.634	
RMSEA	0.086	

[†]Standardized coefficients (coefficients divided by their standard deviations) are used to eliminate the effect of different units. R^2 are squared multiple correlations in the structural model and communalities in the measurement models. CMIN/DF is the minimum sample discrepancy divided by degrees of freedom, NFI is the normed fit index, which varies from 0 to 1 and RMSEA is the root mean square error of approximation. *** (**, *) means statistically significantly different from 0 at the 99% (95%, 90%) confidence level. + means parameter was constrained to 1 before estimation: no significance levels are available.

the main purpose of the SEM analysis was to test hypotheses related to the impact of various variables on the sustainability of relationships, and not necessarily to pursue a 'good fit' in the models (Hair *et al.*, 2006, p. 758).

The path diagram, considering only those variables that were significant in the regression (Table 2), is presented in Fig. 2.

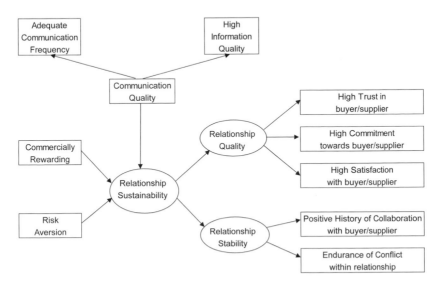

Fig. 2. Path diagram of the SEM for the malting barley-to-beer supply chain.

The results indicate that the two major factors influencing Relationship Sustainability positively in the malting barley-to-beer supply chain were (with coefficients significant at 99%): Communication Quality (with a standardized coefficient of 0.78) and Commercial Reward (0.36). The variable Risk Aversion (–0.15) was significant at the 90% confidence level, indicating that the more a company tries to avoid uncertainty the less sustainable is the relationship. All the other variables (and therefore the hypotheses associated with them) were rejected by the SEM.

Case study of a UK malting barley-to-beer supply chain

This case study examines the operations, supply chain communication and relationships of a UK malting barley-to-beer supply chain. It is centred on a grain cooperative and its members. The cooperative (Camgrain) supplies malting barley via its marketing distributor (Gowlett Grain) to a major maltster (Greencore Malt based at Bury St Edmunds, England), which in turn supplies malt to a brewer (Greene King, also at Bury St Edmunds, England).

The case study endeavours to identify: first, the nature of the relationships and communication between the various chain participants from farmers to the brewer and pub owner (Greene King); second, the benefits of the supply chain relationships to the various participants; and third, the key factors in good supply chain relationships and operations.

The major source of information for the case study was a series of eight face-to-face interviews held with chain participants in October 2007. In addition, this was complemented with available data at the websites of the various businesses and with documentation that the interviewees volunteered.

Each interview lasted about an hour and was assisted by the use of a discussion guide (i.e., the data collection followed a semi-structured interview).

The interviews were with: Malting Barley Growers (3 Camgrain members); Gowlett Grain, a merchant operating Camgrain Malting Barley Pool (1 key staff member); Camgrain staff (Chairman and Managing Director); Greencore Malt (Commercial Manager); Greene King, a Brewer (Head Brewer).

Two topics were explored in the interviews: first, the nature of the relationships and the communication amongst the members of the chain; and second, perceived benefits of the supply chain relationships. The choice of topics was based on the results from the survey and the statistical analysis.

Nature of the relationships and communication

The relationships and the communication between the different segments of the supply chain are summarized below in Fig. 3, which presents both the Camgrain, Greencore Malt, Greene King brewery supply chain together with the relationships between them.

As shown in the figure, it is apparent that there is a strong presence of contractual relationships of one form or another within the chain. These ensure commitment to the relationship from both parties, which is important where financial investment is being made or risk reduction is sought. However, these contractual relationships are reinforced by a high degree of professional regard (embodying technical and commercial competence), trust and in many cases personal acquaintances and friendships.

It should be noted that key points in the business relationships between the parts are: the aims of the supply chain parties are strongly compatible; contractual relationships, which ensure commitment, are backed by professional regard and personal bonds; the professional capabilities and personal qualities of those involved greatly assist the commercial relationships and high levels of trust that exist between the chain participants and there is a general willingness to resolve any problems.

Finally, the chain's communication entails set events complemented by ongoing activity and involves: understanding and communicating the needs of each chain participant; facilitating regular logistical issues; rapid problem resolution where difficulties arise; and maintaining the required quality of service, trust and friendship.

Perceived benefits by the supply chain members

Throughout the chain there is a high level of satisfaction with respect to the nature and performance of the supply chain activities.

Farmers can readily identify a series of benefits from the arrangement, including: the quality, cost and robustness of their storage asset at Camgrain (farmers have to purchase a 'storage entitlement' – expressed as a tonnage per year – in the Camgrain store); the cost-effectiveness and efficiency of the grain-handling operation; greater on-farm flexibility in cropping (larger areas of crops which are harvested at the same time are possible) and

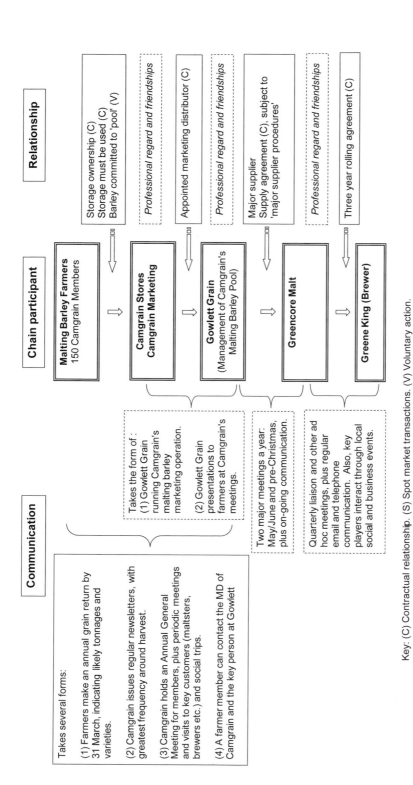

Communication

Takes several forms:

(1) Farmers make an annual grain return by 31 March, indicating likely tonnages and varieties.

(2) Camgrain issues regular newsletters, with greatest frequency around harvest.

(3) Camgrain holds an Annual General Meeting for members, plus periodic meetings and visits to key customers (maltsters, brewers etc.) and social trips.

(4) A farmer member can contact the MD of Camgrain and the key person at Gowlett

Takes the form of :
(1) Gowlett Grain running Camgrain's malting barley marketing operation.

(2) Gowlett Grain presentations to farmers at Camgrain's meetings.

Two major meetings a year: May/June and pre-Christmas, plus on-going communication.

Quarterly liaison and other ad hoc meetings, plus regular email and telephone communication. Also, key players interact through local social and business events.

Chain participant

Malting Barley Farmers 150 Camgrain Members

Camgrain Stores Camgrain Marketing

Gowlett Grain (Management of Camgrain's Malting Barley Pool)

Greencore Malt

Greene King (Brewer)

Relationship

Storage ownership (C)
Storage must be used (C)
Barley committed to 'pool' (V)

Professional regard and friendships

Appointed marketing distributor (C)

Professional regard and friendships

Major supplier
Supply agreement (C), subject to 'major supplier procedures'

Professional regard and friendships

Three year rolling agreement (C)

Key: (C) Contractual relationship. (S) Spot market transactions. (V) Voluntary action.

Fig. 3. The Camgrain, Greencore Malt, Greene King brewery supply chain.

barley husbandry (nitrogen levels are not quite so critical as 'grade segregation' can take place at the store); and more rapid harvesting (larger combines can be used because grain is moved off-farm rapidly). There is a high level of satisfaction with the type and efficiency of the services provided and members generally feel very committed to Camgrain. The membership continues to grow.

Of crucial importance to farmers is the fact that the malting barley pool delivers very good grain prices. The premium over malting barley prices at harvest range from 20 to 40–50%, although higher levels have been achieved, notably in 2006/07. Part of this premium is a return to storage, but a large part of it is due to good marketing and the added value of drying, dressing and delivery to an assured specification, and the large scale of deliveries that Camgrain can engage in.

For its part, Camgrain knows the qualities of grain that it has to handle and can plan its operations accordingly. The good performance of the 'pool' and the storage facilities keeps the membership satisfied. The arrangement with Gowlett ensures access to, and use of, high-quality marketing expertise.

For Gowlett, the Camgrain agreement provides over a quarter of its total malting barley business, and Camgrain is its largest single customer. The success of the relationship and the performance of the 'pool' have encouraged Camgrain to give Gowlett a bonus on the normal trading margin on malting barley of £2 per tonne. This bonus provides an on-going incentive for good performance in securing 'dried-and-dressed' contracts (grain dried and dressed to a set specification). The association with Camgrain also helps to give the merchant a good profile with malting barley buyers.

Greencore Malt derives benefit from the chain in that it provides a reliable and continuous supply of good quality barley, with very quick access to large quantities in the latter part of the year when supplies are more difficult. Greencore Malt sees itself as rewarding Camgrain and Gowlett appropriately for the quality of service they provide. The reliability and scale of the service has enabled Greencore Malt to relinquish its own local malting barley store, thereby effecting major cost savings. Such a store may have had a drying and storage cost of £13–15 per tonne, whereas Camgrain's scale and plant efficiency may achieve a cost of £10 per tonne. Moreover, the value of drying and delivering to specification, without fail, may be worth an extra £5 per tonne. These are benefits that are shared between the businesses with Greencore Malt regarding itself as rewarding Camgrain fairly for the services it provides.

The relationship between Greencore Malt and Greene King offers both parties benefits. Greencore Malt has a significant customer who is taking 10,000–12,000 t of malt a year out of a production of 175,000 t. It is able to make an acceptable margin on that malt. For its part, Greene King has been able to rationalize its supply base because of having a reliable local supplier in Greencore Malt. The low haulage costs (£3 per tonne) for locally produced malt and the savings from dealing with fewer suppliers may give rise to a saving of £20–22 per tonne on malt costing approximately £300 per tonne.

Each party in the chain derives clear benefits from the supply chain relationships. These take the form of both direct financial benefits and improved business service (on the input and/or output side of the business).

Conclusions

The results from the two empirical analyses, the SEM and the case study, point to five factors affecting cohesion of the malting barley-to-beer supply chain: communication, compatibility of partner's aims, contractual relationships backed by professional regard and personal bonds; high levels of trust between the chain participants and a willingness to resolve any problems; and commercial benefit.

As regards communication, this was found to be focused on facilitating regular logistical issues, rapid problem resolution, maintaining the required quality of service and trust and friendship amongst the partners.

The quality of the relationship was maintained/enhanced if the aims of the supply chain parties are strongly compatible; if the existing contractual relationships are backed by professional regard and personal bonds; and if high levels of trust exist between the chain participants and there is a willingness to resolve any problems.

Finally, commercial benefit – direct and indirect financial benefit – was found to be an important determinant of supply chain collaboration and sustainability. Each party has to derive clear benefit from the relationships if businesses are to be readily attracted into a particular set of supply chain arrangements and they are to be maintained. Consequently, market power issues that affect the distribution of rewards amongst the partners, require careful consideration in the maintenance of sustainable chain relationships.

As far as the conceptual framework outlined in the early part of the chapter is concerned, the analyses collectively illustrate that planning and good communication, possibly with investment and certainly good management, enable supply chain benefits to be generated for those involved. Nevertheless, 'soft' factors such as awareness, trust, satisfaction and commitment serve to enable and underpin improved business activity and performance within the context of a supply chain, whilst the other distinctly human factor, that of leadership, provides the direction and vision for improvement and good performance to be achieved, and the wisdom for its fruits to be equitably distributed.

References

Bejou, D., Wray, B. and Ingram, T. (1996) Determinants of relationship quality: an artificial neural network analysis. *Journal of Business Research* 36(2), 137–143.

Bennet, R. and Barkensjo, A. (2005) Relationship quality, relationship marketing, and client perceptions of the levels of service quality of charitable organisations. *International Journal of Service Industry Management* 16(1), 81–106.

Bleeke, J. and Ernst, D. (1993) *Collaborating to Compete*. John Wiley & Sons, New York.

Boles, J., Barksdale, H. and Johnson, J. (1997) Business relationships: an examination of the effects of buyer–salesperson relationships on customer retention and willingness to refer and recommend. *Journal of Business & Industrial Marketing* 12(3/4), 253–264.

Brown, W. (1984) Firm-like behavior in markets. *International Journal of Industrial Organization* 2(3), 263–276.

Coase, R. (1937) The nature of the firm. *Economica*, 4(16), 386–405.

Dorsch, M., Swanson, S. and Kelley, S. (1998) The role of relationship quality in the stratification of vendors as perceived by customers. *Journal of the Academy of Marketing Science* 26(2), 128–142.

Dyer, J. and Singh, H. (1998) The relational view: co-operative strategy and sources of inter-organisational competitive advantage. *Academy of Management Review* 23(4), 660–679.

Fearne, A. (1998) The evolution of partnerships in the meat supply chain: insights from the British Beef Industry. *Supply Chain Management: An International Journal* 3(4), 214–231.

Granovetter, M. (1985) Economic action and social structure: the problem of embeddedness. *American Journal of Sociology* 91(3), 481–510.

Hair, J., Jr, Black, W., Babin, B., Anderson, R. and Tatham, R. (2006) *Multivariate Data Analysis.* 6th edition. Prentice Hall, New Jersey.

Harmsen, H., Grunert, K. and Declerck, F. (2000) Why did we make that cheese? An empirically based framework for understanding what drives innovation activity. *R&D Management* 30(2), 151–166.

Hennig-Thurau, T. and Klee, A. (1997) The impact of customer satisfaction and relationship quality on customer retention: A critical reassessment and model development. *Psychology and Marketing* 14(8), 737–764.

Hobbs, J. (1996) A transaction cost approach to supply chain management. *Supply Chain Management: An International Journal* 1(2), 15–27.

Hobbs, J. and Young, L. (2000) Closer vertical coordination in agri-food supply chains: a conceptual framework and some preliminary evidence. *Review of Agricultural Economics* 5(3), 131–142.

Hughes, D. (ed) (1994) *Breaking with Tradition: Building Partnerships and Alliances in the European Food Industry.* Wye College Press, Wye.

Kidd, J., Richter, F.-J. and Li, X. (2003) Learning and trust in supply chain management. *Management Decision* 41(7), 603–612.

Lagace, R., Dahlstrom, R. and Gassenheimer, J. (1991) The relevance of ethical salesperson behavior on relationship quality: the pharmaceutical industry. *Journal of Personal Selling and Sales Management* 4(1), 39–47.

Lang, B. and Colgate, M. (2003) Relationship quality, on-line banking and the information technology gap. *International Journal of Bank Marketing* 21(1), 29–37.

Leat, P. and Revoredo-Giha, C. (2008) Building collaborative agri-food supply chains: the challenge of relationship development in the Scottish red meat chain. *British Food Journal* 110(4/5), 395–411.

Lewin, J. and Johnston, W. (1997) Relationship marketing theory in practice: a case study. *Journal of Business Research* 39(1), 23–31.

Maltsters' Association of Great Britain (MAGB). http://www.ukmalt.com/index.asp.

Milgrom, P. and Roberts, J. (1992) *Economics, Organisation and Management.* Prentice Hall, New Jersey.

Min, H. and Zhou, G. (2002) Supply chain modelling: past, present and future. *Computers & Industrial Engineering* 43(2), 231–249.

Mohr, J., Fisher, R. and Nevin, J. (1996) Collaborative communication in interfirm relationships: moderating effects of integration and control. *Journal of Marketing* 60(3), 103–115.

Moorman, C., Zaltman, G. and Deshpande, R. (1992) Relationships between providers and users of market research: the dynamics of trust within and between organizations. *Journal of Marketing Research* 29(3), 314–328.

O'Keeffe, M. (1998) Establishing supply chain partnerships: lessons from Australian agribusiness. *Supply Chain Management: An International Journal* 3(1), 5–9.

Peterson, J., Cornwell, F. and Pearson, C. (2000) Chain stocktake of some Australian agricultural and fishing industries, Bureau of Rural Sciences, Canberra. Available online at: http://affashop.gov.au/PdfFiles/PC12761.pdf.

Power, D. (2005) Supply chain management integration and implementation: a literature review. *Supply Chain Management: An International Journal* 10(4), 252–263.

Richardson, G. (1972) The organisation of industry. *Economic Journal* 82(327), 883–896.

Roberts, K., Varki, S. and Brodie, R. (2003) Measuring the quality of relationships in consumer services: an empirical study. *European Journal of Marketing* 37(1/2), 169–196.

Rosen, D. and Suprenant, C. (1998) Evaluating relationships: are satisfaction and quality enough? *International Journal of Service Industry Management* 9(2), 103–125.

Sadler, I. and Hines, P. (2002) Strategic operations planning process for manufacturers with a supply chain focus: concepts and a meat processing application. *Supply Chain Management: An International Journal* 2(4), 225–241.

Sahay, B. (2003) Supply chain collaboration: the key to value creation. *Work Study* 52(2), 76–83.

Storbacka, K., Strandvik, T. and Grönroos, C. (1994) Managing customer relationships for profit: the dynamics of relationship quality. *International Journal of Service Industry Management* 5(5), 21–38.

Tuten, T. and Urban, D. (2001) An expanded model of business-to-business partnership formation and success. *Industrial Marketing Management* 30(2), 149–164.

Wray, B., Palmer, A. and Bejou, D. (1994) Using neural network analysis to evaluate buyer–seller relationships. *European Journal of Marketing* 28(1), 32–48.

Chapter 9

From Transactions to Relationships: the Case of the Irish Beef Chain

Maeve Henchion and Bridin McIntyre

Ashtown Food Research Centre, Teagasc, Dublin, Ireland

Introduction

Traditional issues faced by agri-food producers and processors include small business size and relative bargaining disadvantage, logistical problems, seasonal patterns of production, inconsistency in quality of output and a lack of market orientation due to market interventions (Grunert *et al.*, 2004; Van Tilburg *et al.,* 2007). These issues ensured that transactions were typically short-term in focus and often adversarial in nature.

However, in today's highly dynamic business environment the agri-food sector is faced with a new catalogue of pressures, including: reduction in commodity support and protection, increasing international competition, demands for product differentiation and innovation, increasing consolidation further downstream in the supply chain, increasingly demanding food safety regulations, increasing awareness of animal welfare and environmental issues, and the challenge of ensuring economic, social and environmental sustainability. In a bid to address these challenges, Irish agri-food businesses are considering new options for coordinating agri-food markets. These include non-arm's length relationships as there is evidence that non-arm's length relationships are capable of generating relational rents to partners of a relationship (Claro *et al.,* 2002). Such relationships also offer the possibility of evolving a supply chain that moves away from the free market extreme of 'open commodity-trading' that has been a traditional feature of agricultural markets.

Within the chain, pro-active management of inter-organizational relationships has become a critical source of competitive advantage (Dyer and Singh, 1998) with sustainable business relationships of central importance. Following Chapter 4 we define sustainable business relationships within food chains as:

> longer-term, stable and mutually beneficial interactions and transactions between chain stakeholders. They are based on joint economic interests, such as price stability, profitability of interactions and security of supply. Sustainable business relationships are characterized by high relationship quality (as indicated by mutual trust, satisfaction and commitment) combined with high relationship stability (as indicated by high switching costs, stability and a positive collaboration history). They contribute to achieving the overall

common goal of producing, processing and distributing food effectively, efficiently and consistent with market needs.

These business relationships can be distinguished by their type of governance. Transactions among partners can be organized by spot markets, repeated market transactions with the same partner using either formal or informal contracts, strategic alliances or vertical integration. Given the freedom to choose, only collaborative strategies, which are win–win-oriented and result in benefits for both parties, are likely to be stable under any type of governance. Accordingly, collaborative supplier–customer relationships frequently receive special attention, as they are expected to promote the sustainable development of competitive food chains. As such they are viewed as intangible assets that offer companies a source of long-term competitive advantage and the foundation for long-term success (Duffy, 2005).

The rationale for selecting the beef industry is that while it has previously been a significant recipient of Common Agricultural Policy (CAP) support, it now operates in a highly competitive environment that requires a change from traditional transaction formats. The previous CAP system provided support based on volume (not quality) of output, however support is now decoupled from production, i.e., is independent of output and even whether or not production occurs in some instances, so that market returns will need to provide a strong incentive to motivate production. Thus relationships within the sector encompass a continuum of transaction possibilities, from spot market transactions, which have typically characterized the commodity market, through to collaborative marketing initiatives which provide a point of differentiation and competitive advantage.

This chapter explores the nature of business relationships in agri-food chains and examines options for coordinating agricultural markets, using the Irish beef chain as an example. Following an overview of the relevant structural features of the beef industry, the methodology is outlined and results presented in terms of a discussion of the main types of relationships that prevail in the Irish beef chain, the impact of various factors on the selection of relationship types and the link between relationships and performance. Finally, some conclusions and recommendations are presented.

The Irish beef industry

The Irish beef industry is of considerable importance to the Irish economy due in part to its size and also to its high degree of export orientation. Performance reviews in 2007 highlighted that the output value of the sector was just under €1.5 billion (DAFF, 2008). The Irish beef sector accounted for 8% of all adult cattle slaughtered in the EU-27 in 2007 (Eurostat, 2008), highlighting the sector's significance in EU terms. Imports are limited and are estimated to account for approximately 16% of domestic consumption (Bord Bia, 2009). They are primarily from South America and targeted at the manufacturing and food-service sector.

In total there are about 125,000 cattle herd numbers in the country with beef production carried out on more than nine out of every ten farms. The average size of a cattle farm has increased over time and is now higher than the European average. Production is largely grass based, thus cattle supply can be seasonal with the highest proportion of cattle slaughtered in the autumn to minimize feed cost. It is closely linked to the dairy herd with surplus calves from the dairy herd providing a significant proportion of the raw material for the industry. This has implications for quality as the dairy breeds do not have good conformation from a meat perspective. Production is predominantly based on owner-operator farm enterprises with a significant number of part-time operators.

The processing sector comprises about 40 large-scale export approved plants and over 300 local abattoirs. Most of the abattoirs that are approved for export typically slaughter 50,000 head of cattle per annum, however several of these may slaughter in excess of 100,000 head per annum. Ownership is highly concentrated, with AIBP, Dawn Meats and Kepak accounting for more than half of all slaughterings. However, it is generally acknowledged that there is over-capacity in the slaughtering sector. The high level of concentration, combined with some regional monopolies as well as the low attractiveness of some abattoirs in terms of trustworthiness, terms of payment, etc., may mean the number of perceived market outlets for cattle is limited from a farmer's perspective.

Most plants engage in slaughtering, cutting and deboning activities. However, there are a number of specialized boning halls which purchase carcasses from the export slaughter premises for cutting up for export as vacuum packed or frozen boneless beef. In addition, some Irish processors ship wholesale beef cuts for further cutting and retail packaging to their plants located in the marketplace. Most of the large processors own a central cutting plant in the UK, the main export market, to service these accounts. Ownership tends to be in the form of private or public companies, generally with no ownership by farmers or retailers. Exceptions occur with some independent butchers owning a small-scale abattoir or cutting plant.

It is broadly true to say that the Irish beef industry is under-developed in the area of further processing. Very few of the large meat companies have developed their business with further processed product. However, over the last 10–15 years a new segment of small processing companies has evolved, somewhat independently of the larger slaughtering/deboning plants. These companies usually process both pork and beef but rarely lamb.

Domestic consumption accounts for approximately 15% of total beef available in the market. The domestic retail sector accounts for 70% of Irish beef distribution, 25% is distributed through the foodservice sector and the remaining 5% is sent for further processing. The multiples account for 69% of total retail sales of fresh meat, while independent butchers and grocery independents account for the remaining 31%.

In 2008, 483,000 t of beef were exported, valued at €1687 million. Over 99% of this was destined for EU markets (Bord Bia, 2009). This is the result of strong trade promotion in recent years as the sector used to be very dependent on non-EU markets requiring the support of EU export refunds. The UK is the main destination for Irish beef exports and in 2008 received 261,000 t or 54% of exports. Exports to Continental Europe in that period amounted to 217,000 t (45%). Currently, Irish beef is listed with more multiple retail chains in more EU markets than beef of any other origin (Bord Bia, 2009).

Methodology

Data for this chapter are drawn from a quantitative survey of members of the Irish beef chain and follow-up face-to-face in-depth interviews. For the survey, respondents were interviewed face-to-face where possible with telephone interviews undertaken otherwise. Respondents consisted of 69 farmers, 7 processors and 10 retailers, giving a total of 86 stakeholders in the chain. (Only farmers who sold finished cattle took part in the survey, i.e., farmers selling cattle to other farmers to fatten were not included.) Survey respondents tended to be in upper management positions or (part-) owners of the surveyed businesses. While the number of processors and retailers is lower than anticipated, it is felt that good coverage has been achieved given the concentrated nature of both the beef processing and retail sector in Ireland, and the fact that small- and large-scale operators, from different areas, are included in the sample. Data collection occurred between November 2006 and April 2007.

Depth interviews, using a semi-structured interview schedule, were administered to a purposive sample. The interviews lasted 60 minutes on average. Procurement managers in six beef processing factories were interviewed, as well as two beef farmers, two advisors from the Teagasc beef extension services and one development officer in the organic sector. The processors and farmers were selected on the basis that they represented a range of operations in terms of scale and geographic location. The independent advisors and development officer are believed to have an independent and clear view of the relationships between farmers and processors. All of these interviews were additional to those covered in the survey. In the case of the processors, the same company may have been represented in both the survey and the depth interviews, however different respondents would have taken part.

Analysis of the resultant dataset included bivariate and discrete choice analysis. Discrete choice models were estimated for the beef chain for all stakeholders together, and for farmers as a group, to determine the factors influencing choice of relationship types. While both sets of results are reported, the focus here is on the results for the farmers. All the tested hypotheses regarding the choice of relationships were formulated in terms of stakeholder preference for spot markets versus repeated market transactions, i.e., the reference category was 'repeated market transactions with the same partner'.

How relationships operate

Traditionally, Irish cattle have been sold in livestock auction marts with prices based on supply and demand conditions at the time of ownership transfer, and the identity of the transacting parties being unimportant. Premiums may have been paid for better quality animals based on attributes observable at the time of sale (primarily relating to weight, and visual determination of fat level and conformation, rather than attributes relating to eating quality for example which is of value to consumers). However, paying premiums based on credence-type attributes, such as animal welfare, which are of increasing interest to consumers, was not possible. A major criticism of this traditional system focuses on its inefficiency to communicate customer requirements back to farmers. Spot markets also depend on sufficient competition to arrive at a fair price; however, concentration in the processing sector has meant that there are few buyers relative to sellers, forcing farmers to explore alternative transaction mechanisms and thus relationship types in an effort to secure rewarding outlets for their cattle.

Empirical evidence from this research finds that the nature of relationships has changed over time. Repeated market transactions are now the most favoured relationship type used by farmers. Table 1 shows that 81% of farmers surveyed used repeated market transactions with their main buyer compared to 14% using spot markets.

These relationship types, with a limited number of agents or processors, allow more reliability and lower search costs without loss of switchover options. With only one in seven farmers using spot markets, it is not surprising that the traditional 'mart' is declining in importance in the sale of finished cattle. Multiple retailers, who do not permit cattle that pass through live auction marts into their supply chain, have encouraged this trend. While respondents to the quantitative survey did not include any farmer–processor relationships based on contracts, related research by these authors found that some processors have 'arrangements' which encompass aspects of marketing and production contracts with a limited number of farmers for a specific type of animal. In these circumstances, a base price (or at least a pricing mechanism specifying a pre-determined level above the prevailing market price) is set in advance of the sale of a specific number of animals, produced in line with a set production protocol (encompassing breed, sex, age at slaughter, feed, etc.). However, the actual price may not be finalized until the day of sale when prevailing market conditions are also considered. Thus, in addition to securing market access for the farmer and suitable supplies for the processor, such arrangements shift some of the price risk from farmers towards the processor. Sometimes these arrangements operate within the context of a 'producer club', whereby cattle are produced to meet particular specifications for defined markets. Examples of these include the KK Club, which operates a partnership between Kepak (meat processor), Keenans (a diet feeder supplier), farmers and Italian retailers to produce bull beef to meet the quality requirements of the Italian market. Such arrangements are generally directed at large-scale, specialized farmers rather than small-scale or part-time farmers. The latter may not be able to exploit such opportunities in a profitable way, and may be reluctant to invest in the production of cattle that commit them to certain customer groups. Furthermore, processors may not have sufficient resources to be able to target small-scale farmers. Such results concur with findings by Key (2004) who found that contracting is positively correlated with farm size.

Table 1. Use of different relationship types with main buyer/supplier (% of column total).

Relationship type	Stakeholder				Average %
	Farmers –>	<– Processors	Processors –>	<– Retailers	
Spot markets	14.5	100	0	10.0	31.1
Repeated market transactions	81.2	0	100	70.0	62.8
Formal (written) contracts	0	0	–	20.0	5.0
Financial participation arrangements	0	0	0	0	0
Mixed	4.3	0	0	0	1.1
Total *n*	69	2	5	10	86

Arrows indicate whether the relationship is upstream or downstream, e.g., <– processors refers to processor relationship with farmers.

Financial participation by farmers in the processing sector is not a significant feature of the Irish beef sector, but some processors have feed-lot capacity which can be brought on-stream to off-set poor supply and reduce price volatility. In addition to using increased infrastructure capacity to accommodate uncertainty, another mechanism used is a form of production contract. Processors purchase cattle and supply feed, medication, and nutritional and management advice to farmers in return for the farmer supplying land, buildings, labour, power and equipment. In this instance, the farmer is paid a base price for production on a per-animal basis with incentives for better feed conversion efficiency and lower mortality rates. The farmer remains legally autonomous, but is heavily dependent on the processor who provides all critical resources. The significance of these sources of supply is strongly influenced by market conditions and varies from year to year.

Table 1 belies the importance of third-party agents in the relationship between farmers and processors. In the beef chain, third-party agents account for 50–80% of total cattle sourced by different factories. They are strategically located in areas that best serve the processors and tend to be used by smaller farmers, with larger farmers dealing directly with processors. They have some flexibility in negotiating the price with farmers, however the procurement manager has the final say. Farmers who use agents are thought to have less trust in processors. This may be because of the absence of a personal relationship with the processor and poorer communications between processor and farmer as a result of an additional node in the communication chain. It may also be because farmers in such chains do not identify with the chain *per se*, rather they identify and commit on a personal relationship basis with the agent who markets their cattle. This situation was reported by Gow *et al.* (2005) with regards to the New Zealand wool chain. None the less, agents perform a useful service, to both farmers and processors, in procuring cattle from part-time farmers. They also provide a range of services for farmers other than the purchase of cattle. These include transporting cattle to the factories, purchasing unfinished cattle and providing advice on feed and on when to finish cattle.

Table 1 shows that repeated market transactions and formal contracts are of higher relevance downstream (between processors and retailers) than upstream (between processors and farmers) in the chains, indicating that downstream businesses tend to coordinate and organize their relationships in a more standardized way. This may be due to high switching costs in retailer–processor relationships, as processors generally adapt to retailers' requirements, buy into the brand and make specific investments to maintain and develop the relationship and ensure a strategic fit. Specific investments in this situation usually involve co-branding with supermarket and meat processor insignia, and dedicated meat cutting and packing plants. Thus relationships between processors and multiple retailers are generally quite exclusive and are frequently called 'partnerships' by the parties involved and the trade press. The retail sector in Ireland and in the UK, Ireland's main export destination for beef, is highly concentrated, which means that such partnerships are generally directed by the multiple retailers who act as the chain captain and who generally drive innovation. The benefits to retailers and processors of such an approach are consistent volumes of the required quality and specification for a prolonged time at a competitive price. Advantages to farmers linked into such partnerships include premium prices, protection from low seasonal prices and guaranteed markets.

Factors that influence selection of relationship type

The survey generally found that each of the stakeholders felt free to engage in whatever type of business transaction they chose. Farmers were the only ones to feel somewhat constrained (79% felt free to choose), mainly because they felt they had insufficient bargaining power. This lack of power by farmers is a persistent feature of the sector. In 2000, the Department of Enterprise and Employment (DEE) stated that farmers are located vulnerably towards the upper end of the food chain where they have little bargaining power. They suggested that farmers could increase their power by engaging in cooperatives or producer groups; however the results above indicate that, as suggested by DEE in 2000, while farmers have dabbled in these arrangements, they have never approached them wholeheartedly.

Table 1 also shows that chain stage is a strong influencing factor in determining relationship type, with closer, more formal relationships tending to be more prevalent downstream. However, focusing on factors influencing farmer–processor relationship selection, a number of factors emerged as having an influence. These were identified through the discrete choice analysis (reported in Table 2), descriptive statistics from the survey and key informant observations. These factors include market, economic, social, cultural, political, personal and situational factors.

Considering all chain members as a group and also farmers only as a group, operating in a quality market does not affect the choice of relationship type according to the discrete choice analysis (see Table 2), i.e., farmers/firms are equally likely to use repeated market transactions as spot markets. For farmers this may be linked to the fact that the incentives associated with different qualities do not sufficiently cover the costs associated with achieving them and thus they are not motivated by whether the market is concerned with quality or not when making a choice regarding relationship type. It may also indicate that where farmers are involved in quality-related initiatives, the rewards may not be equitably distributed. Finally, it may indicate that a small proportion of farmers are not commercially aware and continue producing as they have always done regardless of customer requirements.

Schulze *et al.* (2007) argue that quality certification schemes with standards set and audited by independent third parties enforce spot markets between farmers and processors, as they build standards for the whole industry thus reducing the need for company-specific quality approaches on a contract basis. They also support flexibility. However, such schemes operate on a very limited definition of quality allowing scope for retailers to add their own requirements and thus provide a basis for differentiation which may require a more coordinated approach to sourcing. Evidence from this research also does not agree entirely with Schulze *et al.*'s contention, as Irish farmers in quality assurance schemes favour repeated market transactions over spot markets. This may be because the operation of quality assurance schemes requires some specific investments for example relating to improved production practices and communication processes within the chain, thus favouring repeated market transactions over spot markets. It may also be due to the nature of the quality assurance programme in Ireland. Initially processors were responsible for on-farm quality assurance audits; however, responsibility is now shifting to an independent auditor.

The degree of competition in the market place was not found to influence relationship choice (see Table 2). However, anecdotal evidence suggests that the competitive strategies of the multiple retailers are important. Such retailers have tended in recent years to narrow their supply base and build closer relationships with fewer suppliers, not only to control a fully traceable and seamless chain but also in an effort to reduce costs and differentiate themselves from their competitors. When retailers sell beef in pre-packed format, they apply their own private label brand to communicate points of differentiation. This forces a direct relationship between processors and farmers, as retailers will not allow processors to supply them with beef purchased through the auction mart. In the domestic market, some retailers develop point-of-sale information highlighting the particular farm which has supplied the day's beef, as provenance is a primary food choice criterion for Irish beef consumers. Such initiatives allow farmers to visibly appreciate the benefits of relationship development and support a move away from spot markets.

Schulze *et al.* (2007) cited concerns such as fear of losing autonomy and entrepreneurial freedom and becoming dependent on one market partner as reasons against greater integration. Gillespie and Eidman (1998) reported that autonomy dominated risk as the most important attribute in choosing a marketing strategy for pig producers in Minnesota. However, this research found that neither desire for economic independence nor degree of risk aversion influences selection of relationship type for either all stakeholders or for farmers only (Table 2). This may be because the use of repeated market transactions allows businesses to reconcile the dual objectives of business independence and conducting business on a long-term basis.

Table 2. Hypothesis tests related to the choice of business relationship types for the cattle-to-fresh-beef supply chain.

Variables affecting the choice of spot markets[1]	All stakeholders			Only farmers		
	Coefficient	Std error	Wald test[2]	Coefficient	Std error	Wald test[2]
Intercept	−9.98	5.05	3.90	−8.73	6.95	1.58
The firm operates in a quality market	0.70	0.51	1.89	0.97	0.72	1.81
The firm is part of a private and/or non-mandatory QA system with traceability requirements	−3.04**	1.01	9.07	−4.90**	1.72	8.15
The firm operates in highly competitive market	0.14	0.32	0.20	−0.03	0.41	0.00
The firm tries to avoid risks	0.39	0.60	0.42	0.12	0.74	0.03
The firm tries to maintain its independence	1.55	0.88	3.12	1.95	1.19	2.69
The firm tries to establish long-term relationships	−1.15**	0.44	6.78	−1.57*	0.65	5.84
Model fit	Value	Degrees of freedom	Sig.	Value	Degrees of freedom	Sig.
Chi-square[3]	19.98	6	0.00	22.80	6	0.00

[1]The reference category in the regressions is 'repeated market transactions'. [2]Significance test of the coefficient. It has one degree of freedom. [3]It tests the null hypothesis that the explanatory variables do not contribute in terms to explanatory power to a regression considering only an intercept. *(**) means statistically significant at 95(99)% confidence level.

The degree of long-term orientation influences relationship choice for the chain as a whole and for farmers only, i.e. if a firm/farm looks for long-term partnerships then this increases the probability of using repeated market transactions with respect to spot markets. Long-term orientation can strengthen relationships by allowing time for the development of trust and confidence in the other party and may serve to provide the confidence necessary to resolve problems. It can also enhance processors' ability to introduce new technologies (Schulze *et al.*, 2007) as it provides the time necessary to secure benefits from product or process innovation. Some processors, who have committed themselves to new markets, promote the use of producer clubs in an effort to guarantee suppliers and to encourage farmers to also adopt such an orientation. This long-term orientation strengthens relationships between farmers and processors by providing some degree of certainty regarding volumes and prices. The shift towards direct relationships between some processors and farmers builds loyalty and supports a long-term orientation. In addition, some processors provide loyalty bonuses to their farmer suppliers. Further support for the influence of business culture on relationship type is given by the fact that key reasons farmers gave for engaging in various relationship types include the fact that it is a common business practice (see Table 3).

Table 3. Reasons for using repeated market transactions with the same buyer/supplier (degree of agreement expressed on a 7-point Likert scale: 1=strongly disagree, ..., 7=strongly agree).

	Farmers–processors		
	Mean	Std dev.	*n*
Allow us to minimize costs	4.1	1.8	58
Because we have developed personal relationships	4.9	1.7	59
Is a common business practice in our industry	5.6	1.0	63
Are easy/convenient to use	5.6	0.9	63

Std dev. = standard deviation; *n* = no. of valid responses.

The Common Agricultural Policy (CAP) prompted almost one in five farmers to make greater use of repeated market transactions. The CAP has historically had a negative impact on relationships as returns were based on volume, and thereby price, and not quality. It fostered a production-led attitude and farmers engaged in adversarial discrete buyer–seller interactions. However, reform of the CAP which decoupled the traditional links between subsidy payments and production, means that many farmers are looking to the market for production signals, becoming more market aware and engaging more with buyers as they seek to meet the specifications of buyers rather than the requirement of a production-related support system. It is expected that reform of the CAP may ultimately strengthen relationships as commercially oriented farmers seek to maximize returns from the market.

Personal and situational factors are also of importance. For example, in the case of spot markets, important reasons given by farmers for engaging in spot markets were that they provide farmers with flexibility in dealing with buyers and they are easy and convenient to use. Part-time farmers are a significant feature of the Irish beef sector. Such farmers tend to use spot markets, presumably due to the high value they place on convenience and flexibility. Convenience was also an important reason for farmers using repeated market transactions along with the fact that it is a common business practice (Table 3). Flexibility is particularly important in a changing market environment.

Relationships and performance

Relationships in the Irish beef sector are generally enduring with the average length of relationship being 12 years. This suggests that farmers have a high degree of identification with their buyers. From a processor perspective it is important to ensure they identify with the organization as much as buyers representing the organization as individuals who strongly identify with an organization are likely to focus on tasks that benefit the whole organization rather than purely self-interested ones (Gow *et al.*, 2005).

Related to the length of the relationship that farmers have with their buyers they tend to rate their relationship with their main buyer very positively on a number of fronts: 98.5%

rated the relationship as average or better; almost two out of three farmers stated that their trust in their buyer was somewhat good or better; 93% had some level of commitment to this business partner with half of farmers stating that they had a good to very good commitment to them; 50% of farmers reported their satisfaction levels were good or very good; past collaboration experience had been encouragingly good with only 5% stating that they had a poor experience; almost half of farmers stated that they had a good to very good past collaboration experience with this buyer.

Despite a high level of satisfaction with their most important relationship, only 55% of farmers agreed that their most important relationship is commercially rewarding, and of those in agreement only 3% agreed strongly (see Table 4). Furthermore, only three out of five farmers stated that their relationship with their main buyer has a positive effect on their profitability and less than half say it has a positive effect on turnover. These figures could reflect the fact that direct payments (through the CAP) account for a significant proportion of family farm income on beef farms. Market output from cattle farms is sometimes insufficient even to cover production costs according to the National Farm Survey (Teagasc, 2007). Thus, in a difficult market situation, some farmers may find it difficult to attribute positive profitability to supply chain relationships instead ascribing 'profitability' to subsidies and other supports. It may also reflect a belief that performance indicators are influenced by forces outside of the chain relationship. For example with respect to turnover, this is dependent on stock levels which may be influenced by a number of factors including confidence in the market rather than the chain or individual company. Increased participation in quality assurance systems is reflected in two out of five farmers' reponses, stating that their main buyer has had a positive effect on their product or process quality; however, the relationship had a less significant impact on innovation and customer retention in the farmer's business.

Table 4. Effect of relationship with main buyer/supplier on company's performance (% of respondents saying positive effect).

Performance criterion	Stakeholder							
	Farmers -->		<-- Processors		Processors -->		<-- Retailers	
	%	n	%	n	%	n	%	n
Profitability	59.4	41	100	2	60.0	3	72.7	8
Turnover	46.4	32	50.0	1	80.0	4	90.9	10
Cost reduction	29.0	20	0	0	60.0	3	45.5	5
Market share	14.7	10	50.0	1	40.0	2	63.6	7
Customer retention	11.9	8	0	0	50.0	2	63.6	7
Innovation	19.4	13	0	0	60.0	3	54.5	6
Product/process quality	41.2	28	0	0	60.0	3	81.8	9

n = no. of valid responses.

The relatively poor impact of relationships on performance as measured by financial indicators may be due to the fact that relationship performance may be viewed in terms of both financial and non-financial aspects (Gyau and Spiller, 2009), and that farmers see non-financial aspects, such as high levels of satisfaction and information flow, as important. It may also be due to farmers considering other benefits such as identification, tradition and independence in their decision-making process. This argument is supported by McCown *et al.* (2006) who state that private owners in agricultural systems often prefer different types of benefits in addition to profit, and Schulze *et al.* (2007) who claim that trust, attitudes and preferences for entrepreneurial freedom are important but often neglected factors for choosing efficient governance structures.

Again, despite the high satisfaction levels, almost two-thirds of farmers believed they are in an unequal relationship, while only 19% believe they are equal partners. While at face value this may not augur well for sound relationship-building, Hingley (2005) argues that relationship-building is perfectly possible in asymmetric relationships where weaker parties are tolerant of power imbalance. As farmers in this sample generally accept being in business relationships with a stronger partner, this power imbalance may not be an obstacle. For sustainability, rewards and risks do not need to be equally shared, merely equitably distributed.

As noted in Chapter 5, communication in various forms is seen as integral to the improvement of chain performance. For example, it can facilitate superior chain performance by allowing partners to learn about and react to changes in the requirements and expectations of other chain members. However, it appears from the survey that communication is not fulfilling its potential in enhancing chain performance from a financial perspective (see Table 5). For example, less than half of the respondent farmers reported that communication with their buyer had a positive effect on their profitability. Furthermore, while farmers are generally content with communication frequency (94% of respondents stating that the frequency is average or better), the quality of communication with processors in terms of content, relevance and timeliness is perceived as somewhat poor. This is quite surprising as the seasonal nature of beef production militates against frequent communication (many farmers only interact with processors a few times in autumn when their cattle are ready for sale off grass) and because personal communication, i.e., face-to-face and phone, which is expected to be quite interactive, is a central feature of the communication process in the Irish beef chain.

Table 5. Effect of communication with main buyer/supplier on company's performance (% of respondents saying positive effect).

Performance criterion	Stakeholder							
	Farmers -->		<-- Processors		Processors -->		<-- Retailers	
	%	n	%	n	%	n	%	n
Profitability	48.5	32	–	–	60.0	3	60.0	6
Turnover	33.3	22	–	–	80.0	4	60.0	6
Cost reduction	27.7	18	–	–	40.0	2	40.0	4
Market share	10.9	7	–	–	40.0	2	50.0	5
Customer retention	10.9	7	–	–	60.0	3	50.0	5
Innovation	21.5	14	–	–	40.0	2	30.0	3
Product/process quality	37.9	25	–	–	60.0	3	80.0	8

n = no. of valid responses.

It also appears that information may not be being interpreted and used appropriately, e.g., it is not being used as a basis for decision-making on what to produce, or how to produce it. This may be due to the low value that farmers place on such information, a lack of knowledge on how to use it and/or a lack of incentive to use it. Processors appear eager and capable of sending useful information such as kill-out and quality reports to farmers (and frequently have such information available online in real time), however farmers seem to lack the skills to apply an analytical focus to the data available. The poor use of such information may also be as a result of the traditional commodity orientation which emphasized efficiency, high volume, constant quality and economies of scale (Grunert *et al.*, 2004) rather than a market orientation. Limitations with regards to broadband blackspots in more remote/inaccessible areas may also be a factor as well as low levels of information and communication technology skills amongst older farmers in particular.

An issue that emerged from the depth interviews is that information seems to flow between processors and retailers and between farmers and processors but not through the entire chain in all cases. This could reflect the fact that an organization's philosophy about developing long-term relationships with suppliers and customers is pivotal to the nature of the feedback information system adopted (Storer, 2001), i.e., their philosophy could be based on developing relationships with buyers and suppliers rather than developing chain-wide relationships which would result in up-stream and down-stream feedback systems rather than integrated systems. Such a lack of chain-wide communication is an important issue for farmers in particular as their scale/distance from the market place means they may not have sight of end-consumers' demands.

Conclusions and recommendations

The environment within which the Irish beef industry operates has changed over time and the industry has reacted by changing the nature of the prevailing relationships and shifting from primarily spot markets to repeated market transactions. However, a number of indications point to the need for further change. The first of these is the relatively low level of performance associated with the main farmer–processor relationship in terms of financial indicators, despite high reported levels of satisfaction with the relationship overall. Whilst it is acknowledged that farmers consider and value a range of non-financial performance indicators, long-term sustainability will not be achieved without a recognized economic payback. The second is that there continues to be national and EU-wide recognition of the wisdom and benefits of developing collaborative agri-food supply chains (Fearne, 1998; Laurence *et al.*, 2001; Claro *et al.*, 2002; Hornibrook and Fearne, 2005; Schulze *et al.*, 2007), the main argument being that with higher market segmentation the need for processors to define stricter governance structures grows (Schulze *et al.*, 2007). As the Irish beef industry is increasingly competing in highly demanding retailer and high-value food service chains, it is logical to promote more chain integration between farmers and processors to support differentiation and long-term access to these markets.

Such a change does not necessarily imply a shift to contracts or financial participation, so that the disadvantages of such arrangements, such as loss of control and preclusion from other opportunities, are not encountered. Indeed, it is acknowledged that no single type of relationship is 'right' for either buyer or supplier in all circumstances (Ford *et al.*, 2003). Furthermore, it should be noted that considerable challenges exist in increasing vertical integration in the Irish beef chain. The fact that all members of the chain claim to be risk adverse and to avoid uncertainty whenever possible indicates that change will be difficult to achieve. Furthermore, ownership and social structures increase the difficulty due to the need to consider non-financial and intangible objectives.

However, further improvements could be achieved by improving communication within existing relationships. Many authors acknowledge the role of communication between firms to improve performance and satisfaction (Storer, 2001). The current communication process is not fulfilling its potential in enhancing chain performance and this is further exacerbated by the low value some chain members place on using information in the strategic planning process. For example, information related to market-as opposed to production-related competencies needs greater emphasis. In practical terms, this requires a number of steps to be implemented. Firstly, programmes need to be developed that are aimed at increasing awareness of information available within the chain and its potential impact on performance. Secondly, this needs to be followed up with close discussion between chain members to identify information needs at various points in the chain, the format in which the information should be transmitted and with frank discussion on privacy and access issues. Thirdly, mechanisms need to be developed that ensure that relevant information is transmitted along the chain and not just between adjacent chain members. Fourthly, this communication process needs to be linked to an advisory programme so that farmers are in a position to analyse the information sent and use it in their production process. Finally, communication about market requirements should be reinforced by a financial incentive, i.e., the information should be used to support quality-based pricing.

The use of information to improve coordination within the chain will only be realistic when companies share the desire to cooperate and have a common vision. Research for this study provides no evidence of collaborative planning and suggests a commodity orientation still remains in parts of the sectors. None the less, there are some encouraging signs as indicated by greater long-term orientation and the development of producer clubs. There is a need to build on this by whole chain planning with the involvement of all chain members. Partners need to share a common vision of how to work together and to meet their volume and quality requirements. This will require high levels of trust and long-term commitment by all parties. Power asymmetry does not appear to be an obstacle to collaborative planning. Public agencies may have a role to play in promoting dialogue and establishing fora but capacity-building will also be a central aspect of this initiative.

References

Bord Bia (2009) *Meat and Livestock Review, 2008–2009*. Clanwilliam Court, Dublin, Ireland.

Claro, D., Hagelaar, G. and Omta, S. (2002) Selection of suppliers embedded in networks: empirical results on supplier-buyer relationships in the flower supply chain. In: Trienekens, J. and Omta, S. (eds) *Proceedings of Paradoxes in Food Chains and Networks, 5th International Conference on Chain and Network Management in Agribusiness and the Food Industry*. Noordwijk, the Netherlands, Wageningen Academic Press, pp. 615–627.

Department of Agriculture, Fisheries and Food (DAFF) (2008) *Annual Review and Outlook for Agriculture, Fisheries and Food 2007/2008*. Government Publications, Molesworth Street, Dublin 2.

Department of Enterprise and Employment (2000) *Report of the Independent Group into Anti-Competitive Practice in the Irish Beef Industry*. September. Government Publications, Molesworth Street, Dublin 2.

Duffy, R. (2005) Meeting consumer demands through effective supply chain linkages. *Stewart Post Harvest Review* 1(3), 1–15.

Dyer, J. and Singh, H. (1998) The relational view: co-operative strategy and sources of inter-organisational competitive advantage. *Academy of Management Review* 23, 660–679.

Eurostat (2008) *Agricultural Statistics, Main Results, 2006–2007*. Office for official publications of the European Communities, Luxembourg.

Fearne, A. (1998) The evolution of partnerships in the meat supply chain: insights from the British beef industry. *Supply Chain Management: An International Journal* 3(4), 214–231.

Ford, D., Gadde, L.-E., Hakansson, H. and Snehota, I. (2003) *Managing Business Relationships*. John Wiley & Son Ltd, West Sussex.

Gillespie, J. and Eidman, V. (1998) The effect of risk and autonomy on independent hog producers' contracting decisions. *Journal of Agricultural and Applied Economics* 30(1), 175–188.

Gow, H., Stevenson, M., Westgren, R. and Sonka, S. (2005) Farmer identification and commitment responses to institutional change in marketing channel structures. In: *American Agricultural Economics Association Annual Meeting*, July 24–27, Providence, Rhode Island, USA.

Grunert, K., Jeppesen, L., Jespersen, K., Sonne, A.-M., Hansen, K., Trondsen, T. and Young, J. (2004) Four cases on market orientation of value chains in agribusiness and fisheries. Working paper no 83-2004, ISSN 0907 2101.

Gyau, A. and Spiller, A. (2009) An integrated model of buyer–seller relationship performance in agribusiness: the partial least squares approach. *Journal on Chain and Network Science* 9(2009), 25–41.

Hingley, M. (2005) Power imbalanced relationships: cases from UK fresh food supply. *International Journal of Retail and Distribution Management* 19(August), 551–569.

Hornibrook, S. and Fearne, A. (2005) Demand driven supply chains: contractual relationships and the management of perceived risk. In: European Institute for Advanced Studies in Management, Milan, Italy. http://kar.kent.ac.uk/11419/

Key, N. (2004) Agricultural contracting and the scale of production. *Agricultural Resource and Economics Review* 33, 255–271.

Laurence, J., Schroeder, T. and Hayenda, M. (2001) Evolving producer-packer-customer linkages in the beef and pork industries. *Review of Agricultural Economics* 23, 370–385.

McCown, R., Brennan, L. and Parton, K. (2006) Learning from the historical failure of farm management models to aid management practice. Part 1. The Rise and demise of theoretical models of farm economics. *Australian Journal of Agricultural Research* 57, 143–156.

Schulze, B., Spiller, A. and Theuvsen, L. (2007) A broader view on vertical coordination: lessons from German pork production. *Journal on Chain and Network Science* 7(1), 35–53.

Storer, C. (2001) Inter-organizational information feedback systems in agribusiness chains: a chain case study theoretical framework. 2001 International Agribusiness Management Association World Food and Agribusiness Symposium, 25–28 June Sydney. http://www.ifama.org/tamu/iama/conferences/2001Conference/Papers/Area%20VI/Storer_Christine.PDF

Teagasc (2007) *National Farm Survey, 2007.* Teagasc, Oakpark, Co. Carlow, Ireland.

Van Tilburg, A., Trienekens, J., Ruben, R. and van Boekel, M. (2007) Governance for quality management in tropical food chains. *Journal on Chain and Network Science* 7(1), 1–9.

Chapter 10

Reviewing Relationship Sustainability in the Case of the German Wheat-to-Bread Chain

Miroslava Bavorová[1] and Heinrich Hockmann[2]

[1] Martin Luther University Halle-Wittenberg, Germany
[2] Leibniz Institute of Agricultural Development in Central and Eastern Europe (IAMO), Halle (Saale), Germany

Introduction

The concept of sustainability strongly depends upon the context in which it is applied and whether its use is based on an ecological, social or economic perspective. With respect to these different disciplines, most definitions of sustainability have in common that they assume continuity in time (Shearman, 1990). However, Brown *et al.* (1987) argue that any study of sustainability must make both contextual and time assumptions explicit. In a supply chain relationship context, it is useful to distinguish between the length of a relationship and its sustainability. Length as a one-dimensional measure describes the time dimension of a relationship, whereas a two-dimensional concept of sustainability can include time as well as 'profitability'. Sustainable relationships are usually limited in length. Assuming that agents are rational, and thus are able to choose the most 'profitable' relationship, relationships will last as long as expected benefits from cooperation over time exceed the costs of starting a new relationship. Given these conditions, we define relationships as sustainable when they are able to continuously satisfy the needs of the business partners involved.

The principal aim of this chapter is to evaluate whether business relationships in the German wheat-to-bread chain are sustainable, i.e., generate additional rents for chain members and enhance partners' competitiveness. In order to investigate this matter, five hypotheses will be examined:

H1: Business relationships are long-term oriented.

H2: Stakeholders are satisfied with the quality of their business relationships.

H3: Actors are free to choose the most profitable relationship.

H4: A lack of trust is too weak a reason for terminating long-term relationships.

H5: The lower switching costs, the more likely it is that partners are changed.

The first hypothesis concerns the first dimension of sustainable relationships, namely time. The other hypotheses address the profitability dimension, which is difficult to measure directly. Given the possibility to freely choose the most profitable relationship type and partner, we assume that profitability positively correlates with the degree of satisfaction. The two last hypotheses consider the reasons for continuation and termination of relationships. Against this background, we assume that a lack of trust is not sufficient to end long-term relationships in the German wheat-to-bread chain, i.e., trust has no considerable impact on profitability. The fifth hypothesis deals with the role of economic factors on profitability. Only if all hypotheses are supported, chain relationships can be assumed to be sustainable.

The motivation to investigate the German wheat-to-bread chain lies in its significance for the German agri-food sector. In 2004, the share of wheat amounted to about 8% of gross agricultural output in Germany. The first processing level (milling/starch) accounted for about 3% and the secondary level (bread and pastries) for about 10% of total food processing output (BMELV, 2004, 2006).

The chapter starts with a review of the literature on relationship types, quality and development. In the next part, the methodological framework is laid out, and data and results are described and discussed. The final section summarizes the findings and draws some conclusions.

Literature review

As is known from Williamson (1975), different conditions require different types of governance structure. Accordingly, relationships, both short and long-term, help business partners to create sustainable competitive advantages. When considering the whole chain, relationships between partners at different chain levels, however, are likely to show considerable variation. In order to reliably compare the benefits and costs of different relationship types over time, we develop a theoretical concept that allows to determine whether sustainable relationships are likely to continue or rather will be terminated.

Economic factors influencing the quality and longevity of relationships

Applying a relationship marketing approach, several studies found that non-arm's-length relationships are capable of generating relational rents for chain partners. Relational rents are defined as 'profits' (i.e., benefits minus costs), jointly generated in an exchange relationship which cannot be created by either company in isolation but only through joint contributions of alliance partners (Dyer and Singh, 1998). Buyers' interest in long-term relationships is to achieve a competitive advantage through cooperation with a partner who offers the best mix of competitive price, competitive quality and supply continuity. Additionally, public pressure for transparency and traceability also supports long-term cooperation throughout the food supply chain. However, long-term business relations may not only generate additional benefits but also incur extra costs. These costs can be divided into the two categories: (i) costs of starting a relationship; and (ii) costs of maintaining a less beneficial relationship.

The costs of the first category include costs of starting (e.g., cost of locating and contracting new business partners) as well as continuing a relationship. The major part of these costs turns into sunk costs after terminating the relationship; that is why these costs are called 'switching costs' in the marketing literature. According to Press (1997), switching costs are influenced by trust, resource specificity and satisfaction. The higher the value of these three factors, the higher switching costs are, and the stronger commitment to a long-term relationship is. Switching costs are 'the forgone value of investments plus economic penalties and other expenses associated with finding, evaluating, and using a new supplier' (Dwyer and Tanner, 2002). Generally, long-term (contractual) sustainable relationships become more likely when uncertainty regarding product quality is high, products are rather differentiated, the degree of asset specificity is high (e.g., a low number of potential partners) and switching costs are considerable (Williamson, 1975; Kallfass, 1993).

The second type of costs arising in longer-term relationships are maintenance costs and costs induced by cooperation with a business partner who does not offer a competitive mix of price, quality and supply continuity any more. This is likely to occur when partners have developed a high degree of commitment in the long-time relationship or are not flexible enough to terminate a relationship that fails to provide competitive conditions in the long run.

Behavioural quality components and factors influencing the longevity of relationships

One strain of relationship marketing is concerned with the identification of components of relationship quality, such as trust, satisfaction and commitment (see Chapters 4, 7, 8 and 13 of this volume). Other studies look at factors influencing this multidimensional construct and its components (Bejou et al., 1996), as well as similar constructs such as service quality and strength (Donaldson and O'Toole, 2000; Patterson and Smith, 2001) or customers' desired value (Flint et al., 2000). These factors, among others, include performance, structure, personal bonds and communication.

Given that a majority of studies have identified trust as a crucial component of the quality and longevity of relationships, in this investigation, we use trust as a proxy for the behavioural dimension influencing relationship termination. We define trust in accordance with Moorman et al. (1993) as the 'willingness to rely on an exchange partner in whom one has confidence'. This definition is based on the two components personal liking and credibility. Personal liking expresses the inter-personal level towards a particular relationship, i.e., how strong partners feel mutually committed to behaving cooperatively in the future, whereas credibility concerns the belief in partners' willingness and ability to meet competitive conditions (i.e., a competitive mix of price, quality and supply continuity) in the future.

Economic and behavioural factors influencing the termination of relationships

In previous research, the event terminating a relationship was specified by various synonymously used terms such as ending, termination, dissolution, switching, etc. We use the term termination in accordance with Tähtinen and Halinen (2002) who apply the concept of termination 'to refer to an ending where one of the parties, or an outside actor, deliberately ends a relationship'. The term thus accentuates both the intention to end and the decision to (finally) terminate a relationship.

Thus, research identifying and analysing the factors that strengthen relationships by enhancing relationship quality, profitability and satisfaction provides not only significant insights into the mechanisms but also into the understanding of the termination of business relationships. In this context, it has been shown that, for instance, low levels of trust reduce relationship quality (Tähtinen and Halinen, 2002), and thereby increase the probability of termination. We use the concept of trust as a proxy for the behavioural factors influencing the relationships' termination process.

Development of exchange relationships

Most studies have measured the elements of a relationship at a certain point in time without considering that relationships themselves are evolving. Recent research, however, has started to recognize the importance of relationship development over time, especially as the level of certain elements varies in the different phases of a relationship. Another aspect, the development of exchange relationships, has been increasingly put forward during the past decade (Dwyer *et al.*, 1987; Ring and van de Ven, 1994; Jap and Anderson, 2007). Various models have been developed which mostly share the view that a relationship has a beginning, an intermediate phase and an end (Tähtinen and Halinen, 2002).

A framework for developing buyer–seller relationships was proposed by Dwyer *et al.* (1987). Within this framework, buyer–seller exchange is treated as an ongoing relationship, and relationships are assumed to evolve over time and to be temporary. Five typical phases of a business relationship are described: awareness, exploration, expansion, commitment and dissolution.

First, during the awareness phase, a potential partner is identified and his perceived attractiveness is signalled to him. Moreover, initial steps to actively cooperate are undertaken as first transactions are planned jointly. Second, in the exploration phase, first transactions are carried out. Consequently, what follows is a critical assessment of benefits, costs, opportunities and risks involved. Third, in the expansion phase, further transactions are carried out, which in turn leads to increasing mutual dependency. Fourth, the commitment phase is characterized by a high degree of mutual dependency. At the same time, a relationship becomes institutionalized. Finally, in the dissolution phase, business relationships are terminated, a step that is often taken unilaterally.

Relationship length and the sustainable relationship concept

Figure 1 illustrates different aspects of relationship performance. The performance during the individual phases is depicted by the continuous line, the possible performance of conducting transactions with alternative partners by the dashed line. In the first two phases of a relationship start-up costs have to be incurred. Thus, specific 'investments peak early in the relationship, but are consistently lower in every stage thereafter' (Jap and Anderson, 2007). Relationships that are terminated in the awareness or exploration phases induce opportunity costs for the business partners. The costs incurred in the next phases are of a different kind. Compared to those of relationship development, these are costs of maintenance. They can be interpreted as switching costs, and are to be considered as costs of low service quality rather than of market alternatives.

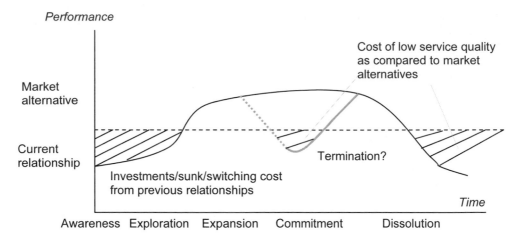

Fig. 1. A business relationship's development, performance and termination (adapted from Dwyer *et al.*, 1987).

When the outcome of a relationship is expected to be higher than the outcome of potential market alternatives, partners gain from a relationship, and thus can reap relational rents. However, to comprehensively assess the net benefits ('profits') of a relationship, it is necessary to consider not only the economic outcome at a single point in time, but also the expected returns aggregated over the expected duration of the relationship. The difference between these approaches can be seen in Fig. 1, considering the case of dropping returns in the expansion phase. Without taking into account the time dimension, a relationship would be terminated before entering the commitment phase. However, as long as expected switching costs exceed expected returns from continuing a relationship, the partnership will prevail. In such a situation, trust can serve as a stabilizing element. Trust encourages business partners to believe in their strength to be able to reduce or resolve problems occurring in the partnership so that they can continue a mutually beneficial relationship.

Data collection

Data were collected as part of the EU-funded FP6 project FOODCOMM in two steps: a quantitative questionnaire survey of supply chain participants in the first, and face-to-face, telephone or written interviews with wheat-to-bread chain stakeholders in the second.

The questionnaire for step one was designed based on information from in-depth telephone or face-to-face interviews with representatives from business associations. A total of 386 questionnaires were mailed to potential respondents (28 farmers, 129 mills, 229 bakeries) along with a cover letter and a postage-paid return envelope. The target population included firms of different sizes and which engaged in different relationship types. The respondents' addresses were obtained from relevant associations and through internet search. The average response rate was low (13.4%), varying considerably between the groups whereby farmers showed a higher response rate than bakeries and mills. The final sample comprised 49 responses (15 wheat farmers, 15 mills and 19 bakeries; Table 1). The information gathered in the survey was used for a first investigation of hypotheses 1 and 2.

Table 1. German wheat-to-bread chain, sample and population.

Business activity	Sample	Whole population
Total farms	15	245,000
– small (5–99 ha)	9	< 239,000
– large (100 and more ha)	6	> 6,000
Total processors	34	18,722
– mills	15	1,315
– bakeries	19	17,407

Sources: BVL (2007), BMELV (2006), AgriMA (2005).

Turning to the second step, interviews with wheat-to-bread chain stakeholders were carried out as outlined in Table 2. Interviewees were asked which relationship type they would prefer, and why and in which phase they had terminated the exchange relationships with their business partners in the past. The interviews mainly collected information that allowed an assessment of the underlying research hypotheses, and in particular 3 to 5.

Table 2. Interview participant details.

Business	Business type	Employees	Method
Producer association	cooperative	n.a.**	phone
Agri-trader	large trader	n.a.**	phone
Mill 1	trade	20	face-to-face
Mill 2	trade	3.5	face-to-face
Mill 3	industry	84	face-to-face
Mill 4	industry	70	written
Bakery 1	trade	4	face-to-face
Bakery 2	trade	4	face-to-face
Bakery 3	trade	10	face-to-face
BÄKO*	purchasing co-op	40	phone
Bakery 4	industry	300	phone

* BÄKO is an important bakery purchasing cooperative; ** n.a. = not available.

The businesses included in the qualitative interviews were chosen to represent a diversity of firm sizes and relationship types. However, for the sake of comparability, interviews were only conducted in the new federal states (NFS) Saxony, Saxony-Anhalt and Thuringia where agricultural enterprises are structurally different as compared to the old federal states (OFS). Vertical coordination, for instance, is more strongly supported by agricultural businesses in the NFS as their desire to remain independent as a company is not as distinct as that of the family farms in the OFS (Lange and Koch, 1995). Because of this bias in the sample, our research results are not valid for other parts of Germany.

Results and discussion

Relationship types and freedom to choose them

Respondents indicated the types of relationships (ranging from spot markets, market transactions with the same buyer/supplier, formal contracts and financial participation) they had with suppliers and buyers in the domestic market. As shown in Table 3, the most typical relationship type in the wheat-to-bread chain is market transactions with the same business partner (= repeated market transactions). Besides these, farmers also use formal (written) bilateral contracts in their downstream relationship. Pure spot markets transactions are preferred by farmers in their downstream business relationships as compared to any other group of participants in the wheat-to-bread chain.

Table 3. Use of different relationship types (% of total).

Relationship types	Stakeholders*		
	Farmers →	← Processors	Processors →
Spot markets	7	0	0
Market transactions with same partner	80	88	87
Formal (written) bilateral contracts	13	8	13
Financial participation arrangements	0	4	0
Total	100	100	100

*Arrows indicate the direction of relationships, i.e., downstream (→) or upstream (←).

A large majority of farmers and processors in their upstream relationships claimed to be free to decide which relationship type they use (Table 4). In the downstream relationships, the discrepancy between farmers and processors was pronounced. Only 50% of processors felt free to choose the relationship type. The reason for this (perceived) lack of freedom lies, as they argued, in their insufficient bargaining power, and customer buying practices.

Table 4. Freedom to choose relationship types (% of total).

Answers	Stakeholders*		
	Farmers →	← Processors	Processors →
Yes	91	100	50
No – member of cooperative	0	0	0
No – not sufficient bargaining power	9	0	25
No – because of customers' buying/suppliers' selling practices	0	0	25
No – because of legal requirements	0	0	0
Total number of responses	11	5	8

*Arrows indicate the direction of relationships, i.e., downstream (→) or upstream (←).

Table 5 provides relevant information on the characterization of the relationship with the most important buyer/seller. For farmers, the most important buyer is located in the same region as the farm. However, this does not necessarily imply that farmers are only regionally oriented as it becomes clear from the upstream relationships of processors. For 25% of the mills, the most important supplier is located outside the region. Moreover, inter-regional relationships become more important for downstream transactions of processors as about 40% of processors stated that their main buyer is located outside their regions.

Table 5. Characterization of the relationship with the most important buyer/supplier.

Characteristics	Stakeholders*		
	Farmers →	← Processors	Processors →
Approximate share in % of total annual turnover/ purchasing value	63 [15]	45 [26]	39 [8]
Average length of relationship in months	160 [15]	255 [26]	237 [8]
Buyer/supplier in same region (% saying yes)	100 [15]	73 [26]	57 [7]

Square brackets give the number of valid responses. *Arrows indicate the direction of relationships, i.e., downstream (→) or upstream (←).

The shares of total annual turnover (purchasing value) generated through the relationship with the most important partner decrease when moving downstream: from 63% for farmers to 39% for processors in their downstream relationship.

Considering the duration of the main business relationship, the results indicate an underlying long-term orientation with an average duration of 160 months between farmers and processors, 255 months between processors and farmer, and 237 months between processors and distributors. These numbers provide support for the first hypothesis, i.e., long-term relationships prevail in the German wheat-to-bread chain.

Perceived relationship quality and stability

Results from the analysis of relationship quality are provided in Table 6. As to overall quality, all respondent groups rated the relationship with their main business partner as good, even though there are slightly lower ratings in the processor downstream relationship. Trust and commitment were perceived in a similar way, i.e., relationships at all levels of the wheat-to-bread chain are characterized by high levels of trust and commitment.

Table 6. Quality of relationship with main buyer/supplier (assessments expressed on a 7-point rating scale: 1 = very poor, …, 7 = very good).

Criteria	Stakeholders*		
	Farmers →	← Processors	Processors →
Overall quality of relationship	6.0 (0.8) [15]	6.1 (0.8) [26]	5.6 (0.9) [8]
Our trust in this buyer/supplier	5.9 (0.9) [15]	6.0 (1.0) [26]	6.1 (0.6) [8]
Our commitment towards this buyer/supplier	5.6 (0.7) [15]	5.7 (1.1) [26]	6.0 (0.9) [8]
Our satisfaction with this buyer/supplier	5.9 (0.7) [15]	6.0 (0.9) [26]	5.4 (1.1) [8]

Numbers in parentheses give standard deviation, in square brackets no. of valid responses. *Arrows indicate direction of relationships, i.e., downstream (→) or upstream (←).

The ratings regarding buyer/seller relationship satisfaction draw a slightly different picture. Farmers are satisfied with their buyer relationships and so are processors in regard to their main supplier. Although processors show less contentment with the buyer relationship, an *F*-test failed to reveal significant differences between the means of the stakeholder statements.

The stability of the relationships with the most important business partners was analysed using a seven-point rating scale for answers to three statements referring to past collaboration experience, dependence and the ability to endure conflicts. The results are listed in Table 7. Actors at all stages have similar perceptions about collaboration experience and the resolution of relationship conflicts. Respondents rated their past collaboration experience as 'good' and considered their ability to endure relationship conflicts with the buyer or supplier as 'rather high'. But the rating was different in regard to the dependency on the main buyer/seller. Farmers and processors in downstream relationships state 'low' dependencies, whereas processors in upstream relationships state moderately high levels of dependency on the main buyer.

Table 7. Stability of relationship with main buyer/supplier (assessments expressed on 7-point rating scales: * 1 = very poor, ..., 7 = very good; ** 1 = very low, ..., 7 = very high).

Criteria	Stakeholders[†]		
	Farmers →	← Processors	Processors →
Our past collaboration experience*	6.0 (0.8) [15]	5.9 (0.7) [26]	5.8 (0.9) [8]
Our dependence on this buyer/supplier**	2.4 (1.5) [14]	2.8 (1.4) [24]	4.3 (1.3) [8]
Our ability to endure relationship conflicts with this buyer/supplier**	5.2 (1.7) [14]	5.5 (1.4) [28]	5.0 (1.0) [8]

Standard deviation in parentheses; no. of valid responses in square brackets. [†]Arrows indicate direction of relationships, i.e., downstream (→) or upstream (←).

In summary, we found strong support for research hypothesis 2. In other words, stakeholders in the German wheat-to-bread chain are rather satisfied with the quality of their business relationships. Besides high quality, interviewees confirmed that in their business relationships problems are solved cooperatively, thus indicating high levels of relationship stability.

Trust in different phases of a relationship

Trust is built when business partners like each other, and when they believe that partners will behave cooperatively in the future and will be able to offer a competitive mix of price, quality and supply continuity. The kind of trust that is based solely on inter-personal liking we define as personal trust.

Personal trust is a crucial factor at the beginning of a relationship, i.e., in the awareness phase, as it encourages one partner to interact with a potential (but largely unknown) partner even though there is no past cooperation experience. Moreover, personal trust also plays a vital role in choosing a new supplier when more suppliers are offering similar price-performance ratios. Farmers and small-trade bakers, especially, said that they would not cooperate with a business when they did not experience a personal liking for the representative. On the other hand, personal antipathy as the opposite of personal liking does not necessarily imply the opposite of fostering a relationship. Indeed, personal antipathy rarely seems to be a reason for terminating long-lasting business relationships in the wheat-to-bread chain in Germany.

In this context, for instance, the respondent of BÄKO (bakery purchasing cooperative) described a situation in which the representative of a long-term business partner had just changed and he, the BÄKO representative, did not like the new representative. Even though personal trust was low, that is to say personal antipathy was high, this did not mean the end of the exchange relationship; instead, he, on behalf of BÄKO, decided to continue the relationship because of the successful past collaboration between the partners. The decision about continuation resulted from the positive past collaboration experience, and thus the conjecture that the partner would further on be able to offer a competitive mix of price, quality and supply continuity.

Bakeries need to meet certain consumer requirements. To be able to produce high-quality bread as demanded by end consumers, bakeries require various kinds and qualities of flour. Therefore, the need for enhanced cooperation and communication between mills and bakeries on technical questions and innovations increases in importance. However, in order to achieve enhanced cooperation, more investments in buyer–supplier relationships (e.g., in information and communication technology) are required; a process that also strengthens relationships as switching costs increase, and thus makes termination more costly.

To satisfy bakery demand for certain wheat qualities, mills have to produce flour from wheat of homogeneous and high quality. However, as there is only a limited number of suppliers that is able to meet these requirements, some German mills which purchase wheat directly from farmers can, to a certain extent, influence farmers' production decisions regarding factors such as choice of varieties, fertilizer use or pest management. Moreover, control of food quality and traceability seems to be less costly in vertically coordinated food chains than in spot markets. The increasing demand for quality ingredients and products thus may foster the development of closer, long-term cooperation between business partners.

Relationship termination

In this section, the results from the qualitative interviews are presented, focusing especially on relationship termination. Three out of four interviewed mills preferred buying wheat directly from farmers in the region to buying from agri-traders. They differentiated the otherwise homogeneous product wheat flour by the special characteristic 'regional origin'. Consumers perceive regional products as being safer and are willing to pay a higher price for them. The mills terminated relationships to farmers primarily because of either differences in price or poor wheat quality (Table 8).

The relationships between agri-traders and mills are usually long-term, i.e., they take the form of repeated transactions with the same partner. Written contracts are often replaced by mutual trust. As for wheat, relationships between buying mills and selling/supplying agri-traders terminate during restructuring processes of the traders or because of price differences in the past. Concentration processes in the sector were primarily responsible for relationship breakdowns. Mills disappear from the market after being bought out by others or in the case of insolvency. Agri-traders also terminate their business relationships when transactions in spot markets secured better prices.

Although relationships between mills and flour buyers are primarily terminated because of price differences, quality and distance also play a considerable role as, e.g., one mill (no. 4) stated: 'long distances reduce flexibility'.

In general, traders who act on behalf of small bakers seek long-term relationships with mills. When a supplier offers competitive prices and quality products, and is able to assure supply continuity, trade enterprises rarely change the supplier. However, a relationship is most likely to terminate when prices are non-competitive or product quality is inferior.

Baker (3): 'we had cooperated for 18 years, but then, some time ago, we broke up because flour quality was not satisfying'.

Baker (4): '12 years ago, we had to cancel cooperation with a mill as the flour quality was not constant'.

Baker (1): 'There were some business relationships which were terminated in the past; once a small company got into financial troubles and could not deliver the agreed volume, so we terminated the cooperation and ordered the products from BÄKO'.

The length of a relationship does not seem to influence the decision to change a partner if the relationship becomes unprofitable. These findings support hypothesis 3 and show that actors in the wheat-to-bread chain are free to choose their most profitable business relationships.

Table 8. Reasons for termination of a supplier relationship in the past.

	After long cooperation – economic reasons				After short cooperation – behavioural reasons
	Price	Quality	Supply continuity	Structural changes	Trust
Mill 1			Termination never happened		
Mill 2		X			
Mill 3	X			X	
Mill 4	X	X			
Bakery 1			X		X
Bakery 2		X			
Bakery 3					X
BÄKO				X	
Bakery (chain) 4		X			

Relationships between large-scale buyers (chain bakeries and industrial bakeries) and suppliers (large-scale mills) are predominantly long-term. When cooperation between large-scale businesses and their suppliers has continued for a while, the partners' attitudes towards written contracts change: the longer a relationship continues, the more unnecessary written contracts are seen, and the more they are substituted by trust. What is more, such relationships are rarely terminated. In addition, as aforementioned, termination seldom results from personal antipathy; rather it is caused by deviations from once-agreed-upon contract terms. Thus, a relationship is most likely to be terminated if a supplier is not able to offer products at competitive prices and/or competitive quality anymore, or to assure supply continuity. Apart from this, another reason lies in the acquisition of the partner's company by other enterprises or its insolvency.

BÄKO, as a large buyer, is a good example to illustrate the just mentioned points; all the more because BÄKO rarely terminates relationships with its suppliers. For BÄKO, the most common reason for losing business partners was either their insolvency or their discontinuation of operations. Termination was also justified when a supplier could no longer offer competitive prices. Even in relationships between large-scale purchasers and suppliers, the previous length of the relationship does not seem to have an impact on the termination decision if a relationship is no longer profitable.

Thus, our fourth hypothesis is supported as the results corroborate that, in the cases considered, a lack of trust was not a sufficient reason to terminate long-term relationships in the wheat-to-bread chain in Germany. Moreover, the results suggest that in relationships in which switching costs are relatively low (e.g., farmers, bakery trade), relationships are usually shorter than in relationships with higher switching costs (e.g., large industrial enterprises like mills or industrial bakeries). Thus, the results highlight the important role of switching costs for the continuation of business relationships, and thereby provide support for research hypothesis 5.

Conclusions

Since we found empirical evidence for all research hypotheses, we argue that sustainable relationships dominate the German wheat-to-bread chains. The results of the survey and the qualitative interviews also revealed that the majority of actors are satisfied with the quality of their business relationships in this chain. In addition, relationship partners feel free to choose a certain relationship type and decide on a relationship's length such as to maximize profit under given environmental conditions. A significant majority of enterprises prefer long-term business relationships; an attitude that clearly helps to strengthen chain relationships. However, the willingness to change relationships seem to depend on switching costs. Relationships with low switching costs (e.g., farmers, bakery trade) are changed more often than those with higher ones (e.g., large industrial enterprises like mills or industrial bakeries).

On the other hand, the desire to remain independent as a company is a key factor for the non-contractual coordination of transactions. Independent businesses prefer to transact without being formally bound to their exchange partners. In the German wheat-to-bread chain, farmers are mainly the ones who seek independence.

For the sake of asserting their independence, even small craft bakeries usually do not enter into contracts, a behaviour consistent with our findings that these small business partners use spot markets or repeated transactions with the same supplier for their purchasing activities. Moreover, as transactions with the same supplier are usually long-term relationships without any contractual obligations, partners not only stay independent but can also reap the benefits of long-term cooperation (i.e., reduce transaction costs) at the same time.

The termination of long-term relationships is typically triggered by price, quality or supply continuity problems, or structural changes – i.e., economic factors. However, low trust seems only to contribute to relationship termination in the awareness and exploration phases.

Two types of trust influencing relationships in the wheat-to-bread chain could be distinguished. The first one is personal trust. This kind of trust builds when business partners have a personal liking for each other. Accordingly, as this type of trust depends above all on the quality of inter-personal communication between business representatives, it is a decisive factor at the beginning of a relationship as it allows the substitution of the lack of past cooperation experience. Moreover, personal trust plays a key role for choosing a partner whenever a number of suppliers are offering similar price performance ratios. Farmers and small-trade bakers especially stated that they would not cooperate with a business when they did not feel a personal liking for the representative.

On the other hand, personal antipathy is rarely seen as a reason for termination of sustainable, long-lasting business relationships in the German wheat-to-bread chain. In such a case, the motivation for continuation can be ascribed to the second type of trust; trust based on the positive past collaboration experience or credibility. This kind of trust develops if partners' objectives were satisfied in the past; a feature that is also strongly connected to inter-organizational communication within relationships, especially in the relationships between mills and industrial bakeries. In these relationships, the required flour quality has to be consulted, product information assuring traceability has to be exchanged, and price and supply conditions have to be negotiated.

References

AgriMA Produkt + Markt (2005) Zielgruppe Landwirte – Bauern, Manager und echte Unternehmer. http://www.lv-h.de/agrarmediaservice/bilder/agrima_pdf/pottebaum.pdf.

Bejou, D., Wray, B. and Ingram, T. (1996) Determinants of relationship quality: an artificial neural network analysis. *Journal of Business Research* 36, 137–143.

BMELV (2004, 2006) *Statistisches Jahrbuch über Landwirtschaft, Ernährung und Forsten der Bundesrepublik Deutschland*. Landwirtschaftsverlag, Münster Hiltrup.

Brown, B., Hanson, M., Liverman, D. and Merideth, R. (1987) Global sustainability: toward definition. *Environmental Management* 11(6), 713–719.

BVL (2007) Struktur und Leistungszahlen des Lebensmittel-Einzelhandels 2005. www.lebensmittelhandel-bvl.de/modules.php?name=Content&pa=showpage&pid=24&cid=7.

Donaldson, B. and O'Toole, T. (2000) Classifying relationships structures: relationship strength in industrial markets. *Journal of Business & Industrial Marketing* 15, 491–506.

Dwyer, F. and Tanner, J. (2002) *Business Marketing: Connecting Strategy, Relationships, and Learning*. McGraw-Hill, New York.

Dwyer, F., Schurr, P. and Oh, S. (1987) Developing buyer–seller relationships. *Journal of Marketing* 51, 11–27.

Dyer, J. and Singh, H. (1998) The relational view: cooperative strategy and sources of interorganizational competitive advantage. *Academy of Management Review* 23, 660–679.

Flint, D., Woodruff, R. and Gardial, S. (2000) Exploring the phenomenon of customers' desired value change in a business-to-business context. *Journal of Marketing* 66, 2–117.

Jap, S. and Anderson, E. (2007) Testing of life-cycle theory of cooperative interorganizational relationships: movement across stages and performance. *Management Science* 53, 260–275.

Kallfass, H. (1993) Kostenvorteile durch vertikale Integration im Agrarsektor. *Agrarwirtschaft* 42, 228–237.

Lange, D. and Koch, H. (1995) *Wettbewerbsfähigkeit durch verstärkte Kooperation: eine Studie zur ostdeutschen Agrar- und Ernährungswirtschaft*. Dt. Landwirtschaftsverlag, Berlin.

Moorman, C., Deshpande, R. and Zaltman, G. (1993) Factors affecting trust in market research relationships. *Journal of Marketing* 57, 81–101.

Patterson, P. and Smith, T. (2001) Modeling relationship strength across service types in an Eastern culture. *International Journal of Service Industry Management* 12, 90–113.

Press, B. (1997) Kaufverhalten in Geschäftsbeziehungen. In: Kleinaltenkamp, M. and Plinke, W. (eds) *Geschäftsbeziehungsmanagement*. Springer, Berlin.

Ring, P. and van de Ven, A. (1994) Developmental processes of cooperative interorganizational relationships. *The Academy of Management Review* 19, 90–118.

Shearman, R. (1990) The meaning and ethics of sustainability. *Environmental Management* 14(1), 1–8.

Tähtinen, J. and Halinen, A. (2002) Research on ending exchange relationships: a categorization, assessment and outlook. *Marketing Theory* 2, 165–188.

Williamson, O. (1975) *Markets and Hierarchies: Analysis and Antitrust Implications*. Free Press, New York.

Chapter 11

Inter-organizational Relationships in the US Agri-food System: the Role of Agricultural Cooperatives

Fabio Chaddad

University of Missouri, Columbia, USA

Introduction

Inter-organizational relationships are increasingly important in agri-food systems. Connecting farmers to food, fibre and bioenergy consumers, however, is a challenging task for two reasons. First, the process of agri-food industrialization leads to vertical coordination challenges as increasingly large, multinational processors and retailers attempt to source raw agricultural inputs from much smaller and dispersed farmers, sometimes across borders. Second, consumers are increasingly demanding agri-food products that are quality, origin and process certified. Responding to these vertical coordination challenges, agri-food system participants are relying less on the price mechanism of spot markets and increasingly adopting hybrids (e.g., inter-organizational relationships) and vertical integration to more efficiently govern transactions along the supply chain (see Chapter 2).

Cook *et al.* (2008) analyse the new institutional approach to contracting and organization in food and agriculture focusing on collaborative or network organizations. In doing so, they observe substantial variety in how these networks are governed and identify the emergence of three stylized network types. The first type comprises networks organized by a lead firm, often a large processor or retailer coordinating a network of farmer-suppliers. These supply chain 'captains' or 'leaders' use different contractual arrangements with a relatively stable network of producers to guarantee consistency and quality of supply, and to provide adequate incentives for producers to stay in the network. Because some of these buyer–supplier relationships tend to rely on frequent and long-term transactions, formal contracts are complemented with trust-based relational agreements. Ménard (2004) categorizes this type of network as 'leadership hybrid' as a firm establishes some degree of authority over partners because of its position in the supply chain or dominance over critical assets, resources or capabilities. Several chapters (e.g., 7, 8, 10) in this book focus on the nature of such vertical, long-term relationships between farmers and a supply chain captain in the agri-food system.

A second type of network is referred to as more 'egalitarian' as participants share similar rights and responsibilities. To deal with the transactional and cooperative interdependencies of inter-organizational relationships described in Chapter 2, these 'egalitarian' networks adopt coordinating structures that shift decision authority to a 'formal government' while retaining a great deal of independence and decentralized decision-making at the participant level (Ménard, 2004; Cook *et al.*, 2008). Differently from the 'leadership' network, participants in 'egalitarian' arrangements willingly and purposefully agree to give up some control to a more centralized, yet non-hierarchical organization. The *label rouge* poultry chain in France is an example of such egalitarian network (Ménard, 1996).

A third type of network or hybrid inter-organizational arrangement is the cooperative. Interestingly enough, despite its importance in the agri-food sector of most advanced agricultural countries, the 'producer-led' cooperative has only recently been the focus of scholarly interest from a new institutional economics perspective (Staatz, 1989; Cook *et al.*, 2004). This paper makes a contribution to this literature by describing and analysing the variety of network configurations organized by US agricultural producers using the netchain approach. Because this chapter deals primarily with contractually tied-together farmer-owned cooperatives and the inter-organizational arrangements they form with agri-food system participants, the concept of 'inter-organizational relationship' is used in a wider sense than in other parts of the book.

In what follows, the chapter describes the organizational architecture of farmer-led, cooperative networks in the US food system using the netchain framework. The next section provides an overview of the economic importance and the evolution of market share gains of cooperatives in the US food system. The third section presents the netchain framework and applies it to the analysis of structural characteristics of producer-led netchains, including federated macro-hierarchies, marketing agencies in common, and alliances. The last section presents a summary with conclusions.

Cooperatives in the US agri-food sector

Despite rapid and fundamental changes in their business environment, agricultural cooperatives in the USA continue to perform an important economic role in agricultural and food chains as farmers' 'integrating agency'. According to USDA Rural Business Cooperative Services (RBS) data, the nation's 2594 agricultural cooperatives generated US$147 billion in sales and accumulated US$57 billion in total assets in 2007 (USDA, 2008). US agricultural cooperatives play a major role in providing production inputs and services to farmers, and in processing and marketing their commodities. In 2002, aggregate cooperative market shares for both farm commodity marketing and purchased inputs reached 27% of total farm output value and input acquisition expenses, respectively (Fig. 1). In the dairy industry, for example, cooperatives collect and market almost 85% of the milk produced by farmers. In the grain and oilseed sector, cooperatives originate and market more than 40% of total US production.

In addition to playing an important role in collecting, originating, storing and marketing agricultural commodities, cooperatives are increasingly involved in upstream and downstream industries. Cooperative involvement in upstream and downstream processing has historically occurred by means of wholly owned subsidiaries. But given the increased capital intensity and competitiveness of farm input manufacturing and value-added food processing and distribution, cooperatives are increasingly forming alliances and joint ventures with other cooperatives, private processors and outside investors. A recent survey of the USDA has found that a sample of 204 US farmer cooperatives had additional sales of at least US$14 billion in 2007 from other, non-cooperative business ventures they have formed or invested in to market value-added food products or sell farm supplies to farmers. These other businesses represent more than US$7.5 billion in assets.

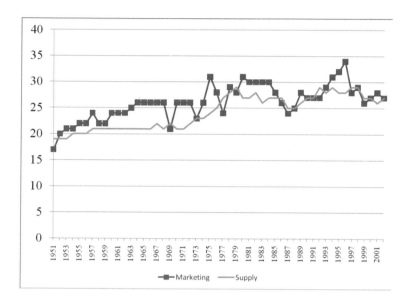

Fig. 1. Market share (%) of cooperatives in farm output marketing and input supply (US Department of Agriculture, 1998/2008).

Cooperative netchains

As agri-food systems become increasingly industrialized and globalized, and hybrid modes of governance emerge to cope with the increased complexities and interdependencies involved in inter-organizational relationships, scholars have sought new ways to analyse the complex organizational arrangements they observe. The 'netchain' approach is a recent attempt to integrate supply chain analysis (SCA) and network analysis (NA) in the study of complex forms of inter-organizational relationships (Lazzarini *et al.*, 2001). A netchain is defined as a set of networks comprised of horizontal ties between firms within a particular industry or group, such that these networks (or layers) are sequentially arranged based on the vertical ties between firms in different layers. Figure 2 shows a generic netchain with four sequentially organized layers (i.e., suppliers, manufacturers, distributors and consumers). Each dot in the figure represents a firm; arrows represent inter-firm linkages or ties.

Netchain analysis explicitly differentiates between horizontal (transactions in the same layer) and vertical ties (transactions between layers), mapping how agents in each layer are related to each other and to agents in other layers. In Fig. 2, the arrows connecting the firms in the supplier layer denote reciprocal interdependencies, while the dotted lines connecting the firms in the distributor layer denote pooled interdependencies. The arrows connecting firms across distinct layers represent vertical ties, which are characterized by sequential interdependencies.

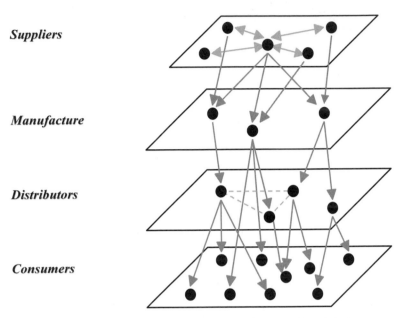

Fig. 2. Example of a generic netchain.

The netchain approach integrates SCA and NA by recognizing that complex inter-organizational relationships embody several types of interdependencies – pooled, sequential and reciprocal (Thompson, 1967). These interdependencies between firms in a netchain are associated with distinct sources of value – that is, strategic variables yielding economic rents – and coordination and control mechanisms to govern such relationships. Lazzarini *et al.* (2001) also examine the applicability of the netchain approach by discussing a diverse set of netchain configurations, which include buyer–supplier relationships, information technology induced inter-organizational collaborations, and the macro-hierarchy organizational structure (which is defined below).

This section applies the netchain approach to describe the configuration of producer-led hybrids in the US food system. These complex farmer-owned and -controlled hybrids include federated macro-hierarchies, marketing agencies in common and strategic alliances with both cooperatives and non-cooperatives.

The federated cooperative structure as a macro-hierarchy

Macro-hierarchies are hierarchies involving organizations – instead of individuals within organizations – that jointly coordinate some of their activities through multiple layers of ownership. One particular example of a macro-hierarchy is discussed in this section: the farmer-owned and -controlled cooperative organized in a multi-layered fashion, which is also known as the federated structure.

Farmers

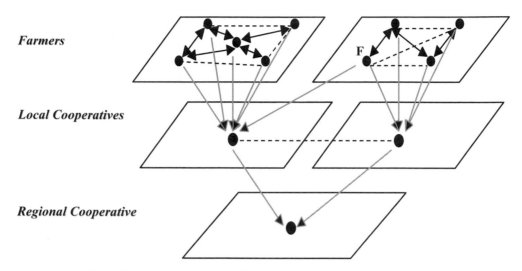

Local Cooperatives

Regional Cooperative

Fig. 3. The federated cooperative structure as a macro-hierarchy.

In a federated cooperative structure, farmers are members of a local, tier-1 cooperative, which in turn is a member of a regional, tier-2 cooperative. Regional cooperatives themselves may also decide to form inter-regional, tier-3 cooperatives, marketing agencies in common, strategic alliances and other forms of inter-organizational relationships upstream or downstream in the agri-food chain (see below for more details). As a result, a federated cooperative is structured by means of sequential layers of ownership. Figure 3 shows an example of a macro-hierarchy with three layers: farmers, local cooperatives and the regional cooperative. Farmers form horizontal ties among themselves characterizing both pooled and reciprocal interdependencies, which are explained below. Local cooperatives, in turn, also pool resources to form the regional cooperative. The vertical arrows connecting firms across layers denote sequential interdependencies.

Cooperatives are characterized by restricted residual claims, i.e., they are owned and controlled by farmer-members. The vertical ties between subsequent layers of a federated cooperative structure therefore entail both a transactional and an ownership relationship. As a result, the transactional and coordination challenges common to all forms of inter-organizational relationships (see Chapter 2 for details) also apply in this context.

The assignment of ownership rights to farmers in a cooperative is often explained as a transaction cost minimization strategy, which is a source of value associated with sequential interdependencies. When farmers own – and thus control – an agricultural cooperative they avoid potential hold-up situations arising primarily from site and temporal asset specificities (e.g., Staatz, 1989; Hendrikse and Bijman, 2002). For example, by forming a cooperative, dairy farmers can more safely invest in equipment for milk production and storage, thereby avoiding downstream pressures to reduce prices given the perishable nature of their product. Additionally, market contracting is costly when upstream farm input providers and downstream processors or retailers have better market information and knowledge than farmers. In case a firm knows more about the quality of the product it sells, it has an incentive to deliver a lower-quality product than originally promised (Barzel, 1982). In this case, customer ownership reduces the firm's incentive to exploit its information advantage. Bonus (1986) and Hansmann (1996) explain the formation of credit, consumer and agricultural input supply cooperatives on the basis of measurement problems of this sort.

There are multiple sequential interdependencies in federated cooperative structures, since on one occasion farmers sell their output to the cooperative (e.g., milk), while on other occasions they acquire production inputs from the cooperative (e.g., feed and fertilizer). These interdependencies are not properly classified as reciprocal because these transactions are not necessarily carried out together, except in some cases where the cooperative 'bundles' services (e.g., product acquisition and technical support) and products (e.g., farm inputs) as in the case of multipurpose local and regional cooperatives. As discussed previously, the contractual hazards created by these sequential transactions determine ownership by farmers, which then acquire rights to control (or 'plan') the allocation of resources through successive stages of the supply chain.

In addition to these sequential interdependencies, federated cooperative structures are also characterized by pooled and reciprocal interdependencies among members within the same horizontal layer. For example, individual members within the same layer keep their independence and decision-making autonomy, but pool financial and other productive resources to create a higher-level structure to jointly develop related businesses thereby characterizing a pooled interdependence. This higher-level structure defines common, standardized rules to commercialize products, purchase inputs, transfer information and divide the residual claims among members (see discussion on hybrids in Chapter 2).

Additionally, Bonus (1986) refers to the local cooperative as a social group with an 'esprit de corps'. The formation of reciprocal interdependencies among farmers in local cooperatives is explained as a consequence of intimate personal knowledge and strong social ties, a distinguishing characteristic of rural communities. In those circumstances, members are likely to employ joint decision-making and problem-solving to coordinate their activities – i.e., mutual adjustments. As a result, the transactional and ownership components of the vertical ties are embedded in a network of personal relationships among members (Granovetter, 2005). These social attachments may foster the emergence of trust, which tends to ameliorate potential internal conflicts and opportunistic behaviour. In other words, reciprocal interdependencies may positively affect vertical transactions between layers (James and Sykuta, 2005). According to Staatz (1987), some cooperatives have an identifiable base of member-patrons who are more inclined to reveal strategic and proprietary information to their cooperative (and vice versa). Farmers can also be members of more than one cooperative (such as farmer *F* in Fig. 3), which tends to facilitate the joint coordination of local cooperatives belonging to a higher-level (regional) cooperative.

In the following sections, the chapter presents a description based on primary and secondary data sources about three configurations of producer-led netchains in the US food system: the federated structure, the marketing agency in common, and the strategic alliance arrangement.

Federated agricultural cooperatives in the US agri-food system

The federated model based on Edwin Nourse's (1922) 'bottom-up' philosophy of cooperation is the cornerstone of the US agricultural cooperative system. In 2000, nearly two-thirds of all US agricultural cooperatives were tier-1, local multipurpose cooperatives that are members of tier-2, regional cooperatives. As explained in the previous section, farmers own the tier-1 cooperatives, which in turn own the tier-2 cooperatives, characterizing such federated systems as macro-hierarchies. More than 75% of all producer memberships held in cooperatives were in these local cooperatives. In 2000, there were 57 federated agricultural associations operating in the USA, some of which were among Fortune 500 companies (Frederick *et al.*, 2002).

In order to describe the actual configuration of these macro-hierarchies, information was collected using a standardized survey instrument about the structure, size and scope of operations of 16 federated cooperative systems. Table 1 shows a summary description of these producer-led macro-hierarchies operating in the US agri-food sector. The federated cooperatives in the sample are among the largest in the country, with combined turnover of over US$45 billion in 2002. These cooperatives operate on a regional and sometimes national and even international scale. The federated cooperative structures shown in Table 1 have business operations in grain marketing, food processing, cotton ginning, financial services, and agricultural input supply manufacturing and distribution.

Based on federated cooperative structural information, heterogeneity in macro-hierarchy architectures is observed. A continuum appears to exist with two polar forms: a purely federated system and a purely centralized system. In a purely federated system, farmers pool resources to form a local, tier-1 cooperative. Tier-1 cooperatives in turn pool resources to form a regional (or national) tier-2 cooperative. Farmers and local cooperatives maintain a high degree of independence, as the regional cooperative has little power or authority in controlling farmers' and local cooperatives' decisions. Farmers and local cooperatives compete amongst themselves and may even compete with the regional. There is a high degree of pooled interdependence governed by standardized rules and mechanisms. In addition, these purely federated structures exhibit some degree of sequential interdependence. Examples include the three federated systems under the Farm Credit System (AgFirst, AgriBank and CoBank), Growmark, Ag Processing Inc. and Florida's Natural.

In a purely centralized system, farmers own and do business directly with the regional cooperative. 'Local' assets are owned by the regional cooperative, such as the case with MFA, Inc. Some centralized cooperatives have adopted this structure since their inception, but there is a trend towards centralization among some federated structures.

Mixed or intermediate macro-hierarchy structures are observed between the two polar forms. These mixed structures include Land O'Lakes (centralized in dairy business and federated in other business units), Dairy Farmers of America (having both tier-1 cooperatives and milk producers as members) and CHS Inc. CHS has a more complex structure – it is also a mixed macro-hierarchy with both centralized and federated business units, and the cooperative has engaged in a 'regionalization' process whereby local assets are owned and operated by CHS but governed by a local board formed by farmers.

Table 1. Description of federated agricultural cooperatives in the USA.

Macrohierarchy	Headquarters	Territory	Business Operations	Volume (2002)	Ownership Structure
AgFirst, FCB	Columbia, SC	15 Eastern States and Puerto Rico	Agricultural lender. Originates real estate, operating and rural home mortgage loans. Offers insurance, equipment leasing, tax and other financial services.	Loans: US$12 billion	Part of Farm Credit System. Owned by 24 Agricultural Credit Associations.
Ag Processing, Inc.	Omaha, NE	16 States (mostly in the Midwest) and Canada	Primary business is soybean processing. Other products include vegetable oil, animal feeds, grain marketing and transportation services, bio-fuels, fuel additives and solvents.	Sales: US$1.8 billion	Owned by 250,000 farmer-members represented through 243 local cooperatives and 8 regional cooperatives.
AgriBank, FCB	St Paul, MN	15 States	Wholesale lending and business services to Farm Credit System (FCS) associations in America's heartland.	Loans: US$19 billion	Part of Farm Credit System. The 7th District Farm Credit Services Associations (19) own AgriBank.
Alabama Farmers Cooperative, Inc.	Decatur, AL	4 States (all of AL and parts of GA, MS and FL)	Fertilizer, feed, grain and farm supply.	Sales: US$300 million	All facilities operated by AFC are governed by local, farmer-owned cooperatives.
CHS, Inc.	Inver Grove Heights, MN	24 States	Grain marketing, food processing, petroleum refineries and pipelines; also markets and distributes energy products, agronomic inputs and feed.	Sales: US$7.8 billion	Owned by 1400 local cooperatives.
CoBank, ACB	Greenwood Village, CO	National	Loan programmes, financial and leasing services.	Loans: US$27 billion	Part of Farm Credit System. Owned by about 2500 stockholders, consisting of agribusinesses, rural utilities and farm credit associations.
Dairy Farmers of America, Inc.	Kansas City, MO	National	Milk collection, hauling, processing and distribution.	Sales: US$8 billion	Owned by 3 tier-1 cooperatives and 18,000 milk producers.

Table 1. Description of federated agricultural cooperatives in the USA (continued).

Macrohierarchy	Headquarters	Territory	Business Operations	Volume (2002)	Ownership Structure
Florida's Natural Growers	Lake Wales, FL	Florida	Orange juice and citrus processor.	Sales: US$ 600 million	1100 member owners in 12 smaller cooperatives.
Growmark, Inc.	Bloomington, IL	6 US States and Ontario, Canada	Regional agricultural supply and grain marketing cooperative.	Sales: US$1.2 billion	Owned by 100 retail and 250 grain marketing cooperatives.
Land O'Lakes, Inc.	Arden Hills, MN	National	Food processing, feed manufacturing and agricultural input supply cooperative.	Sales: US$5.8 billion	Owned by 7000 producer-members and 1300 local cooperatives.
MFA, Inc.	Columbia, MO	Missouri and 3 adjacent States	Agricultural input supply, financial services, livestock and grain marketing.	Sales: US$ 692 million	Owned by 45,000 farmer-members.
PYCO Industries, Inc.	Lubbock, TX	Southern US States	Manufacturer of cotton oil and other cottonseed products.	Sales: US$ 155 million	Owned by over 120 member cotton gins.
Southern States Cooperative, Inc.	Richmond, VA	23 Southern and Midwestern States	Purchases, manufactures and sells agricultural input, animal health, pet food, home and garden supplies. Gins cotton, procures groundnuts and markets small grains, maize, soybeans and fish.	Sales: US$1.5 billion	Owned by 321,000 farmer-members.
Staple Cotton Cooperative Association	Greenwood, MS	8 South-eastern States	Domestic and export cotton marketing.	Sales: US$1.2 billion	Owned by 12,140 member-owners, who sign marketing agreements.
Sunkist Growers, Inc.	Sherman Oaks, CA	California and Arizona	Markets fresh oranges, lemons, grapefruit and tangerines in the USA and overseas. Fruit that does not meet fresh market standards is used in food products.	Sales: US$ 965 million	Owned by 6500 citrus growers who make up its membership. Each is a member of a local association or district exchange (18).
Tennessee Farmers Cooperative	La Vergne, TN	Tennessee	Animal health products, feed, fertilizer, outdoor power equipment, seeds and tyres.	Sales: US$ 400 million	70 member cooperatives own TFC, with 70,000 farmer members.

Marketing agencies in common

In addition to federated structures, US farmer cooperatives have also pursued an alternative netchain structure-strategy configuration by forming marketing agencies in common (MACs) to conduct inter-cooperative coordination of certain facets of their operations, in particular to accomplish specific marketing activities. Members of MACs retain individual asset ownership, while their jointly owned MAC provides various supplementary functions, such as group communications and product selling coordination. A MAC is an example of a horizontal, non-equity inter-organizational relationship defined in Chapter 2.

The term 'marketing agencies in common' is explicitly used in the Capper-Volstead Act as a mechanism for cooperatives to coordinate their marketing functions. MACs may be used by both cooperatives and non-cooperatives. The limited anti-trust exemptions provided by the Capper-Volstead Act to farmer-owned cooperatives have enabled them to use the MAC institutional arrangement more often than other segments of US industry.

The purposes of MACs are, in most cases, related to sales coordination of member products and complementing the marketing programmes that member cooperatives have developed on their own. The marketing coordination functions performed by MACs are sometimes extended to inter-organizational planning and decision-making for certain operational areas that involve mutual or interactive impacts on members.

MACs are a distinct governance alternative for cooperatives seeking the benefits of coordination and economies of size. A comparison of several objectives and alternative organizational forms reveals advantages for MACs when members seek inter-organizational collaboration but have a preference for maintaining their separate identities and independence. MACs have some disadvantages, however, as they generally lack incentives to develop distinctive capabilities and to innovate. This happens because each cooperative in a MAC has a unique set of organization-specific assets and resources that it would like to augment and protect, but not transfer out of its direct and immediate control. An important characteristic with most MACs is that, while their member-cooperatives coordinate some aspects of their marketing programmes, they also compete in the marketplace.

Most studies of cooperative organizations suggest that a MAC is a special type of the federated cooperative structure. Two characteristics, however, distinguish a MAC from the federated structure described in the previous section. Unlike federated cooperative structures, MACs do not serve the purpose of sharing in the acquisition and ownership of financial and physical assets needed for upstream or downstream vertical integration. In other words, MACs are organized by groups of cooperatives to coordinate marketing activities, with each member retaining exclusive ownership over its unique, firm-specific set of physical, financial and human resources. As a result, MACs have relatively low asset-to-sales ratios. In addition to low asset-to-sales ratios, MACs tend to have a smaller number of member cooperatives than other federated structures. The average number of member-cooperatives in MACs is nine, while the average for federated cooperative structures reaches 76 members (Reynolds, 1994). Additionally, most MACs do not take title in marketing transactions with their members; i.e. they solely act as their members' marketing agent.

MACs are very common among cooperatives in the dairy industry. For example, Southern Marketing Agency, Inc. (SMA) is a MAC formed in 2002 to seek efficiencies in supplying the fluid milk needs of the south-eastern US market. SMA is organized as a

Kentucky agricultural cooperative under provisions of the Capper-Volstead Act. It currently has five dairy cooperatives as members – Arkansas Dairy Cooperative Association (Damascus, AK), Dairy Farmers of America (Kansas City, MO), Dairymen's Marketing Cooperative (Mountain Grove, MO), Lone Star Milk Producers (Windthorst, TX) and Maryland & Virginia Milk Producers Cooperative Association (Reston, VA). Strategic goals of SMA include: (i) to promote member cooperation and communication; (ii) to seek cost savings in the purchase of supplemental milk, in farm-to-market milk hauling and in seasonal surplus balancing; and (iii) to preserve over-order prices in the Southeast region. SMA is governed by an Operations Committee made up of senior managers of each member-cooperative. The Operations Committee reports directly to the SMA board of directors, comprised of ten dairy farmers.

Seven MACs in the dairy industry are analysed by Liebrand and Spatz (1993) showing how cooperatives may leverage inter-cooperative relationships to market member products and link farmers to consumers.

Strategic alliances

In the recent past, US cooperatives have been accused of failing to cooperate among them (Torgerson, 1993; Fulton *et al.*, 1998). Interestingly enough, US farmer cooperatives found it easier to collaborate and form alliances with corporations than with other cooperatives. Goldberg (1972) provides an early analysis of corporate-cooperative inter-organizational relationships in the US food system and discusses why they have become a significant factor in many commodity sectors.

It was not until the formation of the Cenex-Land O'Lakes agronomy joint venture in 1986 that US agricultural cooperatives began to seriously consider forming inter-organizational relationships with other cooperatives (Rodriguez-Alcalá, 2000). The recent growth in inter-cooperative alliance formation has been noticed by Merlo (1999). According to the author, there were more mergers, consolidations, acquisitions, joint ventures and alliances among US cooperatives in 1998 than in all the history of cooperative development. Recent USDA data reveal that the number of strategic alliances and joint ventures involving US agricultural cooperatives has increased significantly since the late 1990s.

In order to better understand how cooperatives arc using alliances and joint ventures to advance their strategic goals, data were collected from the SDC Platinum database – a service of Thomson Financial. The search included the 50 largest US agricultural cooperatives and therefore does not comprise alliances involving small, local cooperatives. The search for the period 1980–2003 identified 54 strategic alliances and joint ventures involving at least one US agricultural cooperative. An analysis of this sample of alliances suggests the following initial categorization:

- Forward vertical integration into food processing industries. Examples include Dairy Farmers of America's strategic alliances with several private companies in fresh milk bottling and dairy product processing and distribution; Land O'Lakes' joint venture with Dean Foods in dairy product manufacturing; CHS, Inc.'s joint venture with Mitsui & Co. in edible vegetable oil manufacturing and with Cargill in flour milling.
- Backward vertical integration into agricultural input manufacturing and distribution. Examples include Agrilliance, the largest agronomy company in the USA formed by

CHS, Inc. and Land O'Lakes; CF Industries, a joint venture of seven regional cooperatives in fertilizer and crop nutrient manufacturing; and FS Industries, the joint venture between Farmland and JR Simplot Co in phosphate fertilizer manufacturing.

- Backward vertical integration into agricultural production. Examples include the milk production joint venture between Dairy Farmers of America and Suiza Foods; and MoArk LLC, a joint venture formed by Land O'Lakes in egg production and processing.
- Other alliance ventures, including internet ventures, management consulting, and information and financial services.

Because of the need to become more competitive in the global marketplace, US cooperatives adapted by creating critical mass, achieving economies of scale, and building market share. In addition to consolidation – by means of mergers and acquisitions (Hudson and Herndon, 2002) – cooperatives formed inter-organizational relationships to remain or become more efficient domestically and globally. Cooperatives are not only focusing on domestic partners to achieve strategic objectives, but they have also formed international relationships. Examples include DFA's joint venture with Fonterra in the dairy business; Land O'Lakes' livestock feed venture in Taiwan; and Riceland Foods' venture with two Japanese companies in rice export.

Another factor affecting strategic alliance formation among agricultural cooperatives is the need to acquire risk capital for growth purposes. In general, traditional agricultural cooperatives struggle to acquire equity capital because only members may invest in the cooperative but they lack the necessary incentives to do so (Cook, 1995). Chaddad and Cook (2004) identify alternative capital-seeking strategies for agricultural cooperatives, including joint ventures with other cooperatives, joint ventures with non-cooperatives, trust companies and limited liability companies. In other words, inter-organizational relationships are increasingly utilized as equity capital seeking strategies by US agricultural cooperatives as they seek to link their farmer-members with other supply chain participants and ultimately with consumers.

Conclusions

This chapter described the organizational architecture of an economically relevant network form in the agri-food system – the farmer-led cooperative. Based on data from the USA, the chapter showed that farmer-owned cooperatives play an increasingly relevant role in the agri-food system. Competing with other network forms observed by Ménard (2004) and Cook *et al.* (2008), producer-led networks are organized in complex architectures – including macro-hierarchies, marketing agencies in common, and alliances with participants upstream and downstream in the agri-food chain. The chapter described these architectures using survey and secondary data and analysed them with the netchain approach posited by Lazzarini *et al.* (2001).

Networks, supply chains and alliances have been important in the business world for some time. The same factors that have made these effective strategies for businesses in other sectors of the economy are also coming to bear on agri-food system participants. The question then arises as to how farmers can maintain their independence and specialize in what they do best (i.e., farming) and, at the same time, become more integrated with the rest of the agri-food system. By means of federated cooperative structures, marketing agencies in common, and alliances with sundry partners, US producers have formed complex networks and supply chains (netchains) connecting them with input providers, food processors, distributors, and ultimately consumers. In doing so, they are able to leverage existing resources to achieve economies of scale and scope, to increase operational efficiencies in agricultural input manufacturing and food processing, to access risk capital to finance investment opportunities, and to coordinate agri-food chains. In other words, US agricultural producers have adapted to the evolving business environment by forming cooperatives, which in turn have engaged in several forms of inter-organizational collaborations to leverage their role as farmers' 'integrating agency'.

References

Barzel, Y. (1982) Measurement costs and the organization of markets. *Journal of Law and Economics* 25, 27–48.

Bonus, H. (1986) The cooperative association as a business enterprise: a study in the economics of transactions. *Journal of Institutional and Theoretical Economics* 142, 310–339.

Chaddad, F. and Cook, M. (2004) Understanding new cooperative models: an ownership-control rights typology. *Review of Agricultural Economics* 26(3), 348–360.

Cook, M. (1995) The future of US agricultural cooperatives: a neo-institutional approach. *American Journal of Agricultural Economics* 77(5), 1153–1159.

Cook, M., Chaddad, F. and Iliopoulos, C. (2004) Advances in cooperative theory since 1990: a review of agricultural economics literature. In: Hendrikse, G. (ed) *Restructuring Agricultural Cooperatives*. Haveka, Rotterdam, pp. 65–90.

Cook, M., Klein, P. and Iliopoulos, C. (2008) Contracting and organization in food and agriculture. In: Brousseau, E. and Glachant, J. (eds) *New Institutional Economics: A Guidebook*. Cambridge University Press, Cambridge, pp. 292–304.

Frederick, D., Crooks, A., Dunn, J., Kennedy, T. and Wadsworth, J. (2002) Agricultural cooperatives in the 21st century. *Cooperative Information Report 60*. USDA Rural Business Cooperative Service, Washington, DC.

Fulton, J., Popp, M. and Gray, C. (1998) Evolving business arrangements in local grain marketing cooperatives. *Review of Agricultural Economics* 20(1), 54–68.

Goldberg, R. (1972) Profitable partnerships: industry and farmer co-ops. *Harvard Business Review* (March–April), 108–121.

Granovetter, M. (2005) The impact of social structure on economic outcomes. *Journal of Economic Perspectives* 19(1), 33–50.

Hansmann, H. (1996) *The Ownership of Enterprise*. The Belknap Press of Harvard University Press, Cambridge.

Hendrikse, G. and Bijman, J. (2002) Ownership structure in agrifood chains: the marketing cooperative. *American Journal of Agricultural Economics* 84, 104–119.

Hudson, D. and Herndon, C. (2002) Factors influencing probability and frequency of participation in merger and partnership activity in agricultural cooperatives. *Agribusiness: An International Journal* 18(2), 231–246.

James, Jr, H. and Sykuta, M. (2005) Property right and organizational characteristics of producer-owned firms and organizational trust. *Annals of Public and Cooperative Economics* 76(4), 545–580.

Lazzarini, S., Chaddad, F. and Cook, M. (2001) Integrating supply chain and network analyses: the study of netchains. *Journal on Chain and Network Science* 1(1), 7–22.

Liebrand, C. and Spatz, K. (1993) DariMac: an export marketing agency in common for dairy cooperatives. *Research Report 126*. USDA Rural Business Cooperative Service, Washington, DC.

Ménard, C. (1996) On clusters, hybrids and other strange forms: the case of the French poultry industry. *Journal of Institutional and Theoretical Economics* 152, 154–183.

Ménard, C. (2004) The economics of hybrid organizations. *Journal of Institutional and Theoretical Economics* 160, 345–376.

Merlo, C. (1999) When cooperatives combine. *The Cooperative Accountant* (Spring), 40–47.

Nourse, E. (1922) The economic philosophy of co-operation. *American Economic Review* 12(4), 577–597.

Reynolds, B. (1994) Cooperative marketing agencies in common. *Research Report 127*. USDA Rural Business Cooperative Service, Washington, DC.

Rodriguez-Alcalá, M. (2000) Strategic alliance theory and practice: analyzing the Cenex-Land O'Lakes joint venture. Unpublished MS thesis, University of Missouri.

Staatz, J. (1987) The structural characteristics of farmer cooperatives and their behavioral consequences. In: Royer, J. (ed) *Cooperative Theory: New Approaches*. USDA Agricultural Cooperative Services, Washington, DC, pp. 33–60.

Staatz, J. (1989) Farmer cooperative theory: recent developments. *Research Report No. 84*, US Department of Agriculture, Agricultural Cooperative Service, Washington, DC.

Thompson, J. (1967) *Organizations in Action: Social Science Bases of Administrative Theory*. McGraw-Hill, New York.

Torgerson, R. (1993) Strategic alliances revisited: a cooperative perspective. *American Cooperation*, 151–159.

US Department of Agriculture (1998) Cooperative historical statistics. *Cooperative Information Report 1*, Section 26. USDA Rural Business Cooperative Services, Washington, DC.

US Department of Agriculture (2008) Farmer cooperative statistics 2007. *Service Report 68*. USDA Rural Business Cooperative Services, Washington, DC.

Chapter 12

Guanxi and Contracts in Chinese Vegetable Supply Chains: an Empirical Investigation

Hualiang Lu,[1] Jacques Trienekens,[2] Onno Omta[2] and Shuyi Feng[3]

[1] Nanjing University of Finance and Economics, China
[2] Wageningen University, the Netherlands
[3] Nanjing Agricultural University, China

Introduction

China is one of the world's biggest vegetable producers, accounting for more than one third of world vegetable production (FAO, 2004). China's vegetable industry, however, faces a great number of challenges. The fast growth in production has led to an oversupply of low quality (e.g., high chemical residues) vegetables. Poor quality and food safety issues have subsequently become major constraints for the further development of the Chinese vegetable industry.

Vegetable production in China is characterized as small-scale, with low productivity and inconsistent supply. Vegetable farmers face difficulties in implementing high-quality standards due to technical, managerial and financial constraints. Farmers are not well organized and contract farming is underdeveloped (Lu, 2007). Vegetable transactions are mostly conducted through face-to-face negotiations and cash payments. Collaboration between farmers and buyers is limited and small-scale farmers are largely excluded by most high-value markets including supermarkets.

Vegetable processing and exporting firms, however, have developed quickly in the last decade in China. Nevertheless, most remain uncompetitive in the international market due to inherent disadvantages such as obsolete equipment and technology, untrained and inexperienced labourers and insufficient financial resources (Wang and Yao, 2002). Furthermore, firms also lack legitimacy and government support (Wu and Leung, 2005). Practically, they face difficulties in acquiring high-quality vegetables from farmers and thus face difficulties in selling processed products to markets.

In the Chinese business environment, guanxi provides a basis for information sharing, communication and collaboration, and building trustworthy business relationships (Lu et al., 2008). As a potential mechanism to coordinate markets, guanxi has been extensively researched since the 1990s (Ambler, 1994; Davies et al., 1995). Standifird and Marshall

(2000) reveal that guanxi-based business transactions provide transaction cost advantages. Guanxi networks help to expand markets, contribute to sales growth (Kao, 1993), enhance competitive advantage (Thorelli, 1986) and improve firms' performance (Luo and Chen, 1997). Guanxi networks, however, are by no means always beneficial. Developing and maintaining guanxi is a time-consuming and expensive endeavour, thus inevitably bear costs for those who use guanxi (Yi and Ellis, 2000). Westerners also often view guanxi as related to unethical behaviour and might lead to corruption (Fan, 2002a; Luo, 2008). Despite the drawbacks of guanxi networks, people perceive guanxi networks an efficient way to conduct business in China because the legal systems are far from fully developed (Wong and Chan, 1999).

The ability to achieve high levels of relationship satisfaction has been considered an essential ingredient of business success (Morrissey and Pittaway, 2006), because satisfaction will affect the morale and subsequent intentions of business partners to participate in joint activities (Schul *et al.*, 1985). Therefore, building satisfactory buyer–seller relationships is crucial for both farmers and processing/exporting firms.

Previous research has addressed several aspects of buyer–seller relationships, such as trust (Anderson and Narus, 1990; Zaheer and Venkatraman, 1995), transaction-specific investments (Klein *et al.*, 1990) and networks (Dyer, 1996; Claro *et al.*, 2003), but few systematic attempts have been made to investigate the impacts of guanxi and contracts on relationship satisfaction in China. The objective of this chapter is to explore how Chinese guanxi networks and contracts influence the satisfaction of business relationships in vegetable supply chains. The insights obtained from this study will provide instruments on how to further improve market performance in the Chinese vegetable industry.

The remainder of this chapter is structured as follows. The next section develops the structural model. The research design is then described, followed by the empirical results. This chapter ends with conclusions and some implications.

Theoretical perspectives

In this chapter, we develop a conceptual model to investigate the interactions among guanxi networks, contractual governance, inter-personal trust, transaction-specific investments and relationship satisfaction with a focus on Chinese vegetable supply chains.

Guanxi is commonly defined as special relationships two persons have with each other. Fan (2002b) categorizes guanxi into three types: family, friend and business guanxi. Family guanxi exists among family members and is a relatively permanent and stable social relationship. Friend guanxi also is a rather stable and long-term relationship, and is usually used as an instrument to attain material and/or mutual goals. Business guanxi is defined as the process of finding business solutions through personal connections, which is the focus in this study. Guanxi is transferable from one person to another whereby both direct and indirect personal relationships eventually weave a multilayer guanxi network. The effectiveness of guanxi networks is closely related to the network size, network scope and the strength of the connections (Standifird and Marshall, 2000).

Guanxi is the lifeblood of the Chinese business community and extends into society and politics (Wong and Leung, 2001). Guanxi is first and foremost about the cultivation of long-term personal relationships. Guanxi networks provide certain assurances of exchange partner behaviour (Standifird and Marshall, 2000) because, in a guanxi network, the cost of opportunism is the potential loss of exchange opportunities with all members of the network. Thus, guanxi networks lead to the generation of relationship-sustaining factors such as trust. Empirical research has shown that guanxi networks encourage inter-personal trust and promote trust-based exchanges (Farh *et al.*, 1998; Hill, 1995). Thus, a positive relationship between guanxi networks and the level of trust is expected in Chinese vegetable buyer–seller relationships (Fig. 1).

Guanxi networks also possess the capacity to reduce transaction costs associated with environmental and behavioural uncertainties and opportunism (Standifird and Marshall, 2000). Moreover, the flexible and socially-based nature of guanxi permits the members of a guanxi network to deal with unforeseen contingencies which may arise after agreement has been reached. Under the safeguarding of guanxi networks, business relationships in the presence of specific investments become tighter and more stable. Therefore, we expect that farmers and firms will be more willing to invest when their business relationships are supported by their guanxi networks.

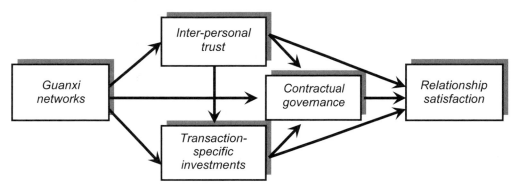

Fig. 1. Conceptual research model.

Contracts, in agriculture, are commonly defined as a written or oral agreement between farmers and companies or other buyers that specify conditions of producing and/or marketing an agricultural product. In addition to specifying quality requirements, contracts can also specify price, quantities, deliver conditions, penalties in case of violation, and so on. Agricultural contracts, mostly yearly-based, are commonly used to regulate the legitimacy of the transactions. In addition, informal governance (guanxi-based transactions) is also widely accepted (Wong and Leung, 2001). Although many researchers have investigated the interaction between contractual and relational governance from a Western perspective (for example see Poppo and Zenger, 2002; Ferguson *et al.*, 2005), the relationships between guanxi networks and contracts in China are still illusive. Braendle *et al.* (2005) argue that Chinese society is currently confronted with new laws and regulations as well as its deeply rooted cultural 'law' of guanxi. In some situations, guanxi can take precedence over legitimate decisions based on law or regulation. Conversely, Guthrie (1998) claimed that guanxi is playing a diminishing role in China and predicted that the legal system (formal governance) would surpass the influence of guanxi in the economic transition. Both studies suggested a substitution effect of guanxi on contracts. Schramm and Taube (2003), however, observed that guanxi still prevails in China and will co-exist with the legal system for a long period. They suggested a complementary effect of guanxi networks on formal governance.

Inter-personal trust, TSIs and contracts in vegetable supply chains

Inter-personal trust and transaction-specific investments (TSIs) are two foci in buyer–seller relationships. Two dimensions of trust, inter-personal and inter-organizational trust, are distinguished (Rousseau *et al.*, 1998). Inter-organizational trust operates as a governance mechanism that mitigates opportunism in exchange contexts characterized by uncertainty and dependence. Inter-personal trust, on the other hand, plays a key role in interfacing with counterparts when implementing marketing strategies (Doney and Cannon, 1997).

Trust is critical in building buyer–seller relationships (Doney and Cannon, 1997), because trust significantly reduces the perception of risk associated with opportunistic behaviour by a partner in a buyer–seller relationship. Thus trust counterbalances the need for costly safeguard mechanisms against opportunism (Rotter, 1980). In considering the role of trust in business transactions, we highlight the personal structure, processes and routines that create a context within which inter-personal trust can develop and persist. This consideration is consistent with the characteristics of Confucian society, where guanxi is a dominant social phenomenon. Under a high level of inter-personal trust, farmers and firms are less inclined to rely on elaborated safeguards for specifying, monitoring and enforcing agreements (Ganesan, 1994). When there is a high level of inter-personal trust in buyer–seller relationships, it is most likely that the sellers and buyers will obey the agreements regarding transaction conditions. Thus, a positive relationship between inter-personal trust and contractual governance can be expected.

In Williamson's (1979) pioneering work, the characteristics of transactions (e.g., asset specificity) were linked to the governance structure from 'classical contracting' (spot markets) at one end of the spectrum, to unified governance (vertical integration) at the other. The level of transaction costs incurred in transactions encourages agents to build closer business relationships.

Contracts are considered as a hybrid form of coordination (Peterson *et al.*, 2001). Contracts are the preferred means of coordination under conditions of high risk and uncertainty. TSI pose a contractual hazard for both buyers and sellers. An opportunistic exchange partner can exploit such assets because they are not re-deployable, or at least will have a reduced value in an alternative exchange relationship. Williamson (1985) argued that in transactions with more one-sided, specific investments, the partners need more formal governance mechanisms, due to the increased dependency of the investing partner on the exchange partners' cooperative behaviour. When the transactions involve a high level of specific investments, detailed transaction conditions (such as the elements of price, volume, quality, payments, and punishment in case of violation, etc.) should be negotiated and agreed upon to reduce risk and uncertainty for both exchange partners. Thus, we expect a positive relationship between TSI and contractual governance in business practice.

TSIs are also very important in achieving closeness in buyer–seller relationships. The decision to create specific transactional assets is a major focus of transaction cost economics (Williamson, 1985). In buyer–seller relationships, TSIs are two dimensional: they are either physical or human (Rindfleisch and Heide, 1997). Physical TSI improve production and handling procedures and therefore lead to a higher level of productivity and efficiency. However, TSIs may also reassure the counterpart about the intention and integrity of the investor. Creating transaction-specific assets signals commitment (Heide and John, 1988). Therefore, the willingness to make transaction-specific investments will be influenced by the level of trust one dedicates to exchange counterparts. Once trust is established, opportunistic behaviour and uncertainty will be much lower in the transaction, and buyers and sellers will be more willing to invest to adhere to their partners' requirements.

Relationship satisfaction in vegetable supply chains

Relationship satisfaction is an essential element of business success (Morrissey and Pittaway, 2006). Satisfied business relationships may be influenced by different factors. The marketing literature suggests a positive relationship between trust and satisfaction (Anderson and Narus, 1990; Dwyer *et al.*, 1987). When one party trusts the other, a relationship is likely to emerge. Trust enhances effective communication and information sharing (Dwyer *et al.*, 1987) and enables partners to manage risk and opportunism in transactions (Nooteboom *et al.*, 1997), thus reducing transaction costs. Trust encourages long-term business relationships such as commitment, a long-term orientation and the propensity to stay in a relationship (Doney and Cannon, 1997). Buyers who trust their exchange counterparts exhibit more integrative bargaining strategies, which leads to mutual benefits (Schurr and Ozanne, 1985). Moreover, trust helps to solve complicated realities quickly and economically (Powell, 1990).

Considering the important requirements for freshness and hygiene, firms should handle vegetables in an efficient and economical way. Therefore standardized handling and processing procedures (e.g., HACCP, ISO) should be applied to produce safe high-quality products which satisfy consumers' requirements. Both human and physical TSIs may help to improve the way of handling products and enhance quality performance, which will inevitably lead to a good business relationship. Thus, a positive relationship between TSIs and relationship satisfaction can be expected in vegetable supply chains.

According to transaction cost theory, the manager's task is to craft governance arrangements at minimal cost to ensure the delivery of the desired products. Therefore, the role of contracts in business relationships has clear managerial implications (Lusch and Brown, 1996). Contracts specify promises, obligations and processes for dispute resolution for both partners. Long-term contracts are also explicitly drafted with a provision to promote the longevity of the exchange (Poppo and Zenger, 2002). If sellers are not able to deliver desired products to buyers in accordance with the contracts, then the buyer–seller relationship will most probably be terminated. As information spreads quickly, once farmers fail to be trusted by one or more members of the business network, they may lose transaction opportunities with other partners. Thus, with a contractual arrangement, farmers have more incentives and obligations to comply with buyers' requirements regarding product quality and delivery conditions, and thus of creating a satisfactory business relationship with buyers.

Research design

Data

The data in this chapter were collected from two major groups of participants in vegetable supply chains in Jiangsu Province: vegetable farmers (i.e., sellers) and vegetable processing and exporting firms (i.e., buyers). Semi-structured questionnaires were used to collect the data. In designing the questionnaires, we made a thorough effort to search and develop the appropriate measures. The initial questionnaire was developed based on previous research and a case study in the field (Lu *et al.*, 2006). The case studies were employed in the research in two ways. First, they were used to discuss the research topics. Discussions with the interviewees helped to identify the main constructs such as guanxi networks, inter-personal trust, TSIs, contractual governance and relationship satisfaction. Second, the case studies were important for formulating the research questions. Interviewees were asked to comment on the questions and layout of the questionnaire. All these measures provided useful suggestions to improve the content validity of the constructs. All measures were operationalized on a rating scale of 1 (not true at all) to 5 (totally true).

We chose the farmers working in the field and the owners of processing and exporting firms as our informants. The actual sample selection, however, was slightly different. Based on the different economic levels in various geographic areas, we first divided Jiangsu Province into three areas: less developed, average and developed areas. We then randomly selected five counties: one each from the less developed and developed area, three from the average area, based on the total number of vegetable farmers in each area. Thereafter, one village was randomly selected from each county. We then randomly selected 30–40 vegetable farming households in each village based on the total number of farmers.

To select the sample firms, based on the information from Jiangsu Agricultural and Forestry Bureau and other sources, we first listed all 250 vegetable processing and exporting firms in Jiangsu Province. After a round of telephone calls, about half of the firms were willing to be interviewed. Interviews with the firms lasted 1 to 2 hours. The interviewees were general managers, CEOs or marketing managers. All interviews were digitally recorded to help clarify any doubts that may arise during data processing.

The actual data collection was based on personal interviews in the field during 2004 and 2005. Excluding those who were not willing to participate or provided incomplete information, a total of 167 farmers and 80 firms were finally interviewed and used for this study.

Methods

The measures chosen were subject to a purification process involving a series of reliability and validity assessments using SPSS (Field, 2005) and SmartPLS (Ringle *et al.*, 2005). Exploratory factor analysis was carried out in SPSS to determine the best items for each latent variable (guanxi networks, inter-personal trust, TSIs, contractual governance and relationship satisfaction).

After purifying the items for the constructs, a partial least squares (PLS) approach was chosen for empirical estimation. Although PLS estimation has some shortcomings, such as the bias and inconsistency of loadings and inner structural coefficients (Fornell and Cha, 1994), our decision was motivated by several considerations. First, the small sample size does not satisfy the assumptions for maximum likelihood estimation (MLE). Second, some theoretical problems such as inadmissible solutions (i.e., negative error) and factor indeterminacy (i.e., non-convergence) have been identified with LISREL's MLE (Fornell and Bookstein, 1982). PLS, however, avoids these two theoretical problems. Third, PLS estimation requires only that the basic assumptions of least squares estimation are satisfied. Fourth, PLS uses jackknife or bootstrap (Efron and Gong, 1983) in combination with the traditional measures of goodness-of-fit (Bagozzi, 1981) to evaluate the model. Furthermore, PLS models both formative and reflective indicators simultaneously (Fornell and Bookstein, 1982). These advantages have encouraged the application of PLS in an increasing number of fields including marketing (Zinkhan *et al.*, 1987) and strategic management (Birkinshaw *et al.*, 1995; Johansson and Yip, 1994). Following Chin (1998), bootstrapping with 500 resamplings was used to show the precision of the PLS estimates.

Measures

The constructs in this study were all measured by multiple items. Guanxi networks imply how vegetable sellers/buyers use their personal networks to facilitate vegetable transactions. We focus on the supportive effects of guanxi networks in business transactions. Four items were used to measure guanxi networks for vegetable sellers and three items for vegetable buyers, such as the extent to which their guanxi networks enabled them to access markets, to improve production technology, and to build trust with suppliers and buyers.

Inter-personal trust refers to the belief that the other party is honest and sincere and in no circumstances will deliberately do anything to damage the relationship. Inter-personal trust is the focus in this study. Previous transaction experience, fairness, reputation and trustworthiness are the major reflective perspectives for inter-personal trust. These, in turn, are used to define inter-personal trust in the current study. Six items were used to measure inter-personal trust for both sellers and buyers. These items were developed from the literature reported by Zaheer *et al.* (1998) and Doney and Cannon (1997).

Transaction-specific investments refer to the sellers' and buyers' perception of the extent to which an investment is made specifically to facilitate the transaction with selected counterparts. We combine the physical and human dimensions of TSIs in this study. This construct, developed from Heide and John (1988), was measured by four items.

Contractual governance refers to transaction arrangements in the vegetable business. The present study defines contractual governance based on the agreements made between the farmers and firms regarding price, quality, quantity and delivery conditions. Statements regarding price and quality arrangements between sellers and buyers were used to measure this construct.

Relationship satisfaction was used to measure the quality of business relationships. People are satisfied when the benefits derived from the relationship are equal to or higher than what they expected. Two items were used to measure relationship satisfaction based on Fornell *et al.* (1996).

Empirical results

Validity and reliability of measures and constructs

Before interpreting the model coefficients, we first check the reliability and validity of the measures. Following common practice (Fornell and Bookstein, 1982; Mathieson *et al.*, 2001), we examine the individual item reliability (factor loading), internal consistency (composite reliability), and discriminant validity (average variance extracted and inter-construct correlations) for each construct.

The acceptability of the measurement model was assessed by first looking at the reliability of the individual items. Individual item reliability was determined by examining the loadings of measures onto their corresponding constructs. In all cases, only factor loadings greater than 0.6 were retained. All loadings were greater than or close to 0.7 for both sellers and buyers, indicating a high degree of individual item reliability.

Internal consistency was analysed using a measure of composite reliability. An internal consistency of 0.7 or greater is reasonable for exploratory research. Composite reliability for all constructs exceeded 0.81 for both sellers and buyers, indicating a good internal consistency.

Discriminant validity was assessed in two ways. First, the square root of the variance extracted should be greater than all construct correlations. Second, the test involves assessing how each item is related to the latent constructs. No item loaded more highly onto other constructs than it did onto its associated construct. Both criteria are satisfied in our case. Therefore, we can confidently rely on the coefficients to interpret the relationships among guanxi networks, inter-personal trust, TSIs and relationship satisfaction.

Empirical results of the research model

In the following, the collected data were analysed in two steps. First, determinants of relationship satisfaction were estimated using the obtained item ratings from vegetable farmers on their downstream customers (processors/exporters, traders/brokers and other buyers). Second, similar determinants were identified for vegetable processors/exporters using their ratings on their upstream suppliers. Hence, the two analyses cover basically the same relationship but are assessed from different chain perspectives.

Determinants of relationship satisfaction for vegetable sellers

The results for vegetable sellers (i.e., farmers) are provided in Fig. 2. Seven out of nine paths are significant at the 5% or 10% level. On average, the model explains 29.1% of the total variance, indicating 'moderate' prediction strength.

As expected, farmers' guanxi networks have direct and positive effects on inter-personal trust (path coefficient = 0.43) and TSIs (path coefficient = 0.28). This implies that vegetable farmers perceive support from their guanxi networks in building inter-personal trust with vegetable buyers and in investing specific assets in vegetable transactions. The results also indicate that the impact that guanxi networks have on inter-personal trust is greater than that on TSIs (0.43 vs. 0.28). However, we did not find a direct impact of guanxi networks on contractual governance.

The results reveal that inter-personal trust significantly improves farmers' relationship satisfaction (path coefficient = 0.40). This implies that vegetable farmers perceive a high level of satisfaction in relationships with their buyers if they trust them. Both inter-personal trust and TSIs improve farmers' willingness to conduct transactions based on formal contracts (path coefficients are 0.45 and 0.18 for trust and TSIs, respectively). This means that when farmers trust their buyers or invest in specific assets, they tend to transact on formal contracts. This is consistent with transaction cost theory. TSIs, however, did not contribute to relationship satisfaction directly for vegetable farmers. Inter-personal trust was positively associated with TSIs for vegetable farmers (path coefficient = 0.43) implying that when farmers trust their buyers, they are more willing to invest in transaction-specific assets in order to better comply with buyers' requirements.

The results also reveal that there was a significant positive relationship between contractual governance and relationship satisfaction (path coefficient = 0.14). This is consistent with our theoretical model. Contractual governance reduces the risk for farmers, which leads to a high level of satisfaction about their business relationships with buyers.

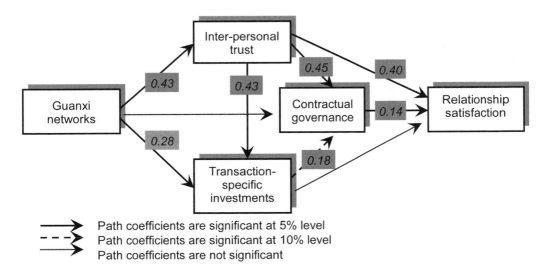

Fig. 2. Path analysis results for vegetable farmers.

To have a comprehensive understanding of the effects of guanxi networks and buyer–seller relationships in Chinese vegetable supply chains, we investigate the total effect (i.e., direct + indirect effect). The total effect provides an indication of the relative magnitude of the effect of each exogenous construct on the endogenous constructs in the research model.[1] The total effect for vegetable farmers is presented in Table 1.

---

[1] Total effects are the sum of all direct and indirect effects of an exogenous variable on an endogenous variable. Indirect effects are calculated by multiplying all path coefficients which lead indirectly from an exogenous to an endogenous variable. For example, the total effects of guanxi networks on TSIs = 0.43 (guanxi -> trust) * 0.43 (trust -> TSIs) + 0.28 (guanxi -> TSIs) = 0.46. Significant levels of total effects were calculated by the estimation software.

Guanxi networks equally affect inter-personal trust and TSIs (total effect coefficients = 0.43 and 0.46 for inter-personal trust and TSIs, respectively). Although the direct effect of guanxi networks on contractual governance is not significant, the total effect is. This implies that with the support of their guanxi networks, farmers are more willing to engage in contract farming. The total effect further reveals that guanxi networks also help to improve relationship satisfaction between farmers and their buyers.

Table 1. Total effects for vegetable farmers.

	Inter-personal trust	Transaction-specific investments	Contractual governance	Relationship satisfaction
Guanxi networks	0.43***	0.46***	0.21***	0.26***
Inter-personal trust		0.43***	0.53***	0.53***
Transaction-specific investments			0.18*	0.15*
Contractual governance				0.14**

Note: *, ** and *** refer to the total effect coefficients being significant at 10%, 5% and 1% statistical level, respectively.

The total effect further confirms that inter-personal trust is important for farmers to make transaction-specific investments, to engage in contractual transactions and improve relationship satisfaction in Chinese vegetable supply chains. Although the direct effect of TSIs on relationship satisfaction is not significant, the total effect shows that TSIs may help farmers to engage in contract farming and improve satisfaction in their relationships with buyers.

Determinants of relationship satisfaction for vegetable processors and exporters

The results for the vegetable buyers (i.e., processing/exporting firms) are listed in Fig. 3. Four out of nine relationships examined are significant at the 5% or 10% level. On average, the PLS model explains 16.4% of the variance. The R^2 for contractual governance and relationship satisfaction would be considered to be of 'moderate' strength. The relative low level of overall variance explained in this model indicates that besides guanxi networks, other factors (e.g., requirements compliance) may also be important for processing and exporting firms to achieve good relationship satisfaction.

Similar to vegetable farmers, the vegetable firms' guanxi networks had a direct and positive effect on inter-personal trust. This indicates that vegetable processing and exporting firms are more willing to trust their vegetable suppliers when the business relationships are supported by their guanxi networks (path coefficient = 0.32). Surprisingly, the firms' guanxi networks had no direct impact on their TSIs. This may be because the firms do not invest much in vegetable purchasing activities. In a buyer-dominated vegetable market in China, the buyers can purchase what they want and can easily find alternative suppliers to fulfill their demand. Therefore, there is less incentive for firms to make TSIs.

The results demonstrate that inter-personal trust significantly improved relationship satisfaction (path coefficient = 0.31) for vegetable processing and exporting firms in the research area. This meant that vegetable processing and exporting firms perceive they have a good relationship if they trust vegetable suppliers. Furthermore, inter-personal trust was positively associated with contractual governance (path coefficient = 0.51). This implies that in a trusting buyer–seller relationship, vegetable processing and exporting firms are more willing to purchase vegetables based on formal contracts. TSIs are positively associated with relationship satisfaction (path coefficient = 0.21). The positive impact indicates that with a high level of TSIs, vegetable firms are able to achieve good relationships with suppliers.

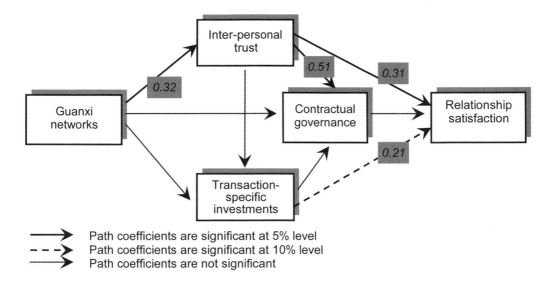

Path coefficients are significant at 5% level
Path coefficients are significant at 10% level
Path coefficients are not significant

Fig. 3. Path analysis results for vegetable buyers.

However, we did not find any significant relationship between TSIs and contractual governance. It is possible that simpler governance structures may be adopted when dealing with trustworthy partners even in the presence of TSIs (Lu *et al.*, 2008). In addition, no significant direct relationship was found between guanxi networks and contractual governance.

We also investigate the total effect for vegetable processing and exporting firms. Same as the direct effect, the firms' guanxi networks have a significant total effect on inter-personal trust (total effect coefficients = 0.32). The total effect also shows that vegetable firms' guanxi networks significantly improve relationship satisfaction with vegetable suppliers (total effect coefficients = 0.16), however this effect was mediated by inter-personal trust (Table 2).

Table 2. Total effects for vegetable processing and exporting firms.

	Inter-personal trust	Transaction-specific investments	Contractual governance	Relationship satisfaction
Guanxi networks	0.32***	0.22	0.23	0.16**
Inter-personal trust		0.18	0.48***	0.38***
Transaction-specific investments			−0.16	0.19
Contractual governance				0.08

Note: ** and *** refer to the total effect coefficients being significant at 5% and 1% statistical level, respectively.

Inter-personal trust shows significant total effects on contractual governance (total effect coefficients = 0.48) and relationship satisfaction (total effect coefficients = 0.38). This implies that if vegetable processing and exporting firms can build trusting relationships with vegetable suppliers, they are more willing to sign contracts with farmers to guarantee a long-term and stable supply of vegetables. This will help to build satisfactory buyer–seller relationships.

Conclusions and implications

Achieving business success in long-term relationships is essential for both farmers and processing and exporting firms in the Chinese vegetable industry. Studying guanxi networks and investigating both the critical elements and the best approaches to achieve satisfactory business relationships will benefit both sellers and buyers. In this study, we examined the effects of guanxi networks and contracts on business relationships in the Chinese vegetable industry using Jiangsu Province as an illustration.

This study shows that guanxi networks positively contribute to trusting relationships and substantially enhance business satisfaction in Chinese vegetable supply chains. However, the approaches that vegetable farmers utilize to achieve good relationships with their buyers is different from that utilized by vegetable processing and exporting firms. For vegetable farmers, with the support of guanxi networks, both inter-personal trust and TSIs help them to build satisfying business relationships with buyers. Their business relationships are further enhanced by contractual governance. For vegetable processing and exporting firms, although guanxi networks and trust are both important in vegetable procurement practices, their business relationships with vegetable suppliers, however, gain little benefit from TSIs and contractual governance.

The critical role of guanxi networks and inter-personal trust for small-scale farmers and processing and exporting firms in managing business relationships with their counterparts suggest that relational governance is advantageous in eliminating transaction costs and in dealing with contingencies in the Chinese vegetable industry. Even though transaction cost economics suggests that contractual governance will be the preferred mechanism in the presence of uncertainty and TSIs (Williamson, 1979), relational governance plays a dominant role in Chinese society (Schramm and Taube, 2003). Although we did not find strong evidence for the direct effects between relational and contractual governance in this study, the total effects of relational governance (based on guanxi networks) on contractual governance underline the importance of trust-building activities within China (Arias, 1998; Wong and Leung, 2001). The complementary effects of guanxi networks on contractual governance imply that a combination of relational and contractual governance in vegetable supply chains may be the best way to improve business relationships and to enhance vegetable marketing performance in China.

The results obtained from this study have significant management implications for the Chinese vegetable industry. Given the prevalence of small-scale vegetable farmers and the relatively low-quality vegetables they produce, the findings in this study imply that implementing guanxi networks in business is very important for vegetable farmers to increase market performance. Farmers will benefit more from increasing the number of connections, their closeness and strength. Farmers can improve their negotiation power in situations where there is good collaboration or organization (Shen *et al.*, 2005; Hu *et al.*, 2004). Collaboration between farmers can further expand and enhance farmers' guanxi networks and trust, and increase farmers' capacity to invest in transaction-specific assets, all of which help to improve farmers' market performance. Government policies aimed at strengthening farmer organizations and facilitating the integration of small farmers into high-value markets (e.g., supermarkets) will play an important role.

Chinese vegetable processing/exporting industries have a low-quality image and lack of competitiveness in international markets. The findings in this study suggest that guanxi networks and inter-personal trust are very important for them in maintaining good relationships, which will help them to acquire consistently good high-quality vegetables from suppliers. Thus, efforts to build and maintain good guanxi networks and to build trusting relationships with suppliers will prove worthwhile, which will further accelerate the development of vegetable processing and exporting industries in China.

Acknowledgements

We gratefully acknowledge the financial support through WOTRO grant W01.65.2001.010 and a research grant from the Mansholt Graduate School of Social Sciences, Wageningen University.

References

Ambler, T. (1994) Marketing's third paradigm: Guanxi. *Business Strategy Review* 5, 69–80.
Anderson, J. and Narus, J. (1990) A model of distributor firm and manufacturer firm working partnerships. *Journal of Marketing* 54, 42–58.
Arias, J. (1998) A relationship marketing approach to guanxi. *European Journal of Marketing* 32, 145–156.
Bagozzi, R. (1981) Evaluating structural equation models with unobservable variables and measurement error: a comment. *Journal of Marketing Research* 18, 375–381.
Birkinshaw, J., Morrison, A. and Hulland, J. (1995) Structural and competitive determinants of a global integration strategy. *Strategic Management Journal* 16, 637–655.
Braendle, U., Gasser, T. and Noll, J. (2005) Corporate governance in China: is economic growth potential hindered by guanxi? *Business and Society Review* 110, 389–406.
Chin, W. (1998) The partial least square approach to structural equation modeling. In: Marcoulides, G. (ed) *Modern Methods for Business Research.* Lawrence Erlbaum Associates Publisher, London, pp. 295–336.
Claro, D., Hagelaar, G. and Omta, O. (2003) The determinants of relational governance and performance: how to manage business relationships? *Industrial Marketing Management* 32, 703–716.
Davies, H., Leung, T., Luk, S. and Wong, Y. (1995) The benefits of 'guanxi': the value of relationships in developing the Chinese market. *Industrial Marketing Management* 24, 207–214.
Doney, P. and Cannon, J. (1997) An examination of the nature of trust in buyer–seller relationships. *Journal of Marketing* 61, 35–51.
Dwyer, F., Schurr, P. and Oh, S. (1987) Developing buyer–seller relationships. *Journal of Marketing* 51, 11–27.
Dyer, J. (1996) Specialized supplier networks as a source of competitive advantage: evidence from the auto industry. *Strategic Management Journal* 17, 271–291.
Efron, B. and Gong, G. (1983) A leisurely look at the bootstrap, then jackknife, and cross-validation. *The American Statistician* 37, 36–48.
Fan, Y. (2002a) Guanxi's consequences: personal gains at social cost. *Journal of Business Ethics* 38, 371–380.
Fan, Y. (2002b) Questioning guanxi: definition, classification and implications. *International Business Review* 11, 543–561.

FAO (2004) *Yearbook of Agricultural Production and Trade.* United Nations Food and Agricultural Organization, Rome.

Farh, J., Tsui, A., Xin, K. and Cheng, B. (1998) The influence of relational demography and guanxi: the Chinese case. *Organization Science* 9, 471–488.

Ferguson, R., Paulin, M. and Bergeron, J. (2005) Contractual governance, relational governance, and the performance of interfirm service exchanges: the influence of boundary-spanner closeness. *Journal of the Academy of Marketing Science* 33, 217–234.

Field, A. (2005) *Discovering statistics using SPSS,* 4th edn. Sage Publications, London.

Fornell, C. and Bookstein, F. (1982) Two structural equation models: LISREL and PLS applied to consumer exit-voice theory. *Journal of Marketing Research* 19, 440–452.

Fornell, C. and Cha, J. (1994) Partial least squares. In: Bagozzi, R. (ed) *Advanced Methods of Marketing Research.* Blackwell, Oxford, pp. 52–78.

Fornell, C., Johnson, M., Anderson, E., Cha, J. and Bryant, B. (1996) The American customer satisfaction index: nature, purpose, and findings. *Journal of Marketing* 60, 7–18.

Ganesan, S. (1994) Determinants of long-term orientation in buyer–seller relationships. *Journal of Marketing* 58, 1–19.

Guthrie, D. (1998) The declining significance of guanxi in China's economic transition. *The China Quarterly* 154, 254–282.

Heide, J. and John, G. (1988) The role of dependence balancing in safeguarding transaction-specific assets in conventional channels. *Journal of Marketing* 52, 20–35.

Hill, C. (1995) National institutional structures, transaction cost economizing and competitive advantage: the case of Japan. *Organization Science* 6, 119–131.

Hu, D., Reardon, T., Rozelle, S., Timmer, P. and Wang, H. (2004) The emergence of supermarkets with Chinese characteristics: challenges and opportunities for China's agricultural development. *Development Policy Review* 22, 557–586.

Johansson, J. and Yip, G. (1994) Exploiting globalization potential: US and Japanese strategies. *Strategic Management Journal* 15, 579–601.

Kao, J. (1993) The worldwide web of Chinese business. *Harvard Business Review* 71, 24–36.

Klein, S., Frazier, G. and Roth, V. (1990) A transaction cost analysis model of channel integration in international markets. *Journal of Marketing* XXVII, 196–208.

Lu, H. (2007) *The Role of Guanxi in Buyer–Seller Relationships in China: A Survey of Vegetable Supply Chains in Jiangsu Province.* Wageningen Academic Publishers, Wageningen.

Lu, H., Trienekens, J. and Omta, S. (2006) Does guanxi matter for vegetable supply chains in China? A case study approach. In: Bijman, J., Omta, S., Trienekens, J., Wijnands, J. and Wubben, E. (eds) *International Agri-food Chains and Networks: Management and Organization.* Wageningen Academic Publishers, Wageningen, pp. 31–47.

Lu, H., Trienekens, J., Omta, S. and Feng, S. (2008) The value of guanxi for small vegetable farmers in China. *British Food Journal* 110, 412–429.

Luo, Y. (2008) The changing Chinese culture and business behaviour: the perspective of intertwinement between guanxi and corruption. *International Business Review* 17, 188–193.

Luo, Y. and Chen, M. (1997) Does guanxi influence firm performance? *Asia Pacific Journal of Management* 14, 1–16.

Lusch, R.F. and Brown, J.R. (1996) Interdependency, contracting, and relational behavior in market channels. *Journal of Marketing* 60, 19–38.

Mathieson, K., Peacock, E. and Chin, W. (2001) Extending the technology acceptance model: the influence of perceived user resources. *The Data Base for Advances in Information System* 32, 86–112.

Morrissey, W. and Pittaway, L. (2006) Buyer–seller relationships in small firms. *International Small Business Journal* 24, 272–298.

Nooteboom, B., Berger, H. and Noorderhaven, N. (1997) Effects of trust and governance on relational risk. *Academy of Management Journal* 40, 308–338.

Peterson, H., Wysocki, A. and Harsh, S. (2001) Strategic choices along the vertical coordination continuum. *International Food and Agribusiness Management Review* 4, 149–166.

Poppo, L. and Zenger, T. (2002) Do formal contracts and relational governance function as substitutes or complements? *Strategic Management Journal* 23, 707–725.

Powell, W. (1990) Neither market nor hierarchy: network form of organization. In: Cummings, L. and Shaw, B. (eds) *Research in Organizational Behavior*. JAI Press, Greenwich, CT, pp. 295–336.

Rindfleisch, A. and Heide, J. (1997) Transaction cost analysis: past, present, and future applications. *Journal of Marketing* 61, 30–54.

Ringle, C., Wende, S. and Will, A. (2005) SmartPLS 2.0 (M3) Beta. Hamburg, http://www.smartpls.de.

Rotter, J. (1980) Inter-personal trust, trustworthiness and gullibility. *American Psychologist* 35, 1–7.

Rousseau, D., Sitkin, S., Burt, R. and Camerer, C. (1998) Not so different after all: a cross-discipline view of trust. *Academy of Management Review* 23, 393–404.

Schramm, M. and Taube, M. (2003) On the co-existence of guanxi and a formal legal system in the P.R. China – An institutionalist approach. The 7th Annual Conference of International Society for New Institutional Economics (ISNIE) on 'Institutions and Change'. 11–13 September 2003, Budapest, Hungary.

Schul, P., Little Jr, T. and Pride, W. (1985) Channel climate: its impact on channel members' satisfaction. *Journal of Retailing* 61, 9–38.

Schurr, P. and Ozanne, J. (1985) Influences on exchange processes: Buyers' preconceptions of a seller's trustworthiness and bargaining toughness. *The Journal of Consumer Research* 11, 939–953.

Shen, M., Rozelle, S. and Zhang, L. (2005) Farmer professional associations in rural China: state dominated or new state-society partnerships? In: Sonntag, B., Huang, J., Rozelle, S. and Skerritt, J. (eds) *China's Agricultural and Rural Development in the Early 21st Century*. Australian Centre for International Agricultural Research, Canberra, pp. 197–228.

Standifird, S. and Marshall, R. (2000) The transaction cost advantage of guanxi-based business practice. *Journal of World Business* 35, 21–42.

Thorelli, H. (1986) Networks: Between markets and hierarchies. *Strategic Management Journal* 7, 37–51.

Wang, Y. and Yao, Y. (2002) Market reforms, technological capabilities and the performance of small enterprises in China. *Small Business Economics* 18, 197–211.

Williamson, O. (1979) Transaction cost economics: the governance of contractual relations. *The Journal of Law and Economics* 22, 233–261.

Williamson, O. (1985) *The Economic Institutions of Capitalism: Firms, Markets and Relational Contracting*. The Free Press, New York.

Wong, Y. and Chan, R. (1999) Relationship marketing in China: guanxi, favoritism and adaptation, *Journal of Business Ethics* 22, 107–118.

Wong, Y. and Leung, T. (2001) *Guanxi: Relationship Marketing in a Chinese Context*. The Haworth Press, Inc., Binghamton.

Wu, W. and Leung, A. (2005) Does a micro-macro link exist between managerial value of reciprocity, social capital and firm performance? The case of SMEs in China. *Asia Pacific Journal of Management* 22, 445–463.

Yi, L. and Ellis, P. (2000) Insider-outsider perspectives of guanxi. *Business Horizons* 43, 25–30.

Zaheer, A. and Venkatraman, N. (1995) Relational governances as an interorganizational strategy: an empirical test of the role of trust in economic exchange. *Strategic Management Journal* 16, 373–392.

Zaheer, A., McEvily, B. and Perrone, V. (1998) Does trust matter? Exploring the effects of interorganizational and inter-personal trust on performance. *Organization Science* 9, 141–159.

Zinkhan, G., Joachimsthaler, E. and Kinnear, T. (1987) Individual difference and marketing decision support system usage and satisfaction. *Journal of Marketing Research* 24, 208–214.

Chapter 13

Inter-organizational Relationships as Determinants for Competitiveness in the Agri-food Sector: the Spanish Wheat-to-Bread Chain

Azucena Gracia, Tiziana de Magistris and Luis Miguel Albisu

Agri-Food Research and Technology Center of Aragón (CITA), Zaragoza, Spain

Introduction

The establishment, development and maintenance of inter-organizational relationships between partners are crucial and it is increasingly important that partners build stronger and longer-term relationships in the supply chain to remain competitive because of the ever changing competitive environment (Parsons, 2002). Moreover, Holmlund and Kock (1996) stated that relationships between buyers and suppliers tend to be closer and last longer because a firm might not become and stay competitive on its own.

Results from a survey on inter-enterprise relations, in selected European countries, conducted by Eurostat (Schmiemann, 2007) suggest that firms strongly believe that longer-term relationships have a positive impact on their competitiveness, but more importantly, they expect that longer-term relationships would increase their competitiveness over the next years.

According to Hobbs and Young (2000, 2001) companies are moving towards closer vertical coordination which may improve the relative competitiveness of business and result in an outward shift of the demand function, because firms are better able to offer consumers the specific products they demand.

The aim of this chapter is to analyse the link between longer-term inter-organizational relationships and the stakeholders' competitiveness. In particular, it is investigated whether higher-quality relationships in the Spanish wheat-to-bread supply chain positively influence the competitiveness of small- and medium-sized enterprises (SMEs). Although this aim seems to be very relevant, as far as we know, no empirical work has been undertaken with this specific goal. Only few studies, mentioned in the next section, have been conducted with a similar aim, to analyse the impact of more integrated relationships on stakeholders' performance.

The Spanish wheat-to-bread supply chain is very fragmented all along the chain, with a high number of wheat farmers, millers and bakers. One of the main characteristics of wheat producers in Spain is the lack of a classification system for different wheat qualities preventing the supply of homogeneous quality. In the past, exchanges in the sector were mainly carried out in the open market but there has been a sharp tendency to maintain high-quality relationships with suppliers to assure the required levels of product and process quality. Emerging characteristics of the wheat-to-bread sector include the increasing demand for a greater variety of breads. A few years ago, bread in Spain was a homogeneous product consisting mostly of 'baguettes' made of white flour. However, in the last few years, the market has demanded a greater variety of bread (i.e., whole grain, multi-cereals) and more convenience (i.e., fresh bread available all day long). Therefore, the demand for new bread types and subsequent product innovations have induced that the most common relationship type in the wheat-to-bread chain are 'repeated market transactions with buyers/suppliers' and 'spot markets',[1] but the former is by far used more often.

We focus on SMEs because most firms in the Spanish bread chain are of small and medium size and they play an important role in the development of rural areas. Moreover, we only analyse SMEs because as Morrisey and Pittaway (2006), and Cambra-Fierro and Polo-Redondo (2009) pointed out, the different characteristics of the SMEs lead to different ways to build the business relationships along the supply chain compared to larger enterprises, although the latter are also oriented towards long-term relationships.

To achieve the main objective, a model is developed in which the main components that define competitiveness, relationship quality and the main factors explaining supply chain relationships are defined.

The next section presents the theoretical framework and the specification of hypotheses. The third section deals with the model development, followed by a section describing the methodology, data collection and variable definition. The fifth section presents the empirical application and the results as well as the verification of hypotheses. The final section offers concluding remarks about research findings and their implications.

Theoretical framework: literature review and hypotheses formulation

Few empirical papers have been conducted with a similar aim to ours. The existing scarce literature has mostly been published within this decade as only two significant works were produced earlier (Palmer, 1996; Hobbs *et al.*, 1998), both based on the supply chain management approach. Palmer (1996) discussed the key factors to develop effective business linkages in the meat supply chain, while Hobbs *et al.* (1998) analysed whether an appropriate supply chain management in the Danish pork sector could create the right conditions to compete in international markets.

Most of the papers developed during this decade focus on the analysis of the factors that reinforce more integrated relationships in the agri-food chain and the establishment and maintenance of these higher-quality relationships. Han *et al.* (2007) tried, in addition, to gain insights into the impact of supply chain integration on the performance of the pork chain in China. They arrive at the conclusion that, in order to improve the quality of the

---

[1] 'Repeated market transactions with buyers/suppliers' means repeated exchange of goods and services with the same buyer/supplier and 'spot market' are immediate exchanges of goods and services at current prices.

offered products and reduce uncertainty in the pork supply chain, companies should develop more integrated relationships with their suppliers, as higher integration was expected to have a positive impact on firm performance. This result is in line with the outcome of Hobbs *et al.* (1998), who suggested that closer vertical coordination results in lower product and transaction costs and produces high-quality products tailored to specific demands. Both of these results are the essence of the competitiveness in the Danish pork sector. Consequently, the first hypothesis of our model is defined as follows:

H1: Higher relationship quality should positively influence stakeholders' competitiveness.

Batt (2003) analysed buyer–seller relationships in the Australian fruit and vegetable supply chain and, in particular, the key factors in building long-term relationships. He found that the aspects that would maintain a relationship in the long term are trust, satisfaction and the sharing of similar goals, although the major reason for continuing a relationship is the expectation of higher returns. More recently, some studies on relationship quality for specific food supply chains have been conducted in Europe. Schulze *et al.* (2006) developed a model for the pork and dairy sectors in Germany to assess the impact of improved relationship quality on stakeholder willingness to switch from some buyers to others. Findings indicate that relationship quality is a construct shaped by satisfaction, commitment and trust. The improvement of relationship quality depends on several factors such as communication quality and personal bonds/friendships between partners. Matopoulos *et al.* (2007) analysed supply chain relationships for SMEs in the agri-food industry. They identified the importance of elements such as trust, power, dependence and risk/reward sharing in establishing and maintaining supply chain relationships. Fischer *et al.* (2007) analysed the role of trust and economic relationships in selected agri-food chains in four EU countries. Their results indicate that trust is more pronounced among SMEs which are characterized by the existence of personal relationships between business partners. Fischer *et al.* (2009) further analysed the role of inter-enterprise relationships and communication in selected European agri-food chains. Their findings show that effective communication is the most important factor determining the sustainability of relationships in the agri-food supply chain. The existence of personal bonds and equal power distribution between buyers and suppliers are the second-most important determinants of relationship sustainability. They defined relationship sustainability using four dimensions: trust, commitment, satisfaction and positive collaboration history. Based on these findings, three hypotheses can be defined as follows:

H2: Higher communication quality along the chain should positively influence the quality of chain relationships.

H3: Stronger personal bonds among partners should positively influence the quality of chain relationships.

H4: A more equal distribution of power among partners should positively influence the quality of chain relationships.

Model development

As a first step it is essential to define and identify the specific dimensions of the variables involved in the model: the quality of the inter-organizational relationships, the factors potentially explaining them (communication, personal bonds and equal power distribution) and the concept of competitiveness.

Relationship quality

Few definitions of relationship quality have been offered in the literature. As Henning-Thurau (2000) stated, this deficiency is due to the fact that researchers assume that everyone has some kind of intuitive understanding of the factors that shape a high-quality relationship. Crosby *et al.* (1990), in a study of the insurance service industry, defined relationship quality as a state in which the 'customer is able to rely on the salesperson's integrity and has confidence in the salesperson's future performance because the level of past performance has been consistently satisfactory'. In the context of business purchasing literature, Smith (1998) defined quality relationship as 'an overall assessment of the strength of a relationship and the extent to which it meets the needs and expectations of the parties based on a history of successful or unsuccessful encounters or events'. No consensus has been reached about the dimensions that comprise the quality of the relationship. Crosby *et al.* (1990) and Parsons (2002) suggested that relationship quality consists of at least two dimensions, trust and satisfaction. However, Smith (1998) suggested that relationship quality is composed of, at least, trust, satisfaction and commitment. Other studies have extended the structure of relationship quality by including more dimensions. For instance, Dorsch *et al.* (1998) considered the importance of trust, satisfaction, commitment, minimal opportunism, customer satisfaction and ethical profile, while Naude and Buttle (2000) considered commitment, trust, satisfaction, social bonds and conflict resolution.

Based on the insights obtained from the literature review we assume that relationship quality is a multidimensional construct with the most important dimensions being trust, satisfaction and commitment, which are considered in this chapter to measure relationship quality.

Factors determining relationship quality

According to Mohr and Nevin (1990), communication can be defined as the glue that holds together a channel of distribution. They built a theoretical model where communication strategies are linked to relationship quality defined by coordination, satisfaction and commitment. Schulze *et al.* (2006) found that communication, quality and quantity, positively influenced relationship quality in the German pork chain. Fischer *et al.* (2009) suggested that effective communication, defined by adequate information quantity and high information quality, has a positive and highly statistically significant impact on relationship sustainability. In this chapter, communication is also defined by these two components, adequate information quantity and high information quality. Fischer *et al.* (2009) also found that factors affecting relationship sustainability are the existence of personal bonds and equal power distribution between buyers and suppliers. Moreover, Rodríguez and Wilson (2000) analysed how personal or social bonds influence relationship-building. They defined personal bonds characterized by familiarity, friendships and personal confidence which are incorporated in the relationship. These inter-personal ties are a form of social capital that enhances the maintenance of relationships. Equal power distribution among partners is preferred by stakeholders because it eliminates the possibility that benefits may be distributed unfairly along the chain. For instance, Boger (2001) shows that in the Polish pig chain, farmers prefer to conduct business with buyers who are on a more equal power level.

Competitiveness

In the economic literature, several definitions of competitiveness exist. However, we focus on those more specific because they can provide the key indicators of competitiveness. Competitiveness, at the firm level, is the ability to consistently and profitably deliver quality products and services, which customers are willing to purchase in preference to those of competitors (National Competitiveness Council, 1998). This definition raises the importance of profitability, product quality and customer loyalty in achieving competitiveness. The following definition also points out the importance of product quality; a firm is competitive if it can produce products and services of superior quality and lower costs than its domestic and international competitors (National Competitiveness Council, 1998). Finally, agribusiness competitiveness has been defined as the sustained ability to profitably gain and maintain market share (Martin *et al.*, 1991). These definitions outline the importance of profitability and market share on the competitiveness concept. As Murphy *et al.* (1996) pointed out, in their revision of performance measures, no single measure of competitiveness is appropriate but distinct dimensions of performance exist. They suggested that future studies should use multiple dimensions of performance. Then, the different dimensions of competitiveness mentioned in the different definitions should be taken into account to measure competitiveness.

Following the key concepts found in the previous definitions of competitiveness and the indicators used in some empirical papers (Tan *et al.*, 2002; Kim, 2006; Han *et al.*, 2007) the dimensions selected are: profitability, turnover, market share, customer loyalty and product quality.

Methodology

Structural equation modelling

A structural equation modelling (SEM) approach is used to empirically test the influence of relationship quality on SMEs' stakeholder competitiveness in the wheat-to-bread supply chain in Spain. This approach is selected because the analysed concepts, competitiveness and relationship quality, cannot be directly observed, but can be considered latent variables measured by one or more indicators. Moreover, SEM allows the analysis of simultaneous relationships between dependent and independent variables affecting firm-level competitiveness. Structural equation modelling encompasses an entire family of models where the multiple and interrelated dependence relationships are estimated, and unobserved concepts are represented in these relationships (Hair *et al.*, 2001).

The data analysis procedure consists of a confirmatory factor analysis (CFA) to assess the measurement model and the SEM analysis to examine the overall relationships among the constructs (Hair *et al.*, 2001).

Data gathering

Data were obtained in a survey conducted with farmers, processors and retailers in the wheat-to-bread chain in Spain (the region of Aragon) from November 2006 to April 2007. The final questionnaire was administrated to a total of 175 stakeholders, in particular 104 wheat farmers, 45 bread processors and 26 small independent bread shops. In order to approach farmers, there was a first contact with advisory extension services in the region as well as cooperatives and veterinary services. Twelve counties were selected because they are the ones producing wheat. The directors of these services contacted the farmers as they had to approve voluntarily to respond to the questionnaire. The directors of the extension services were requested to select the wheat producers in order to obtain a sample representative of the entire population.

Farmers were interviewed face-to-face at the extension services facilities. In the case of agri-food processors, due to the difficulty to reach company managers, a mixed strategy was adopted. The first step was to send letters with questionnaires and there were some responses through a pre-paid return envelope. The second step was to contact them through the telephone, or, if they did not wish to response on the phone, to arrange a meeting to conduct the interview face-to-face. Retailers were approached directly through face-to-face interviews in their own stores. Table 1 describes the final sample in terms of its composition by region, size and business age. Respondents are located in the region of Aragon, and many of them, in Zaragoza. The size of interviewed businesses is quite small, mainly for farmers and retailers, corresponding with the small size of the entire business population. Around 60% of farmers and retailers have only one employee and 22% and 15%, respectively, have two employees. Processors are also quite small although they are larger than the other actors in the chain. Around 25% of bread processors have more than ten employees. Hence, our sample mainly consists of SMEs.

Table 1. Sample characteristics.

Farmers		Processors		Retailers	
1 employee	68 (66%)	1 to 3 employees	13 (28.9%)	1 employee	15 (57.7%)
2 employees	22 (22%)	4 to 10 employees	21 (46.7%)	2 employees	4 (15.4%)
3 or more employees	13 (12%)	11 or more employees	11 (24.4%)	3 or more employees	7 (26.9%)
Average number of years in business	17.7		22.4		16.1

The questionnaire was developed based on previous expert interviews conducted during summer/autumn 2005 (Fischer *et al.*, 2007). The final questionnaire contains three groups of questions. The first group consisted of questions related to the type, nature and quality of the chain relationship. The second group included questions on information and communication strategies with the main buyer/supplier. Part three consisted of questions related to the effect of relationship quality on competitiveness and to the factors that might influence relationship quality. Finally, some questions on specific actor characteristics were included. At the beginning of the questionnaire, stakeholders were asked to focus on their main buyers/suppliers to whom the following questions were related. This questionnaire was pre-tested using a small sample of stakeholders.

Definition of variables

The definition of the dimensions and factors in the model of relationship quality and competitiveness is shown in Table 2. As a first step, the competitiveness construct was measured by asking respondents whether the main relationship with their buyer/supplier affects one of the competitiveness indicators: profitability, turnover, market share, customer loyalty and product quality (F3 in Table 2).

The relationship quality construct was measured by asking participants to rate the quality of different aspects that define the relationship with their main buyer/supplier: trust in the supplier/buyer, commitment towards the supplier/buyer and the satisfaction with the supplier/buyer, using a scale from one to seven with the latter as the maximum level (F2 in Table 2).

Table 2. Measurement of used exogenous and endogenous variables.

Factors	Observed variables	Rating scale	Name
Communication quality (F1)	High information quantity or adequate frequency of communication	1 = very poor, ..., 7 = very well	FREQ
	High information quality	1 = very poor, ..., 7 = very well	QINFO
Relationship quality (F2)	Trust in this supplier/buyer	1 = very poor, ..., 7 = very well	TRUST
	Commitment to this supplier/buyer	1 = very poor, ..., 7 = very well	COMMIT
	Satisfaction with this supplier buyer	1 = very poor, ..., 7 = very well	SATIS
Competitiveness (F3)	Profitability	1 = negative effect 2 = no effect 3 = positive effect	PROF
	Turnover	1 = negative effect 2 = no effect 3 = positive effect	TURN
	Market share	1 = negative effect 2 = no effect 3 = positive effect	SHARE
	Customer loyalty	1 = negative effect 2 = no effect 3 = positive effect	LOYAL
	Product quality	1 = negative effect 2 = no effect 3 = positive effect	QUAL
Personal bonds (F4)	This relation is based on strong personal bonds	1 = strongly disagree, ..., 7 = strongly agree	PERSONAL
Equal power distribution (F5)	We are equal partners in this relationship	1 = strongly disagree, ..., 7 = strongly agree	POWER

The communication quality construct was measured asking participants to rate how good or adequate is the information quantity and the quality of the information shared with their main buyer/supplier, using an increasing scale from one to seven where the latter was the maximum level of goodness (F1 in Table 2).

The other variables in the model (personal bonds and equal power distribution) were also obtained from the questionnaire asking respondents to what extent they agree or disagree with the statement 'The relationship with our main buyer/supplier is characterized by strong personal bonds' and 'We are equal partners in this relationship' using a scale from one to seven where seven indicates the highest level of agreement (F4/F5 in Table 2).

Results

Structural equation modelling was employed to examine the general fit of the proposed model and to test the hypotheses. The data analysis procedure consists of CFA to assess the measurement model and SEM analysis to examine the overall relationships among the constructs (Anderson and Gerbing, 1988; Hair *et al.*, 2001).[2]

Testing the measurement model

To see whether the fit between the proposed measurement model and the data is satisfactory, a set of indices have been calculated for the measurement model ($\chi^2_{(46)} = 82.1$, $\chi^2/46 = 1.785$, Bentler-Bonett normed fit index (NFI) = 0.918, comparative fix index (CFI) = 0.961, and root mean square error of approximation (RMSEA) = 0.067). Taking into account that all of these fit indices exceed the recommended acceptable levels (Bagozzi and Yi, 1988) we can conclude that there is a satisfactory fit between the proposed measurement model and the data.

Reliability refers to the consistency of the measurement. Table 3 shows that the Cronbach's Alpha values, measuring scale reliability of communication quality (0.907), relationship quality (0.89) and competitiveness (0.80) exceed the recommended level of 0.70.

Table 3. Confirmatory factor analysis results: standardized parameter estimates for the used measurement model.

Variable	Cronbach's α	Standardized factor loading	*t*-value
Communication quality (F1)	0.91		
FREQ		0.92	(0.000)[a]
QINFO		0.89	11.665
Relationship quality (F2)	0.89		
SATIS		0.92	(0.000)[a]
TRUST		0.91	16.343
COMMIT		0.73	11.724
Competitiveness (F3)	0.80		
PROF		0.89	(0.000)[a]
TURN		0.76	10.248
SHARE		0.54	7.071
LOYAL		0.56	7.363
QUAL		0.54	7.026
Personal bonds (F4)	1		
PERSONAL		1	(0.000)[a]
Equal power distribution (F5)	1		
EPOWER		1	(0.000)[a]

[a]The value was not calculated because loading was set to one to fix construct variance

---

[2] Only the best model in terms of statistical measures and goodness-of-fit is presented.

Convergent validity is evaluated by the *t*-ratio tests of the factor loadings. In Table 3 it can be observed that for each variable the *t*-value associated with each of the loading exceeds the critical value, at the 1% significance level. It means that all variables are statistically significant in their specified constructs, verifying the hypothesized relationships between indicators and constructs. Thus, it can be concluded that the fit of the measurement model is reasonable.

Finally, discriminate validity is achieved if the correlations between different constructs, measured by their respective indicators, are relatively weak. The χ^2 difference test is used to assess the discriminate validity of two constructs by calculating the difference of the χ^2 statistics for a constrained and an unconstrained measurement model. The constrained model is identical to the unconstrained model, in which all constructs co-vary, except that the correlation between the two constructs of interests is fixed at one. Since every pair of five constructs needs to be tested, the Bonferroni method was used because it allows multiple comparisons, assuring that the overall 0.001 significance level with the critical values of the $\chi^2_{(1)}$ of 10.82, is maintained. Table 4 displays the χ^2 difference tests for the different pairs of constructs. Because all of them exceed 10.82, the discriminate validity of the model is successfully achieved.

Table 4. Discriminate validity for the measurement model.

Construct pair	Standard measurement model $\chi^2_{(46)} = 82.1$ ($p < 0.001$)	
	Uni-dimensional model $\chi^2_{(47)}$	χ^2 difference
(F1, F2)	111.2	29.1
(F1, F3)	125.3	43.2
(F1, F4)	109.6	27.5
(F1, F5)	114.6	32.5
(F2, F3)	126.2	44.1
(F2, F4)	96.4	14.3
(F2, F5)	111.6	29.5
(F3, F4)	106.6	24.5
(F3, F5)	124.9	42.8
(F4, F5)	112.5	30.4

Testing the structural model

The structural coefficients in the model have been estimated using the maximum likelihood estimation procedure with the AMOS 5.0 computer software. Figure 1 presents the standardized parameter estimates for the structural model and shows that all coefficients are statistically significantly different from zero at the 0.1% significance level.

The assessment of the overall fit of the proposed model which ensures that it is an adequate representation of the entire set of casual relationships, is shown in Table 5, where absolute fit, incremental fit and parsimonious fit measures are used.

Absolute fit measures determine the degree to which the overall model (structural and measurement models) predicts the observed covariance or correlation matrix. The likelihood ratio χ^2 (52 degrees of freedom) is 111.912 indicating that it is statistically significant at the 5% significant level. Since the chi-square statistic is sensitive to sample size, other measures are examined to assess the model goodness-of-fit. In fact the $\chi^2/52 = 2.152$, which is smaller than 3, and the RMSEA value (0.08) indicate a reasonable error of approximation. Thus, these results mean that there is a good correspondence between the resulting model-implied covariance matrix and the empirical or data-based covariance matrix. Regarding the comparisons to a baseline model, the TLI, NFI, IFI and CFI values were calculated, and they are at 0.90 (Table 5), indicating that the proposed model can be considered as acceptable. Finally, the parsimonious fit measures represent the degree of model fit per estimated coefficient. These measures attempt to correct for any 'over-fitting' of the model and evaluate the parsimony of the model compared to the goodness-of-fit (Hair *et al.*, 2001). The results indicate that the model is parsimonious because the PRATIO value, although not high, is good enough. Therefore, the proposed model is statistically reasonable.

Table 5. Goodness-of-fit measures of the structural model.

Measures	Acceptable level	Estimated model
Absolute fit measures		
x^2	–	111.912
Degrees of freedom (df)	–	52
p	> 0.05	0.000
RMSEA	0.05–0.08	0.081
Incremental fit measures		
TLI	> 0.90	0.902
NFI	> 0.90	0.899
IFI	> 0.90	0.937
CFI	> 0.90	0.935
Parsimonious fit measures		
x^2/df	1–3	2.152
PRATIO	Close to 1	0.68

Verification of hypotheses

The path diagram for the estimated model is shown in Fig. 1. This figure represents the latent variables as ellipses and the indicators as rectangles. Moreover, the single-headed arrows are the causal relations. Path coefficient values are placed on the arrows from latent variables to indicators, or from one latent variable to another one. In addition, the standardized values (between zero and one) of the estimated coefficients for each indicator and for each latent variable are presented.

Competitiveness is a multidimensional concept where profitability is the first indicator followed by turnover at a second level. However, customer loyalty, market share and product quality are in a third level. This result indicates that competitiveness is a broad concept that takes into consideration not only firms' economic indicators, but also the firms' ability to increase market share through product quality, in order to establish customer loyalty or preference for those higher-quality products.

The causal relationship among relationship quality and competitiveness is statistically significant and positive. This means that hypothesis 1 is confirmed. Then, in the Spanish wheat-to-bread chain, as the quality of relationships improves, stakeholder competitiveness increases. The high path coefficient values between the latent variable relationship quality and their indicators prove that, as expected, relationship quality is a multidimensional construct of trust, satisfaction and commitment. This means that the basis for relationship quality in the wheat-to-bread supply chain in Spain is the trust, satisfaction and

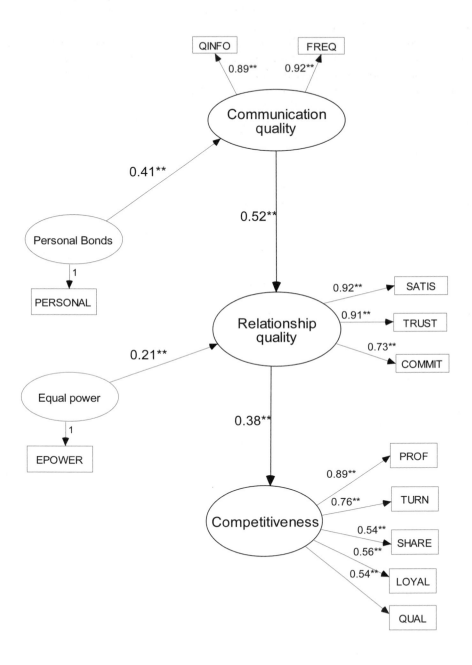

Fig. 1. Path diagrams of the estimated model (** significant level 0.1%).

commitment between buyers and sellers. Similar results were obtained by Schulze *et al.* (2006), who stated that trust, satisfaction and commitment build a one-dimension construct of relationship quality for the pork and dairy chain in Germany. Fischer *et al.* (2009) assessed relationship sustainability in selected agri-food chains in Europe measured by trust, satisfaction, commitment and positive collaboration history.

The positive and statistically significant coefficient estimate between the quality of communication, measured by quality and quantity of shared information, and the quality of relationship indicates that hypothesis 2 is verified. This result means that as communication quality in the wheat-to-bread supply chain increases, the quality of the relationship improves. Similar results were obtained by Schulze *et al.* (2006) and Fischer *et al.* (2009) for specific agri-food chains in different European countries. Then, the quality of communication also has a positive effect on stakeholder competitiveness through relationship quality.

The second more relevant conclusion is that only one of the hypothesized factors that might explain relationship quality is, in fact, a relevant factor. This factor is equally distributed power between stakeholders along the chain. However, personal bonds do not influence directly the quality of relationships but through communication quality. Then, hypothesis 3 has been only partially verified while hypothesis 4 has been confirmed. This result indicates that if negotiation power is equally distributed in the wheat-to-bread chain, this has a positive effect on the quality of relationships. Moreover, personal bonds positively influence the quality of communication indicating that communication improves as personal bonds are closer along the chain. This last result was expected because of the small size of the stakeholders in the sample.

Conclusions

Our results confirm that in the Spanish wheat-to-bread chain, as the quality of relationships improves, stakeholder competitiveness increases. Moreover, the quality of relationships in the wheat-to-bread supply chain in Spain is based on trust, satisfaction and commitment between buyers and sellers. In addition, competitiveness is a broad concept that comprises firm-level economic indicators, such as profitability and turnover, but also the firms' ability to increase market share, through product quality, in order to establish customer loyalty or preference for those higher-quality products. Second, as communication quality in the wheat-to-bread supply chain increases, the quality of relationships improves. In addition, the quality of communication has a positive effect on stakeholder competitiveness through relationship quality. Thus, in order to increase competitiveness, stakeholders in the chain have to implement quality communication systems with their buyers/suppliers and to build higher trust, commitment and satisfaction among them. The extent to which the different actors in the chain, such as farmers, processors and retailers succeed in creating an improved communication system, building higher trust and commitment as well as higher levels of satisfaction among them, will increase their own competitiveness. However, we have to take into account that in our model we have only included some of the factors that might affect stakeholder competitiveness. This is a shortcoming of our work and constitutes the aim of future research.

Finally, only one of the hypothesized factors that might explain relationship quality, equally distributed power among stakeholders, influences the quality of relationships. In the same way, the only factor that positively influences communication quality is personal bonds. Then, whether negotiation power is equally distributed along the chain, and the establishment of strong personal relations with their buyers/suppliers have positive impacts on their competitiveness through the improvements of both the quality of relationships and communication.

References

Anderson, J. and Gerbing, D. (1988) Structural equation modeling in practice: a review and recommended two-step approach. *Psychological Bulletin* 103(3), 411–423.

Bagozzi, R. and Yi, Y. (1988) On the evaluation of structural equation models. *Journal of the Academy of Marketing Science* 16(1), 74–98.

Batt, P. (2003) Building trust between growers and market agents. *Supply Chain Management: An International Journal* 8(1), 65–78.

Boger, S. (2001) Quality and contractual choice: a transaction cost approach to the Polish hog market. *European Review of Agricultural Economics* 28(3), 241–261.

Cambra-Fierro, J. and Polo-Redondo, Y. (2009) Long-term orientation of the supply function in the SME context. Reasons, determining factors and implications. *International Small Business Journal* 26(5), 619–646.

Crosby, L., Evans, K. and Cowles, D. (1990) Relationship quality in services selling: an interpersonal influence perspective. *Journal of Marketing* 54(3), 68–81.

Dorsch, M., Swanson, S. and Kelley, S. (1998) The role of relationship quality in the stratification of vendors as perceived by customers. *Journal of the Academy of Marketing Science* 26(2), 128–142.

Fischer, C., Gonzalez, M., Henchion, M. and Leat, P. (2007) Trust and economic relationships in selected European agri-food chains. *Food Economics* 4(1), 40–49.

Fischer, C., Hartmann, M., Reynolds, N., Leat, P., Revoredo-Giha, C., Henchion, M., Albisu L. and Gracia, A. (2009) Factors influencing contractual choice and sustainable relationships in European agri-food supply chains. *European Review of Agricultural Economics* 36(4), 541–569.

Hair, J., Anderson, R., Tatham, R. and Black, W. (2001) *Análisis multivariante*. V Edición. Prentice Hall. Iberia, Madrid.

Han, J., Omta, S. and Trienekens, J. (2007) The joint of supply chain integration and quality management on the performance of pork processing firms in China. *International Food and Agribusiness Management Review* 10(2), 67–98.

Henning-Thurau, T. (2000) Relationship quality and customer retention through strategic communication of customer skills. *Journal of Marketing Management* 16(1–3), 55–79.

Hobbs, J. and Young, L. (2000) Closer vertical co-ordination in agri-food supply chains: a conceptual framework and some preliminary evidence. *Supply Chain Management: An International Journal* 5(3), 131–143.

Hobbs, J. and Young, L. (2001) Vertical linkages in agri-food supply chains in Canada and United States. Agriculture and Agri-Food Canada, Publication 2083/E, Ottowa.

Hobbs, J., Kerr, W. and Klein, K. (1998) Creating international competitiveness through supply chain management: Danish pork. *Supply Chain Management: An International Journal* 3(2), 68–78.

Holmlund, M. and Kock, S. (1996) Buyer dominated relationships in a supply chain – a case study of four small-sized suppliers. *International Small Business Journal* 15(1), 26–40.

Kim, S. (2006) Effects of supply chain management practices, integration and competition capability on performance. *Supply Chain Management: An International Journal* 11(3), 241–248.

Martin, L., Westgren, R. and Van Duren, E. (1991) Agribusiness competitiveness across national boundaries. *American Journal of Agricultural Economics* 73, 1456–1464.

Matopoulos, A., Vlachopoulou, M., Manthou, V. and Manos, B. (2007) A conceptual framework for supply chain collaboration: empirical evidence from the agri-food industry. *Supply Chain Management: An International Journal* 12(3), 177–186.

Mohr, J. and Nevin, J. (1990) Communication strategies in marketing channels: a theoretical perspective. *Journal of Marketing* 50, 36–51.

Morrisey, W. and Pittaway, L. (2006) Buyer–supplier relationships in small firms. *International Small Business Journal* 24(3), 272–298.

Murphy, G., Trailer, J. and Hill, R. (1996) Measuring performance in entrepreneurship research. *Journal of Business Research* 36, 15–23.

National Competitiveness Council (1998) Annual Competitiveness Report. Available from http://www.forfas.ie/media/ncc980311_competitiveness_1998.pdf (accessed 2007).

Naude, P. and Buttle, F. (2000) Assessing relationship quality. *Industrial Marketing Management* 29, 351–361.

Palmer, C. (1996) Insights from industry building effective alliances in the meat supply chain: lessons from the UK. *Supply Chain Management: An International Journal* 1(3), 9–11.

Parsons, A. (2002) What determines buyer–seller relationship quality? An investigation from the buyer's perspective. *The Journal of Supply Chain Management* (Spring), 4–12.

Rodríguez, C. and Wilson, D. (2000) Relationship bonding and trust as a foundation for commitment in US–Mexican strategic alliances: a structural equation modeling approach. *Journal of International Marketing* 10(4), 53–76.

Schmiemann, M. (2007) Inter-enterprise relations in selected economic activities. *Statistics in Focus* 57. Eurostat.

Schulze, B., Wocken, C. and Spiller, A. (2006) Relationship quality in agri-food chains: supplier management in the German pork and dairy sector. *Journal on Chain and Network Science* 6, 55–68.

Smith, J. (1998) Buyer–seller relationships: similarity, relationship management and quality. *Psychology and Marketing* 15(1), 3–21.

Tan, K., Lyman, S. and Wisner, J. (2002) Supply chain management: a strategic perspective. *International Journal of Operations & Production Management* 22(6), 614–631.

Chapter 14

How Buyer–Supplier Relationships can Create Value: the Case of the Australian Wine Industry

Lynlee Hobley and Peter J. Batt

Curtin University of Technology, Perth, Australia

Introduction

In 2005, Australia was the fourth largest wine exporter in the world (Winetitles, 2005). Over many years, Australia has achieved strong sustained growth and is internationally competitive across all price points on quality-for-money (AWBC, 2003). Nevertheless, the change in the product value mix is of concern as export sales are increasing primarily at lower price points. Furthermore, there has been a significant reduction in the premium and super premium segments (Van der Lee, 2004) and a considerable decline in the average dollar per litre return on sales. While Australian wine exports increased 13% by volume in 2003/04, over the same period, the value increased by only 2.9% (ABS, 2004a, b). Similarly, while domestic sales increased 3% by volume, they declined 6% by value. Reasons for the change have been attributed to greater consumer interest in lower-price-point wine, consolidation in the retail sector, excess wine in international and domestic inventories and less favourable exchange rates (Stanford, 2005).

Declining returns in the Australian wine industry now threaten the financial viability of all but the major wine producers (Deloitte, 2005). The average rate of return has declined from 7.6% in 1997–1998 to 4.2% (EBT/total assets) in 2000–2001 (AWBC, 2003). Returns on vineyard operations show a similar decline due to falling prices for both red and white wine grapes as a result of the current global and domestic oversupply (ABARE, 2005). As a result, many wineries and wine grape suppliers find that although their sales revenues are growing, it is often at the expense of profit (Deloitte, 2005).

In order to compete, most wineries are adopting a value-driven relational marketing and purchasing approach to profitably grow the demand for Australian wine (Centre for International Economics, 2004). Wineries are developing closer relationships with their contracted grape suppliers in an integrated effort to enhance the more efficient production of grapes to produce wine styles that meet market demands (Beuman and McLachlan, 2000; Osborn, 2000). Generally, contracted growers cooperate with wineries in the adoption of new technologies and often sacrifice yields in the pursuit of superior quality grapes for their customers (DeGaris, 2000; Swinburn, 2000). As directed by their customers, grape suppliers also manage grape sugar, colour, berry size, pH levels, titratable acidity and vineyard pests and diseases (Clancy, 2005). In the export market particularly, wine grape growers need to comply with mandatory requirements for agrochemical use and application. The growers' willingness to deliver grapes that wineries want has given Australian wine exports a quality advantage over international competitors at most price levels (Donald and Georgiadis, 2000).

From a winery perspective, the potential benefits of a long-term relationship with wine grape suppliers include: better access to a more reliable supply of high-quality grapes, on-going improvements in the quality of wines produced, a higher level of technical interaction in the form of information exchange, and greater support from suppliers in introducing new wine grape varieties or new wine styles (Wilson, 2000).

From a supplier perspective, contract wine grape growers can potentially achieve greater customer loyalty, an increase in the production of higher-quality grapes that consistently meet the wineries specifications, a higher level of technical assistance and potentially higher returns. Both parties also benefit from being able to better plan and forecast production schedules, to optimize operational processes and coordinate deliveries.

Irrespective, in order to prosper, these collaborative customer–supplier relationships must achieve the desired benefits and a fair appropriation of value for both parties. This does not imply that value appropriation is equally shared or that it is even linked to the relative contribution each partner makes towards the creation of value (Pardo *et al.*, 2006). Essentially, each party must be clear about their specific value focus in the relationship in order to optimize activities and resource allocation. If the relationship is to be sustainable, each party must achieve some value beyond the level which they would receive if they were to operate independently (Wilson, 1995).

This study seeks to examine customer and supplier perceptions of the factors that are instrumental in the optimization of relationship value in the grape and wine industry and to identify the extent to which the relational constructs are consistent within the winery and grape supply sectors.

Theoretical background

A steady shift has been observed in trading relationships from a transaction-orientation to a relationship-orientation as firms strive to strengthen their position in today's highly competitive and dynamic business environment (Ryssel *et al.*, 2000). In business-to-business markets, customers rely on the products and services they buy from their suppliers to improve their own market offering and to increase the overall profitability of their firm (Ulaga, 2001). Thus the main reason for firms to engage in business relationships is to create value (Walter *et al.*, 2001; Anderson and Narus, 2004; Pardo *et al.*, 2006).

Anderson and Narus (2004) define value in business markets as the net worth in monetary terms of the economic, technical, service and social benefits a customer firm receives in exchange for the price it pays for a market offering. While the construct remains elusive, relationship value is most often described in terms of benefits and sacrifices. Value is attained when the negative aspects of the exchange, such as costs and sacrifices, are exceeded by the positive aspects, i.e., benefits (Werani, 2001; Walter *et al.*, 2001). Not unexpectedly, as value perceptions are influenced by personal values and individual expectations, any attempts to measure or to evaluate value are highly subjective.

At the relationship level, value creation is viewed as a dynamic process which concerns both parties. Close long-term relationships provide a means for the co-creation of value where the competitive abilities of each trading party are enhanced by being in the relationship (Anderson, 1995; Wilson, 1995).

Relationships between customers and suppliers are important in the grape and wine industry because a good-quality wine starts in the vineyard. Many industry experts agree that at least 60% of the work in making good wine is done in the vineyard (Scales *et al.*, 1995). Value creation depends on the ability of the winery and their grape suppliers to deliver on those quality attributes that are important to the consumer at predetermined price points. For this reason, the market for wine grapes is unlike markets where anonymous buyers and sellers meet to conclude transactions on the spot. Rather, the typical situation is for a winery to establish a contract for 3 to 5 years duration with growers (Scales *et al.*, 1995), with an understanding that, subject to meeting grape quality specifications, the winery will accept all the grapes produced from the designated vineyard. In turn, the grape supplier will deliver a specified volume of grapes to the winery (Allen, 2003).

A great number of behavioural variables have been implicated in the development of sustainable trading relationships including trust (Ganesan, 1994; Morgan and Hunt, 1994; Doney and Cannon, 1997), cooperation (Morgan and Hunt, 1994; Cannon and Perreault, 1999; Cannon *et al.*, 2000), performance satisfaction (Anderson *et al.*, 1994; Wilson, 1995), communication (Morgan and Hunt, 1994; Cannon and Perreault, 1999), power asymmetry (Anderson and Weitz, 1989; Cannon and Perrault, 1999; Cannon *et al.*, 2000) and conflict resolution (Werani, 2001; Anderson and Narus, 2004).

Cooperation

Cooperation in a working relationship implies a joint effort, team spirit and collaboration towards achieving both intra-firm and inter-firm goals (Cannon and Perreault, 1999). For the exchange partners, there is an expectation of a balanced exchange, reciprocity and mutuality over time (Morgan and Hunt, 1994; Leonidou, 2004).

Cooperation is a key dimension for coordinating the activities and resources between parties involved (Morgan and Hunt, 1994; Håkansson and Snehota, 1995). This construct is fundamental to closely linked relationships where the importance of supply is high and purchase requirements are complex (Cannon and Perreault, 1999).

The extent to which customers and their suppliers will cooperate depends on: (i) the degree to which the parties believe that they can simultaneously achieve their goals; (ii) the existence of mutual agreement between the parties concerning their actions in achieving individual goals; (iii) the perceptual clarity of the information processed by the interacting parties; (iv) the establishment of mutually accepted norms upon which the achievement of individual goals are approved and disapproved; and (v) the acceptance of norms of exchange, which protect the exchange parties from opportunistic and self-centred behaviour (Childers and Ruekert, 1982).

Trust

Trust is a fundamental building block in most long-term relationships (Ganesan, 1994; Morgan and Hunt, 1994; Wilson, 1995; Doney and Cannon, 1997).

Four frequently cited definitions of trust include: (i) a willingness to rely on an exchange partner in whom one has confidence (Moorman *et al.*, 1992); (ii) when one party believes that its needs will be fulfilled in the future by actions taken by another party (Anderson and Weitz, 1989); (iii) a firm's expectation that another firm will fulfil obligations and will put their weight into the relationship (Dwyer *et al.*, 1987); and, (iv) the belief that a firm's word or promise is reliable and a firm will fulfil their obligations within an exchange relationship (Wilson, 1995).

Trust signifies an attitude by one party to have confidence in, attach credibility to, and show benevolence towards the other party in a working relationship (Moorman *et al.*, 1992; Morgan and Hunt, 1994; Doney and Cannon, 1997). Credibility exemplifies a common belief by one trading partner that the other partner is honest, dependable, reliable and will honour its word. Furthermore, credibility is based on the belief that the other partner has the necessary expertise to perform the task effectively and reliably (Dwyer *et al.*, 1987).

The benevolent component of trust involves a belief, attitude or expectation that relationship partners will act in the best interests of the other partner (Wilson, 1995). Benevolent trust has been defined as 'the firm's belief that another company will perform actions that will result in positive actions for the firm, as well as not take unexpected actions that would result in negative outcomes for the firm' (Anderson and Narus, 2004, p. 407). These characteristics in relationships reduce the tendency for firms to take advantage of each other when the possibility for opportunism arises (see Chapter 3).

According to Anderson and Narus (2004), once trust is established, firms learn that coordinated joint efforts will lead to outcomes that exceed what the firm would achieve if it were to act solely in its own best interests. Hence trust will have a positive association with cooperation.

Performance satisfaction

Wilson (1995, p. 338) defines performance satisfaction as 'the degree to which the business transaction meets the performance expectations of the partner ... it includes both product specific performance and non-product attributes'. From a customer viewpoint, it is important for suppliers to deliver on the basic elements of the business transaction (superior quality, reliable delivery, competitive price). At the same time, the customer must satisfy the supplier's business needs or they risk becoming marginalized.

Performance satisfaction in an on-going trading relationships is defined most frequently as a positive affective state resulting from the appraisal of all aspects – economic and non-economic – of a firm's working relationship with another firm (Frazier *et al.*, 1989; Geyskens *et al.*, 1999). Over time, on-going satisfaction with past outcomes builds equity (Ganesan, 1994), which in turn gives a trading partner confidence that they are not being taken advantage of in the relationship and that both parties are concerned about the other's welfare. When a firm finds that their trading partner is willing and able to satisfy their requirements in a reliable and predictable manner, they are more likely to trust the trading partner (Ganesan, 1994; Geyskens *et al.*, 1999). Therefore, it is hypothesized that performance satisfaction will have a positive association with trust.

Firms are predominantly concerned about functionality and performance in business markets (Anderson and Narus, 2004). A thorough understanding of a partner's requirements and preferences will enable a firm to know where their resources and capabilities have the greatest potential to create and deliver superior value. Value for a firm depends on the 'promise' of a partner's offering, but more importantly, value depends on the partner's ability to fulfil that promise to the firm's satisfaction. Therefore, it is hypothesized that performance satisfaction will have a positive association with relationship value.

Communication

Communication is the basis of interaction between all suppliers and customers. It is 'the formal and informal sharing of meaningful and timely information between firms' (Dwyer *et al.*, 1987). Communication processes underlie most aspects of inter-firm activity for they provide a means to better understand a partner's expectations, to solve problems, to build trust and to demonstrate commitment.

Communication quality refers to the extent to which interaction between customers and their suppliers is frequent, formal/explicit, bidirectional (to include positive and negative feedback) and non-coercive (Mohr *et al.*, 1999). Cooperative business relationships imply intensive mutual coordination through high frequency two-way interaction (Werani, 2001).

The systematic availability of information allows people to coordinate and complete tasks more effectively and is associated with mutually fulfilled expectations and increased levels of performance satisfaction. Hence, it is hypothesized that communication will have a positive association with performance satisfaction.

Communication in trading relationships is critical in building trust (Mohr and Spekman, 1994). The quality of information transmitted and joint participation by partners in planning and goal setting sends signals to trading partners. Joint participation in decision-making ensures that both parties understand the strategic choices facing each other. Furthermore, the disclosure of confidential information to an exchange partner exposes one's vulnerability. As timely communication fosters trust by resolving disputes and aligning perceptions and expectations (Morgan and Hunt, 1994), it is hypothesized that communication will have a positive association with trust.

Anderson and Narus (1990) propose that communication leads to greater cooperation in dyadic exchange relationships. Consequently, it is hypothesized that communication will have a positive association with cooperation.

Power asymmetry

Power is an integral component of customer–supplier relationships (Hingley, 2005). Power resides in the ability of a firm to make another firm undertake actions it wouldn't undertake on its own. However, the presence of power does not necessarily mean that it will be explicitly exercised, as a firm's possession of power is separate from the way power is applied (Ogbonna and Wilkinson, 1996).

Hingley (2005) proposes that power asymmetry exists in the vast majority of dyadic relationships. While the imbalance most often favours the customer, this does not mean that weaker partners cannot benefit in such relationships. Often, more powerful partners assume responsibility for the inter-firm division of labour, monitoring outcomes, linking discrete activities between actors, establishing and managing relationships between the various actors and organizing logistics. Nevertheless, when there is an inordinate power imbalance, there may be an unwillingness on the part of the stronger party to respond to the other firm or to participate in joint problem resolution. Therefore, it is hypothesized that power asymmetry will have a negative association with conflict resolution.

Dwyer (1993) found suppliers were satisfied in their relationship with more powerful buyers for as long as the norms of stewardship were nurtured and tendencies to centralize and threaten were restrained. Therefore, it is hypothesized that power asymmetry will have a negative association with performance satisfaction.

Researchers have argued that inter-firm relationships with more asymmetric power are more dysfunctional because there is greater opportunity to exploit the weaker party (Geyskens *et al.*, 1996; Kumar *et al.*, 1995). Such circumstances can reduce or eliminate feelings of trust on the part of the more vulnerable party. This leads to the hypothesis that power asymmetry will have a negative association with trust.

Power can be used as a mechanism for achieving cooperation from exchange partners (Hingley, 2005). The more dominant firm can either use coercive, negative types of power to achieve immediate cooperation or, it can exercise more positive, collaborative types of influence with the aim of increasing cooperation over the long term. When the dominant firm chooses to exercise its power collaboratively, the more dependent partner may interpret this as a signal that its partner wishes to work together to promote long-term joint goals and valuable outcomes. Therefore, it is hypothesized that power asymmetry will have a positive association with cooperation.

Conflict resolution

Conflict refers to the general level of disagreement between customers and suppliers (Anderson and Narus, 2004). Although some level of conflict is normal in every relationship, if the conflict gets out of hand it may be harmful to the relationship or even cause its demise. Conflict is either attitudinal or structural (Leonidou, 2004). Attitudinal conflict may be due to ill-defined and poorly performed roles, different expectations about potential outcomes, different opinions about the relationship or the capabilities of the parties involved. Structural conflict usually occurs in the pursuit of different or even opposite goals by participants, the need to protect and maintain autonomy in the relationship and competition between the two parties for the same resources.

Firms in cooperative relationships are inclined towards joint problem solving since integrated outcomes satisfy more fully the needs and concerns of both parties (Mohr and Spekman, 1994). Although partners may attempt to persuade each other to adopt particular solutions, this approach is generally more constructive than the use of coercion.

The benefits of conflict resolution in dyadic relationships include: (i) more frequent and effective communication between the parties and the establishment of mechanisms to express complaints; (ii) a more equitable review of relationship resources; (iii) a more balanced distribution of power; and (iv) standardization of modes of conflict resolution (Assael, 1969). Consequently, it is hypothesized that conflict resolution will have a positive association with communication.

Furthermore, as firms that lower the overall level of conflict in their relationship experience greater satisfaction (Anderson and Narus, 1990), it is hypothesized that conflict resolution will have a positive association with performance satisfaction.

Effective conflict resolution produces feelings of procedural justice and trust. According to Ganesan (1994), when two parties have successfully resolved critical problems in the relationship, mutual trust will strengthen. Therefore it is hypothesized that conflict resolution will have a positive association with trust.

Research design

For this study, an eight-page structured questionnaire was developed containing comparable measures for customers and suppliers to permit direct comparisons of wine grape suppliers and the wineries. It was developed from a comprehensive literature review and the findings of a qualitative field study involving 16 wineries and their grape suppliers in Western Australia (Hobley, 2007). The items (Table 1) were measured on a seven-point Likert scale.

Relationship value, as the dependent variable, was assessed by a four item construct that had been developed and used by Walter *et al.* (2001, 2002a, b).

Cooperation was evaluated by a four-item construct developed from the literature reported by Cannon and Perrault (1999) and Cannon *et al.* (2000).

Trust was evaluated by an eight-item construct derived from Doney and Cannon (1997), Ganesan (1994), Morgan and Hunt (1994) and Walter *et al.* (2000, 2002a).

Performance satisfaction was derived from a five-item construct derived from Anderson *et al.* (1994), Cannon and Perrault (1999) and Geyskens *et al.* (1999).

Communication was evaluated by a seven-item construct developed from the literature reported by Mohr and Spekman (1994), Mohr *et al.* (1999) and Morgan and Hunt (1994).

Conflict resolution was evaluated by a four item construct derived from Ford (1984) and Werani (2001).

Similarly, power asymmetry was assessed by a four-item construct derived from Frazier *et al.* (1989) and Wilson (2000).

The two survey populations were independently selected: (i) wine producers with an annual wine grape crush of 50 t or more that sourced wine grapes from independent grape suppliers; and (ii) independent wine grape suppliers currently supplying wine grapes to a winery.

As statistics on the number and size of wineries that outsource wine grape growing are not readily available, all 859 wineries that crush more than 50 t were contacted using the Australian and New Zealand Wine Industry Directory (Winetitles, 2005). Unlike the wine industry, the wine grape industry does not have a database and contact with the target population was only possible through the Phylloxera and Grape Industry Board, State Departments of Primary Industry, national government industry organizations and regional grape and wine associations. By necessity, the survey had to rely on a non-probability sample.

The main survey method was a two-step combination: (i) person-to-person contact over the telephone to identify an appropriate respondent; then (ii) self-completion of the questionnaire either by email, post or fax. The identified key winery participants typically held the position of CEO, winemaker, grower liaison officer or winery vineyard manager. Key grape supplier participants were usually either the vineyard owner or manager. Questions were designed to seek information about a trading relationship with one specific partner at a firm level. The surveys resulted in responses from 175 wineries (a 20.4% response rate) and 400 wine grape suppliers in South Australia, Victoria, New South Wales and Western Australia.

The structural equation modelling (SEM) process involved two main steps: (i) validating the measurement model; and (ii) testing the hypothesized structural model (Anderson and Gerbing, 1988). Validation of the measurement model was achieved mainly through confirmatory factor analysis (CFA), while testing the structural model was achieved through path analysis with latent constructs using AMOS 6.0 software to confirm the models. Both measurement and structural models were performed in two-group analyses for cross-validation, as well as to compare the two groups of wineries and wine grape suppliers in a cross-sectional sample.

Empirical results

One-factor congeneric measurement models were used to assess item reliability, determine scale reliability and to generate factor score regression values for relationship value, communication, trust, cooperation, conflict resolution, power asymmetry and performance satisfaction with the recommended number of multiple observed items (between four and eight) associated with each of the latent constructs (Hair *et al.*, 2006).

CFA highlighted the unidimensionality of the constructs (Jöreskog and Sorbom, 1998) with high correlations among indicators and high proportions of variance explained by each factor. With the exception of cooperation, the reliability of each construct (Cronbach's alpha) was above 0.7 and all items had factor loadings greater than 0.50, supporting the convergent validity of the constructs (Hair *et al.*, 2006). Discriminant validity was assessed in two ways: (i) by testing if correlations between constructs were significantly different from 1, comparing a constrained model with the unconstrained model (χ^2 difference value with $p < 0.05$ supported the discriminant validity criterion for all constructs); and (ii) by comparing the variance extracted for any two constructs with the square of the correlation between constructs (confirmed in 85% of cases). There were no cross-loadings between the indicators for any construct (see Appendix).

An analysis of the individual constructs revealed that: (i) the wineries' perceptions of relationship value were significantly higher than the grape suppliers'; (ii) perceptions of trust for each of the measures were significantly higher for wineries than grape suppliers; (iii) perceptions of performance satisfaction were significantly higher for wineries than grape suppliers; (iv) while both groups had positive perceptions about the communication in the relationship, responses were significantly higher for wineries; (v) while the wineries were significantly more confident about their ability to resolve conflicts arising in the relationship, grape suppliers were more likely to suggest that disagreements and problems had yet to be resolved; (vi) while both groups had positive perceptions of cooperation and the need to 'work together to be successful', the wineries believed that they were more flexible than the grape suppliers in 'putting aside contractual terms' and 'going along with this trading partner'. Similarly, the wineries believed that they and their grape suppliers pursued compatible goals to a higher degree than the reality of the situation would suggest; (vii) with respect to power asymmetry, all item measures showed wineries to hold the dominant position. Grape suppliers often 'had no choice other than to adhere to the customers' demands', 'the customer controlled all the information', 'the customer had all the power' and 'the customer exerted a strong influence'.

A two-group model was estimated that was judged to provide acceptable goodness-of-fit, despite a chi-square value (41.439) that was statistically significant. Similar to Anderson (1987), this judgement was made on the basis of meaningful interpretability of the model from a content and theoretical viewpoint and a value of 0.98 for both the NFI (Bollen, 1989) and GFI (Jöreskog and Sorbom, 1984). This judgement was further supported by an IFI value of 0.99 (Bollen, 1989).

Figure 1 depicts the estimated model. The line weights of the arrows indicate the size of the obtained effects (the thicker the line the stronger the effect).

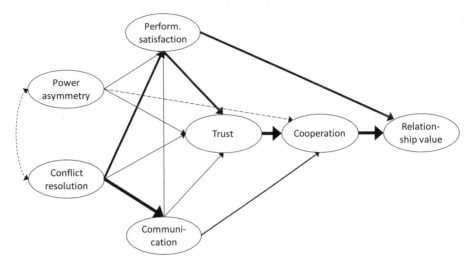

Fig. 1. Estimated causality links between analysed constructs.

As hypothesized, the relationships between performance satisfaction, conflict resolution, communication and trust were positive and significant. Power asymmetry was negatively related to performance satisfaction and trust. However, while power asymmetry was positively related to cooperation for the wineries it was not significant for grape suppliers. Conversely, while power asymmetry and conflict resolution was negative and significant for grape suppliers, it was not significant for the wineries (Table 1).

Table 1. Results of the two-group structural equation model.

	Wineries			Grape growers		
Cooperation → Relationship value	0.550	(4.331)	A	0.469	(3.089)	A
Trust → Cooperation	0.413	(4.634)	A	0.616	(9.053)	A
Performance satisfaction → Trust	0.222	(9.025)	A	0.628	(12.942)	A
Performance satisfaction → Relationship value	0.134	(4.102)	A	0.258	(2.178)	A
Communication → Performance satisfaction	0.643	(2.700)	A	0.375	(5.325)	A
Communication → Trust	0.124	(1.790)	A	0.158	(2.956)	A
Communication → Cooperation	0.234	(2.811)	A	0.180	(2.503)	A
Power asymmetry ↔ Conflict resolution	0		R	−0.388	(−7.242)	A
Power asymmetry → Performance satisfaction	−0.478	(−2.687)	A	−0.211	(−5.259)	A
Power asymmetry → Trust	−0.187	(−3.516)	A	−0.101	(−3.321)	A
Power asymmetry → Cooperation	0.299	(3.814)	A	0		R
Conflict resolution → Communication	0.690	(7.970)	A	0.805	(19.158)	A
Conflict resolution → Performance satisfaction	1.054	(3.700)	A	0.430	(5.665)	A
Conflict resolution → Trust	0.257	(2.880)	A	0.164	(2.789)	A

Unstandardized parameter estimates with critical ratios in brackets; A = hypothesis is accepted; R = hypothesis is rejected.

Of the two direct contributors to relationship value, cooperation had the biggest impact ($\beta = 0.45$ wineries/0.40 grape suppliers), while performance satisfaction was somewhat less (0.35/0.27) for both the wineries and grape suppliers, respectively.

Both trust (0.52/0.76) and communication (0.29/0.19) were antecedents to cooperation, while power asymmetry facilitated cooperation (0.36), but only for wineries.

Four constructs directly contributed towards trust. The strongest and most direct effects were achieved through performance satisfaction (0.56/0.62), and to a lesser extent by conflict resolution (0.23/0.15) and communication (0.12/0.16). In contrast, power asymmetry had a significant negative impact on trust (–0.17/–0.10).

Good communication had a strong positive impact on performance satisfaction (0.25/0.36), as did conflict resolution (0.38/0.41). However, power asymmetry detracted from performance satisfaction (–0.17/–0.20), more so for the wine grape suppliers who, in the current market, risk losing their market if they do not follow the advice and recommendations provided by the winery.

The strongest direct effect in the model arises from the positive contribution of conflict resolution on communication (0.63/0.80), highlighting the importance of being responsive to a partner's complaints and working on joint problem solutions to keep communication channels working efficiently.

In summary, the model for both groups did not vary in direction, but there were differences in perceptions for key relationship constructs between wineries and grape suppliers. There were two hypothesized pathways for which relationships were significant for one group, but not for the other. In the wine grape suppliers' model, the regression weights corresponding to power asymmetry and cooperation were not statistically significant and for the wineries, the regression weights corresponding to power asymmetry and conflict resolution were not significant.

Concluding comments

The Australian grape and wine industry was selected for the research because of the importance of the trading relationships between buyers and sellers of wine grapes. Wineries rely on the quality of the wine grapes they buy from their grape suppliers to improve their own market offer and to increase the overall profitability of their firm. Value outcomes call for the two parties to work together to deliver on those quality attributes that are important to the consumer at predetermined price points. For this reason, the relationships between wineries and their grape suppliers tend to be close, communicative and cooperative. Most of these trading arrangements are on-going, with wineries seeking to establish contracts with preferred grape suppliers for an average of 3 to 5 years' duration (Scales *et al.*, 1995).

Conflict resolution was a core antecedent which directly drove communication and performance satisfaction to increase trust and cooperation, leading to superior relationship value for customers and suppliers. Joint conflict resolution, in order to develop meaningful joint solutions to problems, would seem to be a worthwhile mechanism for bringing these essential constructs together in practice. Such an approach would also entail a quick response to a partner's complaints and resolving disagreements and problems as they arise. Even firms in successful trading relationships readily acknowledge that disagreements or conflicting views on important issues are inevitable from time to time. Rather than trying to ignore problems or allowing them to run their course capriciously, a more effective management strategy is to develop mediating mechanisms to defuse and settle differences in a timely manner.

For grape suppliers, their ability to resolve disagreements and problems was impeded by the power held by their customers. Readily acknowledged as the dominant party, wineries exert a strong influence over their grape suppliers. As a result, grape suppliers were not completely confident that wineries would not take advantage of their strong bargaining positions. Sentiments of this nature can increase stress in the working relationship which reduces a firm's perceptions of trust and overall satisfaction with their partner's performance.

Identifying trust as a principal mediating variable was critical in achieving a better insight into the process of making customer and supplier relationships work. Good communication between firms enhanced performance satisfaction and increased the level of benevolence and credibility in their relationship, while at the same time it reduced their partners' perceptions of opportunistic behaviour.

Cooperation was a strong predictor of relationship value, along with performance satisfaction. The winery model supports the view that in close, more collaborative relationships, suppliers support the wineries legitimate right to specify their wine grape requirements, particularly where the importance of supply is high and purchase requirements are complex. A more collaborative approach motivates firms to cooperate because they realize they must work together to be successful. There was also some expectation of reciprocity in situations where flexibility in trading arrangements was shown towards the other partner. Irrespective, it was evident from the model that the two direct pathways from performance satisfaction and cooperation were currently realizing higher relationship value for wineries than grape suppliers.

Clearly, for the wine grape suppliers, winery power was an important issue, with some negative relational connotations for both parties in terms of reducing the perceptions of performance satisfaction and trust in the relationship. Wineries may use various reward and coercive powers and legitimate authority to cajole and coerce cooperation from suppliers, however, it was the use of non-mediated power that inevitably built social bonds and close relationships. While it is not unreasonable for wineries, as the customer, to stipulate purchase specifications and to expect supplier compliance and cooperation, providing it is beneficial for the supplier to do so, some wine grape suppliers found that it was more difficult to resolve problems at a higher management level.

From a temporal perspective, while grape suppliers were looking for immediate gratification for the contribution they made to the wineries' value creation process, for the wineries, it may take several years before the value is realized. Furthermore, while social satisfaction may enhance the wine grape suppliers' relationship with the winery, declining returns in the Australian wine industry threaten the financial viability of all but the major wine producers. As prices for both cool and warm climate wine grapes decline, wine grape suppliers are not being rewarded for producing superior quality as the wineries simply cannot afford to pay higher prices. No doubt, as the supply of wine grapes diminishes and supply and demand return to some equilibrium, the balance of power will shift and thus the very nature of the collaborative relationships established within the Australian wine grape industry will evolve.

References

Allen, W. (2003) Winegrape assessment in the vineyard and at the winery. Retrieved: 11 Aug 2005, from www.dtftwid.qld.gov.au/Documents/Wine-Research+Data/Winegrape+Assessment.pdf.

Anderson, E. and Weitz, B. (1989) Determinants of continuity in conventional industrial channel dyads. *Marketing Science* 8(4), 310–323.

Anderson, J. (1987) An approach for confirmatory measurement and structural equation modeling of organizational properties. *Management Science* 33(4), 525–541.

Anderson, J. (1995) Relationships in business markets: exchange episodes, value creation and their empirical assessment. *Journal of the Academy of Marketing Science* 23(4), 346–350.

Anderson, J. and Gerbing, D. (1988) Structural equation modeling in practice: a review and recommended two-step approach. *Psychological Bulletin* 103(3), 411–423.

Anderson, J. and Narus, J. (1990) A model of distributor firm and manufacturing firm working relationships. *Journal of Marketing* 54(1), 42–58.

Anderson, J. and Narus, J. (2004) *Business Market Management: Understanding, Creating and Delivering Value*. Prentice Hall, New Jersey.

Anderson, J., Håkansson, H. and Johanson, J. (1994) Dyadic business relationships within a business network context. *Journal of Marketing* 58 (October), 1–15.

Assael, H. (1969) Constructive role of interorganizational conflict. *Administrative Science Quarterly* 14, 573–582.

Australian Bureau of Agricultural and Resource Economics (ABARE) (2005) Wine outlook to 2009-10: returns to improve in the medium term. Retrieved: 13 April 2006, from http://www.abareonlineshop.com/PdfFiles/ac06.1_part_a.pdf.

Australian Bureau of Statistics (ABS) (2004a) Australian wine and grape industry, Cat No. 1329, Canberra. Retrieved: 20 April 2006, from http://www.abs.gov.au/AUSSTATS/abs@.nsf/allprimarymainfeatures/890C00E5719D0AFDCA25710100165162?opendocument.

Australian Bureau of Statistics (ABS) (2004b) Fruitful year for the wine industry, Canberra. Retrieved: 20 May 2006, from http://www.abs.gov.au/ausstats/abs@.nsf/mediareleasesbytitle/AE40B4D83279F6C1CA256F95007765FD?OpenDocument.

Australian Wine and Brandy Corporation (AWBC) (2003) Australian wine sector scorecard, Winefacts statistics. Retrieved: 1 May 2006, from https://www.awbc.com.au/winefacts/data/category.asp?catid=8.

Beuman, K. and McLachlan, E. (2000) Grower feedback and developing relationships – Orlando Wyndam's perspective. In: Davies, C., Dundon, C. and Hamilton, R. (eds) *Modern Viticulture – Meeting Market Specifications*. Australian Society of Viticulture and Oenology, pp. 50–51.

Bollen, K. (1989) *Structural Equations with Latent Variables*. John Wiley & Sons, New York.

Cannon, J. and Perreault, W. (1999) Buyer–seller relationships in business markets. *Journal of Marketing Research* 36, 439–460.

Cannon, J., Achrol, R. and Gundlach, G. (2000) Contracts, norms and plural forms of governance. *Journal of the Academy of Marketing Science* 28(2), 180–194.

Centre for International Economics (CIE) (2004) A national wine-grape growers association: a discussion paper. Department of Agriculture, Canberra.

Childers, T. and Ruekert, R. (1982) The meaning and determinants of cooperation within an interorganizational marketing network. In: Bush, R. and Hunt, S. (eds) *Marketing Theory: Philosophy of Science Perspectives*. American Marketing Association, Chicago, pp. 116–119.

Clancy, P. (2005) Time to bridge the great divide. *The Australian and New Zealand Wine Industry Journal* 20(2), 4.

DeGaris, K. (2000) Targeting and achieving quality improvements. In: Davies, C., Dundon, C. and Hamilton, R. (eds) *Modern Viticulture – Meeting Market Specifications*. Australian Society of Viticulture and Oenology, pp. 33–35.

Deloitte (2005) Annual financial benchmarking survey for the Australian wine industry – Vintage 2004, Deloitte and the Winemakers' Federation of Australia. Retrieved: 2 February 2006, from http://www.deloitte.com/dtt/cda/doc/content/Final%20Benchmarking%20Survey%20Vintage%202004.pdf.

Donald, F. and Georgiadis, P. (2000) Setting quality categories for particular markets. In: Davies, C., Dundon, C. and Hamilton, R. (eds) *Modern Viticulture – Meeting Market Specifications*. Australian Society of Viticulture and Oenology, pp. 15–17.

Doney, P. and Cannon, J. (1997) An examination of the nature of trust in buyer–seller relationships. *Journal of Marketing* 61(2), 35–51.

Dwyer, R. (1993) Soft and hard features of interfirm relationships: an empirical study of bilateral governance in industrial distribution. ISBM Report 6-1993. Institute for the Study of Business Markets, Pennsylvania State University. University Park, PA.

Dwyer, R., Schurr, P. and Oh, S. (1987) Developing buyer–seller relationships. *Journal of Marketing* 51(2), 11–27.

Ford, D. (1984) Buyer–seller relationships in international industrial markets. *Industrial Marketing Management* 13, 101–112.

Frazier, G., Gill, J. and Kale, S. (1989) Dealer dependence levels and reciprocal actions in a channel of distribution in a developing country. *Journal of Marketing* 53(January), 50–69.

Ganesan, S. (1994) Determinants of long-term orientation in buyer–seller relationships. *Journal of Marketing* 58(2), 1–19.

Geyskens, I., Steenkamp, J., Scheer, L. and Kumar, N. (1996) The effects of trust and interdependence on relationship commitment: a trans-Atlantic study. *International Journal of Research in Marketing* 13, 303–317.

Geyskens, I., Steenkamp, J. and Kumar, N. (1999) A meta-analysis of satisfaction in marketing channel relationships. *Journal of Marketing Research* 36(May), 223–238.

Hair, J., Black, W., Babin, B., Anderson, R. and Tatham, R. (2006) *Multivariate Data Analysis*. 6th edition. Pearson Prentice Hall, Upper Saddle River, New Jersey.

Håkansson, H. and Snehota, I. (1995) *Developing Relationships in Business Networks*. Thomson International Press, London.

Hingley, M. (2005) Power to all our friends? Living with imbalance in supplier-retailer relationships. *Industrial Marketing Management* 34, 848–858.

Hobley, L. (2007) The value of trading relationships between buyers and sellers of wine grapes in Australia. PhD thesis. Curtin University of Technology.

Jöreskog, K. and Sorbom, D. (1984) *LISREL VI: Analysis of Linear Structural Relationships by the Method of Maximum Likelihood*. National Educational Resources, Chicago.

Jöreskog, K. and Sorbom, D. (1998) *LISREL 8: Structural Equation Modeling with the SIMPLIS Command Language*. Scientific Software International, Chicago.

Kumar, N., Scheer, L. and Steenkamp, J. (1995) The effects of perceived interdependence on dealer attributes. *Journal of Marketing Research* 32, 348–356.

Leonidou, L. (2004) Industrial manufacturer–customer relationships: the discriminating role of the buying situation. *Industrial Marketing Management* 20, 731–742.

Mohr, J. and Spekman, R. (1994) Characteristics of partnership success: partnership attributes, communication behaviour and conflict resolution techniques. *Strategic Management Journal* 15, 135–152.

Mohr, J., Fisher, R. and Nevin, J. (1999) Communicating for better channel relationships. *Marketing Management* 8(2), 38–45.

Moorman, C., Zaltman, G. and Deshpande, R. (1992) Relationships between providers and users: the dynamics of trust within and between organisations. *Journal of Marketing Research* 29, 314–328.

Morgan, R. and Hunt, S.D. (1994) The commitment–trust theory of relationship marketing. *Journal of Marketing* 58(3), 20–38.

Ogbonna, E. and Wilkinson, B. (1996) Inter-organisational power relations in the UK grocery industry: contradictions and developments. *International Review of Retail Distribution and Consumer Research* 6(4), 395–414.

Osborn, C. (2000) Grower feedback and developing relationships. In: Davies, C., Dundon, C. and Hamilton, R. (eds) *Modern Viticulture – Meeting Market Specifications*. Australian Society of Viticulture and Oenology, pp. 52–53.

Pardo, C., Henneberg, S., Mouzas, S. and Naude, P. (2006) Unpicking the meaning of value in key account management. *European Journal of Marketing* 40(11/12), 1360–1374.

Ryssel, R., Ritter, T. and Gemunden, H. (2000) The impact of IT on trust, commitment and value-creation in inter-organizational customer–supplier relationships. 16th Annual IMP Conference, Bath, UK.

Scales, W., Croser, B. and Freebairn, J. (1995) Winegrape and wine industry in Australia: a report by the committee of inquiry into the winegrape and wine industry. Australian Government Printing Service, Canberra.

Stanford, L. (2005) Australian wine industry: demand assessment 2004. *The Australian and New Zealand Wine Industry Journal* 20(1), 52–55.

Swinburn, G. (2000) Achieving and maintaining winegrape quality improvements. In: Davies, C., Dundon, C. and Hamilton, R. (eds) *Modern Viticulture – Meeting Market Specifications.* Australian Society of Viticulture and Oenology, pp. 50–51.

Ulaga, W. (2001) Customer value in business markets: an agenda for inquiry. *Industrial Marketing Management* 30(4), 1–7.

Van der Lee, P. (2004) The next frontier in export performance. *The Australian and New Zealand Wine Industry Journal* 19(6), 54–61.

Walter, A., Ritter, T. and Gemuenden, H. (2001) Value creation in buyer–seller relationships: theoretical considerations and empirical results from a suppliers perspective. *Industrial Marketing Management* 30(4), 365–377.

Walter, A., Holzle, K. and Ritter, T. (2002a) Relationship functions and customer trust as value creators in relationships: a conceptual model and empirical findings for the creation of customer value. 18th Annual IMP Conference, Dijon.

Walter, A., Mueller, T., Helfert, G. and Wilson, D. (2002b) Delivering relationship value: key determinant for customer's commitment. Institute for the Study of Business Markets, Pennsylvania State University, Pennsylvania.

Werani, T. (2001) On the value of cooperative buyer–seller relationships in industrial markets. Institute for the Study of Business Markets, Pennsylvania State University, Pennsylvania.

Wilson, D. (1995) An integrated model of buyer–seller relationships. *Journal of Academy of Marketing Science* 23(4), 335–345.

Wilson, H. (2000) Long-term buyer–seller relationships in the Western Australian wine industry. Curtin University of Technology.

Winetitles (2005) *The Australian and New Zealand Wine Industry Directory.* 23rd edition. Wine Titles and Hartley Higgins, Adelaide.

Appendix. Measures and construct reliabilities (winery/grape supplier).

Construct Eigenvalue Cronbach's alpha	Measures
Relationship value 2.637 / 2.875 0.826 / 0.869	This supplier/customer relationship has a high value for our firm The value of the relationship with this supplier/customer is very high in comparison with alternative suppliers/customers Considering all benefits and sacrifices associated with this supplier/customer relationship, how would you assess its value? How do you rate the value of all performance contributions that your firm gains from this supplier/customer
Trust 4.893 / 5.198 0.839 / 0.879	We have confidence in this supplier/customer When problems arise, this supplier/customer is honest about these problems We can count on the promises this supplier/customer makes to our firm We can count on this supplier/customer to do what is right This supplier/customer performs its tasks competently When making important decisions, this supplier/customer is concerned about our welfare This supplier/customer is knowledgeable about viticulture
Performance satisfaction 2.954 / 4.020 0.825 / 0.939	Working with this supplier/customer puts less strain on our organization than working with other suppliers Generally, we are satisfied with our overall relationship with this supplier/customer My firm usually gets at least a fair share of the rewards and cost savings from our relationship with this supplier/customer The benefits achieved from our relationship with this supplier/customer have greatly exceeded our expectations The financial returns our firm obtains from this supplier/customer are better than we envisaged
Communication 4.147 / 4.557 0.786 / 0.697	Our firm and this supplier/customer keep each other well informed This supplier/customer keeps me well informed on technical matters There is excellent communication between our firms so there are never any surprises that might be harmful to our working relationship This supplier/customer communicates his expectations of our firm This supplier/customer frequently informs me of any information or change that could affect the expected grape quality or yield There is frequent face-to-face contact with this supplier/customer It is relatively easy to contact this supplier/customer
Conflict resolution 2.276 / 2.695 0.707 / 0.767	Our relationship with this supplier/customer enables joint conflict resolution This supplier/customer is quick to handle complaints We work on solutions together to solve problems so they do not happen again In the past, disagreements and problematic issues with this supplier/customer have not been resolved
Cooperation 1.960 / 2.235 0.576 / 0.584	I feel that by going along with this supplier/customer, I will be favoured on some other occasion We are willing to put aside contractual terms in order to work through special circumstances or difficult problems with this supplier/customer This supplier/customer and our firm have compatible goals We must work together with this supplier/customer to be successful
Power asymmetry 2.820 / 3.155 0.820 / 0.728	This supplier/customer exerts a strong influence over us This supplier/customer has all the power in our relationship This supplier/customer controls all the information in our relationship We have no choice other than to adhere to this supplier's/customer's demands

Results are presented on left/right side for wineries/grape growers. All factor loadings are significant at the 99.9% confidence level.

Part III

Implications and Outlook

Chapter 15

Best Practice in Relationship Management: Recommendations for Farmers, Processors and Retailers

Hualiang Lu,[1] Peter J. Batt[2] and Christian Fischer[3]

[1] Nanjing University of Finance and Economics, China
[2] Curtin University of Technology, Australia
[3] Massey University, New Zealand

Introduction

The previous chapters of this book have analysed theoretically (Part I) and empirically (Part II) the relevance and nature of agri-food chain relationships. While there is still no consensus in the academic literature on how to define and measure inter-organizational relationships, there is little doubt that good, sustainable relationships – when competently managed – can create value for, and enhance the competitiveness of, all chain partners.

Relationships are the glue that hold partnerships together. Without effective interaction, coordinated, mutual value creation is not possible. In a business context, relationships are driven by mutual commercial benefits; in the same way as love or friendship are the engines of inter-personal relationships. Once the engine dies, relationships slowly wither and eventually die. Given that relationships need constant care, it is best to work proactively and continuously to stabilize and to strengthen them, rather than to address problems only when they emerge. Because relationships fail easily and it usually takes great effort to repair them, it will be more satisfying for everyone involved if these relationships are managed competently and professionally. Hence, there is a need for some simple but reliable rules which can be used to build and maintain strong agri-food chain relationships.

'Best practice' describes procedures and methods that are time-proven, science-based solutions to common problems, which can be relied on to yield desirable results. Because building and managing business relationships is usually a complex process, the challenge lies in implementing many small actions at the right time and in the right order. Having good management 'recipes', while not automatically guaranteeing desired outcomes, often works well in the hands of skilled leaders. Needless to say, knowing what to do is one thing, successful implementation is quite another. Yet, without a proven 'compass', most leaders responsible for managing any group of people or project are more likely to fail.

In this chapter we describe how to create and maintain good, sustainable business relationships between agri-food chain buyers and sellers. The approach is practice-oriented. That is, less weight is given to theoretical or numerical analysis and more to real-world implications and recommendations. Consequently, we discuss what chain managers can and should do in their daily work practices when interacting with their customers and/or suppliers. The first section covers the necessary foundations of relationship management in agri-food chains. The second section gives general instructions of how to 'grow' and 'nurture' agri-food chain relationships. The third section provides more specific advice for individual chain stakeholders. Section four concludes.

Foundations of agri-food chain relationship management

In managing any long-term trading relationship within an agri-food chain, it is important to appreciate that exchange partners will only enter into a relationship when they expect to receive returns that are greater than those they can obtain elsewhere.

The importance of prices

At the producer level, the primary motivation for maintaining any relationship with a downstream customer is commercial benefit and most frequently price. However, in making the decision to sell, astute producers will evaluate the price offered on a number of criteria such as the price per unit, the number of units that conform to buyers' specifications, and the terms of payment.

However, achieving the highest possible price is not always the principal objective, for high prices are often achieved for only a small proportion of a producer's output. If producers go out of their way to grade their products to meet buyers' specifications, they may experience some difficulty in finding an alternative market that is willing to pay a fair price for that proportion of the harvest that is rejected. In transacting with modern retailers and food manufacturers, producers may be forced to accept lower prices in the knowledge that they have an assured market for that product which meets buyers' specifications. Furthermore, it is widely acknowledged that higher prices are sometimes paid for inferior quality product on the basis that it is out-of-season or inclement weather has unexpectedly reduced the quantity of product available. On other occasions, producers may find to their dismay that a buyer's promise to pay a higher price does not materialize. Where credit has been extended, the returns that producers receive are on occasions no better than those they could have obtained on the spot market. In other instances, attempts to extract the amounts owing fail because buyers no longer have the funds or buyers have simply disappeared.

Given the importance of price in commodity markets, producers need to know not only the price that they received (for each grade), but also the prevailing market price (for each grade). This enables producers to compare their net returns with other producers, enabling them in the first instance to determine if they have been treated fairly and equitably, and in the second, to identify where improvements can be made to improve their offer quality. It also provides an opportunity for buyers to manage their producers' expectations and where appropriate to offer technical assistance and advice.

Building a trust basis

Such scenarios as those described above occur on a daily basis, both in developed and transitional economies. In order to accommodate the risk which is inherent within any exchange, producers prefer to transact with those buyers that they have come to trust. Trust is derived both from producers' knowledge that they have been treated fairly and equitably and buyers' willingness to invest in a relationship. These investments may take many forms including the exchange of price information, technical information, the development of new products and product lines, and particularly in the transitional economies, the provision of credit to farmers so that they can purchase inputs.

If buyers are willing to assist producers to meet either their own or their downstream customers' specifications, producers must first be provided with written specifications which clearly describe what is, and what is not, acceptable. In many cases, not only must the product be defined, but the processes producers are expected to follow must also be clearly outlined. Inevitably, in order to respond to society's increasing concerns about food safety and traceability, such process control may lead to the implementation of quality assurance systems. In this respect, downstream customers have a pivotal role to play in working with their upstream suppliers to facilitate compliance. Such activity may in part result in buyers actively supporting quality training workshops. In other instances, it may require buyers to facilitate meetings with their downstream customers so that producers fully understand and appreciate both the need for, and the benefits they can derive from, quality assurance systems. However, almost invariably, it requires an element of patience and a willingness to accommodate producers during the transition. Clearly, both parties will only enter into such a relationship where there is an element of trust and some temporal commitment, which provides an assurance that the investment in time and capital (where applicable) will pay dividends in the future.

With greater resources and greater access to information, buyers are often in a better position to facilitate the adoption of new product and process innovations to the mutual benefit of both parties. Indeed, in many food processing operations, the provision of agronomists and technical support staff is normal practice, whereby producers are encouraged to adopt innovations which will not only improve their marketable yields but also reduce their costs of production, which, in turn, greatly enhances food processors' competitiveness in the market. In some instances, it may even extend to facilitating credit, although more often than not, downstream market intermediaries are reluctant to go this far. The reasons for their reluctance are self-evident: the risks are too high. Not only must they assume the risk of crop failure, and thus the need, should the crop fail, to reinvest again so that producers can repay their initial loans, but also the risk that as the harvest approaches, producers may opt to sell to other buyers who offer higher prices.

Opportunism therefore potentially exists on both sides of the exchange transaction. While producers may choose to forgo existing contractual arrangements (whether implicit or explicit) in the pursuit of higher prices, buyers may at their discretion adjust quality specifications and/or agreed prices in response to changing market conditions. While it is unrealistic and non-sustainable in the long term to expect buyers to pay more than the prevailing market price, from time-to-time it may be necessary for downstream customers to sacrifice their profit margins so that, in effect, there is some risk-sharing arrangement. While such behaviour is rarely seen in modern agri-food chains, it is a sure fire way of building and enhancing trust.

However, on occasions, rather than to enhance trust and a long-term commitment to relationships, it is the propensity for either party to engage in opportunistic trading that threatens to undermine them. Here it is important to realize that an exchange partner's satisfaction with an exchange, and the trust that they place in their respective exchange partners, is often re-evaluated after each transaction. Two variables, which are to some extent interrelated, act to constrain opportunistic behaviour: the desire to protect reputation and the socially embedded nature of relationships (Chapter 3).

An organization's (farm or firm) capacity to do business is very much defined by its reputation. Not unlike trust, reputation provides a signal of past behaviour which potential and existing exchange partners utilize to predict or to anticipate future behaviour. Organizations that have developed a reputation for treating their suppliers and downstream customers fairly and equitably will find it relatively easy to attract new business. Conversely, those organizations which seek to take advantage of their exchange partner will find it more difficult to conduct business.

Within all agri-food supply chains, communication occurs not only between the different actors (producers–market intermediaries–consumers), but also between those actors on the same level (or node). In this respect, information about opportunistic traders and/or producers will diffuse very quickly within the network, damaging reputations and thus the ease with which that party which has behaved opportunistically will be able to enter into new relationships.

Evidence has been presented (Chapters 6 and 12) to suggest that in transitional economies, personal relationships are of much greater importance, for in the absence of appropriate vehicles for legal redress, people prefer to transact with those people they already know. Such relationships may emerge from religious or political affiliations, ethnicity, or from the long-term relationships that have been built up over many generations as evidenced by the presence of guanxi in much of Asia. This is often referred to as social capital, the shared knowledge, understandings, norms, rules and expectations about patterns of behaviour which reside within a network and facilitate interaction (Ostrom, 2000). In this context, trust is an expectation that arises within a community of regular, honest and cooperative behaviour, based on commonly shared norms (Fukuyama, 1995). Those who choose to ignore the rules are ostracized and their position within the network compromised.

Dealing with power imbalances

A variable which most often leads to the demise of long-term relationships in agri-food chains is the frequent use of coercive market power. While both marketing theory and the supply chain literature acknowledge the legitimate right of a commercial customer to lead or to control the supply chain, the manner in which this is done will have a profound impact on the relationship. Where power is overtly exercised and exchange partners frequently threatened or coerced into taking action which is not in their best long-term interest, not only will trust diminish, but the aggrieved party is less likely to collaborate and may even withdraw to pursue self interests.

When a channel member controls resources that another channel member wants, various power relations will emerge that enable the party controlling those resources to exert some influence or power (El-Ansary and Stern, 1972). Power resides in the ability of one party to make another do what s/he would not have otherwise done (Gaski, 1984).

According to French and Raven (1959), power is derived from the more dependent organization's perception of the dominant organization's ability to mediate rewards, mediate punishment, its legitimate right to prescribe behaviour, to communicate some specific knowledge or expertise and the extent to which the more dependent organization identifies with the dominant one. More recently, Johnson *et al.* (1993) have classified power as either mediated power (reward, coercion and legal legitimate power) or non-mediated power (expertise, information and traditional legitimate power). This dichotomy reflects whether the source does or does not control the reinforcements that guide the target organization's behaviour (Brown *et al.*, 1995).

While the power to coordinate is the prerogative of the dominant organization (Achrol, 1997), the subsequent use of that power will impact on the exchange partner's perception of relationalism (Brown *et al.*, 1995). Channel leaders have often used various reward and coercive powers and legitimate authority to cajole and coerce cooperation among channel members. However, it is the use of non-mediated power that inevitably builds social bonds and close relationships. Granovetter (1985) describes how even in the presence of substantial power imbalances, exchange partners may prefer to operate in ways that are informed by habit, custom and practice. Relationships may be so embedded that the more powerful organization risks the loss of its reputation should it be found to have engaged in opportunistic trading.

It is widely accepted that the frequent use of mediated power is likely to damage relational norms, cooperation and accommodation between channel partners (Brown *et al.*, 1995). Overt attempts to directly influence weaker parties through the use of mediated power are generally viewed with considerable disfavour. Not only will this lead to conflict, but the relative attractiveness of alternative exchange partners will increase. Not unexpectedly, most of the conflict that arises in agri-food chains will relate to price, and with the continual erosion of trade barriers under the WTO, competition is increasing, putting more downward pressure on price. Hence, if producers are to remain competitive, they must look towards ways to reduce costs, to add value or to differentiate their product in what is already a saturated market.

In providing that competitive edge and an assured market, downstream customers have a critical role to play in facilitating the investment which will ultimately be of benefit to both parties. Hence, on the one hand, while it is necessary to retain firm control and to provide appropriate incentives, on the other, it is the willingness to share sensitive market information and to provide technical support that ultimately builds enduring long-term relationships.

Best practice for 'growing' strong chain relationships

Most natural and man-made processes depend on evolution and repetition. Starting with a decision to achieve a certain goal and making a plan of how to do this, a second phase involves the administration of resources and a steering process before results can be reaped, and learning from the achievement and optimization of the next process can start.

Agricultural production processes follow the same pattern. At the start of the growing season, farmers/growers have to make a decision on what, where and how much to plant. After sowing, plants need to be nurtured by providing necessary resources in the right amount and at the right time. Harvesting requires some work. After the harvest, preparation for the next production cycle begins. 'Growing' strong, successful business relationships is quite similar. Figure 1 illustrates the process.

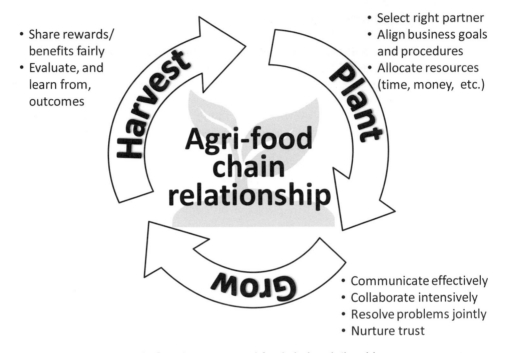

Fig. 1. Growing strong agri-food chain relationships.

Planting

Selection of right partners

The first step in the business relationship growing process involves finding a suitable partner. A relationship with a business partner is a longer-term venture. Partners will be together in the same boat. One should think hard about with whom one will spend one's time, share risks and hopefully reap rewards. There are many criteria which can and need to be considered when selecting suitable business partners. Competence, trustworthiness, sympathy are but a few. Which ones are considered most important depends on the individual decision-makers, their values, experiences and feelings.

Alignment of business goals and procedures

After both sides have agreed that they want to form a business partnership, their individual business goals need to be aligned. Often, of course, a decision on whether to enter into a partnership with someone will be the result of an agreement on shared goals. That is, forming partnerships is an interactive process and may take some time and even require previous successful interaction episodes (see Chapter 4) before a final decision is made on closer collaboration. Whatever the process might be to arrive there, at some stage business partners have to agree on what to do together, and how. This may then be the 'official' starting point of their relationship.

Allocation of appropriate resources to a key relationship

Allocating appropriate resources to a key relationship is the next step. In a business venture, often investments have to be made. Not only intangible ones, such as time and commitment, but commonly tangible ones as well, such as regular meetings, data exchange interfaces, or even the employment of specialized staff. In many larger corporations, it is common practice today to assign dedicated 'relationship managers' to look after key customers or suppliers. In smaller companies, senior managers will need to allocate appropriate amounts of time and dedication to build and maintain relationships with their business partners.

Growing

Effective communication

Effective communication has been shown to be one of the most important determinants of good, sustainable chain relationships (see, e.g., Chapters 7 and 9). Both components count: good information quality and an adequate communication frequency. The right amount of each component has to be identified by relationship partners. It is probably true that there is no such thing as too much communication, as long as it is relevant. Overall, effective communication can be seen as the water that plants need on a regular basis to grow and flourish.

Intensive collaboration

Ongoing collaboration, or some combination of cooperation, coordination and collaboration ('C3 behaviour', Humphries and Wilding, 2004), is the basis for a business partnership's success. Whatever the exact name and nature of 'working together', in practice it means to only engage in actions that are in the best interest of all partners. True collaboration strengthens relationships, while paying mere lip service to it weakens them.

Joint problem resolution

The emergence of problems in business partnerships can be considered as normal. The issue is not whether they are there but how to deal with them. The main point is to resolve each problem jointly, as this avoids undermining trust. Moreover, having effective conflict resolution mechanisms in place is also vital. Successful relationship managers need to be skilled in the identification of potential relationship hazards, and in resolving conflicts. Adequate inter-personal skills, such as the mastering of effective communication and negotiation techniques, are important. As in the agricultural growing process, there will be periods of storm, flooding or drought. It is the job of farmers to look after their plants and to compensate for the vagaries of nature. Growing strong business relationships requires no less care and dedication.

Nurturing trust

Trust is a necessary component of sustainable business relationships (see Chapters 5, 6, 8, 9, 10). Since risk is a common characteristic of most business ventures, trusting commercial partners is a cost-effective strategy to cope with such risks. However, trusting in itself – like hoping – is not enough. Trust can be easily exploited and only a fool would blindly rely on it. Due diligence is required in addition to trust. That is, making maximum efforts to ensure that things happen in the right way. In the absence of trust, contracts can be used. Given that contracts are rarely complete and their enforcement involves costs, trust, in some instances, is necessary in addition to a contract. Seen from another perspective, even if good trust between business partners exists, they may choose to close a contract precisely to protect this trust in the case of contingencies.

Contracts can be an effective conflict resolution framework, specifying a priori what to do when things go wrong. Hence, trusting relationships may be the most satisfying ones but building, and protecting, trust requires continuous efforts such as effective communication, real commitments, and certain premises such as favourable personal characteristics (honesty, integrity etc.) which lead to accepted reputations. Like crop rotation practices in the agricultural growing process, trust in relationships is the ingredient that does not come cheap, nor are its benefits immediate. However, without such practices, crops will suffer in the long run.

Harvesting

Fair allocation of rewards

If relationships are to be sustained, the reaped rewards or benefits from them need to be shared equitably (i.e., in a fair way). Finding an acceptable (or even better, a motivating) benefit allocation mechanism requires effort and compromise. Similar to agriculture, 'harvesting' relationship rewards is hard work. However, given that commercial rewards are the engine of business relationships, it pays to get the allocation process right and to devote adequate time and attention to it. Even if contracts, specifying expected rewards, have been designed at the start of a business venture, it is probably useful to reconsider original agreements and to make suitable adjustments if necessary. It is well known that maintaining an existing relationship is overall less expensive than building a new one. Making sure that everyone is, and stays, 'happy' may require some effort but it could pay off later on.

Evaluation of relationship outcomes

The performance of relationships needs to be reviewed from time to time. This procedure fulfils two purposes and answers two key questions. First, whether the expected outcomes have been achieved; and, second, whether the relationship can be improved. In inter-personal relationships, satisfaction is an immediate result and no great analytical exercise needs to be undertaken to know whether a relationship is worthwhile or not. In inter-organizational relationships, with potentially many people from both organizations involved, the assessment is more difficult. A more systematic and analytical approach will be necessary to find out whether a relationship is performing or not.

Finding out whether and how relationships can be improved may be even more difficult. It usually requires all relationship partners to come together and to review past interaction episodes, to analyse past problems and their resolution, and to define future goals. Despite the required efforts involved, it is an exercise which should be done on a regular basis, whether this is after one interaction episode or a certain number of them. As in agricultural production cycles, after the harvest is finished and before a new growing season starts, some thought needs to be given to the question of what to plant next, and how yields can be improved, based on the experiences gained in previous seasons.

Recommendations for individual agri-food chain stakeholders

Building and maintaining good business relationships requires care and effort, just like growing agricultural products. However, the emphasis of the elements of managing business relationships differs at various chain stages (i.e., producing, processing and retailing stages) and with different partners (i.e., upstream and downstream partners). Therefore, in this section, some practical recommendations for individual agri-food chain stakeholders, such as farmers/growers, processors and distributors/retailers, are provided. These serve as guidelines for their relationship management processes.

Farmers/growers

Most farmers/growers, especially those in developing countries, are rather small, and thus have less capacity to implement advanced crop management technology due to technical, managerial and financial constraints. Hence, they often deliver products in an unstable manner and with inconsistent quality, all which may hinder them from being attractive to buyers. Furthermore, a highly competitive market environment makes it even more difficult for farmers/growers to sell products effectively and at the desired profit. By following some of the points mentioned hereafter, farmers/growers may improve significantly their marketing practices.

Staying in regular contact with buyers

Farmers/growers are the starting point of agri-food supply chains. In order to effectively market their products, farmers should from time to time consult with buyers in order to collect information about the price and market demand for their products, so that they may know in advance what to plant and how much to produce. In order to reduce market risk and achieve satisfactory profitability, they should especially make regular contact with 'good' buyers who can pay a fair price. A fair price, on the other hand, signals the quality of, and the desire to sustain, their relationships.

To build strong business relationships, farmers/growers also should keep in regular contact with, and show a sincere attitude towards buyers, and always deliver the best products to markets. Personal visits, telephone communication and eating out together are some of the common practices to keep in contact with buyers. In the case of problems arising, farmers/growers should sit down together with buyers immediately and solve a problem jointly. Regular contact helps them to achieve this.

Investing in relationship-specific assets

Research also suggests that the willingness to invest in relationship-specific assets, such as human capital or transaction-specific assets, will enable farmers/growers to be treated as insiders and therefore be selected as preferred suppliers, which may help to strengthen their business relationships.

Investing in transaction-specific assets is an important mechanism for achieving closeness in agri-food chain relationships. Bounded rationality and opportunism are the two key assumptions of transaction cost economics. Bounded rationality implies that human actors, as well as organizations, are incapable of perfect contracting. As such, certain environmental and behavioural uncertainties inevitably arise. Opportunism is the assumption that decision-makers may act in their own interests, and that it is difficult to know in advance who is trustworthy and who is not. By investing in transaction-specific assets, farmers/growers show commitment or pledges to their buyers, which will help to build trusting business relationships. Once a trusting relationship is established, opportunistic behaviour and uncertainty (i.e., relational risk) will be lower, which will, in turn, enhance the relationship.

The agri-food processing industry often requires considerable investment such as cool storage and transport facilities as well as standardized handling and processing procedures. As farmers may not be able to undertake large investments in physical assets, they may compensate for this by investing more in relational assets. Closer relationships can help farmers/growers to acquire advanced production techniques or managerial skills and thus improve handling processes. Typically, good agricultural practice is facilitated by higher levels of physical as well as human assets (such as knowledge training and know-how). Hence, good business relations may eventually lead to quality enhancements which better comply with buyers' requirements.

Building strong personal relationships

From research conducted in Europe (as described in Chapters 5 and 7) and Asia (Chapter 12) it has become clear that for farmers/growers personal relationships are more important than for other chain stakeholders, such as retailers. Moreover, research suggests that a wide network of personal relationships in Asian countries ('wa' in Japan, 'guanxi' in China, 'inhwa' in Korea, etc.) are important in commercial practice, because there is a long tradition of this in their business cultures.

Commercial practices are personally and socially oriented (Chapter 3). Relationships emanate from the individual and are more inter-personal and person-specific. Social interaction is typically necessary prior to business relationships. Social interaction will determine whether a relationship improves, deteriorates or stays the same. Personal relationships may help farmers/growers to access scarce resources, to contact new buyers, to receive new orders, to participate in new marketing initiatives, etc. In a highly competitive market environment, personal relationships can help open new market opportunities for farmers/growers.

Food processors

Food processors are typically positioned in the middle of the chain. They add value by buying and processing commodities, which in many instances they sell on as consumption-ready products to retailers. So they are heavily dependent on both input suppliers (farmers) and distributors (retailers). Maybe for this reason, food processors need to put most effort into building and maintaining good business relationships upstream as well as downstream.

Providing technology to upstream and downstream partners

In order to purchase high-quality inputs from suppliers, food processors should select the right partners to make both tangible and intangible investments. Processors can nurture their business relationships with suppliers by providing advanced production technology, transferring know-how, and offering training programmes. By doing so, processors are able to purchase products that fulfil their specific specifications. Suppliers (farmers), on the other hand, can improve their capacity to produce high-quality products, and show commitment and loyalty to processors, by delivering their best products.

Offering fair contracts

The business relationships between processors and suppliers may be further enhanced by preferred supplier arrangements or contractual agreements. Although it is widely agreed that a contract is an efficient way to reduce opportunistic behaviour and to lower marketing risk, relational arrangements are complementary to contracts (see, e.g., Chapter 12). That is, making and relying on business relationships still makes sense for processors even in the presence of contractual arrangements.

Contracts between processors and their suppliers imply specific transaction agreements, whose terms may include a specified price, quantity, quality and timing. The more complex a contract, the greater the specification of promises, obligations and processes for dispute resolution, and the less risk and uncertainty for ongoing transactions. Long-term contracts are also explicitly drafted with a provision to promote the longevity of exchange. A fair contract between processor and supplier will improve the sustainability of business relationships since it signals the fair distribution of the benefits and value which will motivate suppliers to maintain existing relationships.

Providing relevant market information to suppliers

Providing market information to suppliers is another way to foster relationships. Farmers/growers are eager to know what and how much to produce. In general, they welcome help from processors. It is even better if processors can make decisions jointly with producers about product variety and volume and transaction conditions. Processors who share market information and are willing to work with suppliers closely not only signal their desire to enter into a long-term relationship, but may also signal a greater willingness to pursue mutually beneficial goals, which may help processors engage in closer business relationships with suppliers.

Investing in relationship-specific assets

In relationships with downstream partners (distributors/retailers), on the other hand, processors should show sincere attitudes and honesty towards them, and they should also invest in transaction-specific assets. In order to minimize losses when transporting perishable agri-food products, the processing industry requires investment in refrigerated vehicles to deliver products to customers. Processors also should be able to comply with retailers' requirements in terms of quality standards and special requirements. Therefore, investments in sorting, processing, packaging and labelling know-how are also necessary.

Appropriate resources, such as time and commitment, should be allocated because staff training and the retention of specialized staff in trading positions is important for food processors. The employment of supply chain staff who culturally and/or socially get along with those they transact with may facilitate relationships. This is especially important for processors when marketing their products internationally or in otherwise dynamic and competitive marketing environments.

Distributors/retailers

Effective communication with suppliers

Distributors/retailers are at the end of business-to-business supply chains and have close contacts with consumers. To satisfy consumer requirements, distributors/retailers need to deliver the right products to the right place at the right moment to the right person. Effective communication with their suppliers may help retailers to achieve this. From research conducted in Europe (Chapters 5, 7 and 9) it has become clear that for retailers, effective communication is more important than for farmers, because companies are usually larger and the importance of personal relationships smaller. Retailers should share information about consumer requirements, market trends for their products and major competitors with their supply chain partners.

The need for more professional management support

Retailers generally run larger businesses than farmers and there is a need to use more professional, systematic management approaches. Retailers often use customer relationship management (CRM) software when dealing with suppliers and loyalty card schemes (or market research panels) when trying to understand end customers (consumers). Farmers will rarely be in a position to use such professional data collection and analysis tools. Nevertheless, no matter how advanced the technology or tools used, one goal retailers always want to achieve is letting consumers know that they care about them. Improving consumer satisfaction levels is an important aspect that will also enhance supplier business relationships because suppliers recognize that their products are 'in good hands'.

Compensating power asymmetries by nurturing trust with suppliers

It is widely recognized that for building satisfying agri-food business relationships, good communication and trust are key points. However, there are problems for suppliers to communicate effectively with distributors/retailers, and to build trusting relationships. This is because of the fact that suppliers, compared to distributors/retailers, are usually much smaller in terms of both scale and power. Thus, distributors/retailers should tackle the possible mistrust with suppliers arising from the fact that they are more powerful than suppliers. In order to nurture trust in suppliers, distributors/retailers should avoid over-exercising their power and threatening or coercing suppliers to take action that is not in their best interests.

Conclusions

It is widely agreed that building and maintaining good, sustainable business relationships can lower transaction costs and result in competitive advantages, which will benefit both sellers and buyers in agri-food chains. The management of successful business relationships, however, should take into account several aspects, such as sharing equitably commercial risks, avoiding abuse of coercive market power, making relationship-specific investments, creating trusting partnerships, sharing commercial information, building reputation, fostering personal relationships, etc. Similar to growing agricultural products, there is a best-practice approach to managing agri-food chain relationships.

The major ingredients for 'planting' business relationships are selecting the right partner, aligning business goals and procedures, and allocating appropriate resources. To 'grow' a relationship, effective communication, intensive collaboration, joint problem-solving, and the nurturing of trust are pivotal. For a successful 'harvest' one should share rewards/benefits fairly and evaluate all relationship outcomes. However, there is no simple rule or identical routine to achieve successful business relationships for all actors. Different stakeholders in agri-food chains should instead focus on the most important aspects of their business relationships. Thus, successful implementation practice differs for farmers/growers, processors and distributors/retailers respectively, as well as across cultures and market development stages.

References

Achrol, R. (1997) Changes in the theory of interorganizational relations in marketing: toward a network paradigm. *Journal of the Academy of Marketing Science* 25(1), 56–71.
Brown, J., Lusch, R. and Nicholson, C. (1995) Power and relationship commitment: their impact on marketing channel member performance. *Journal of Retailing* 71(4), 363–392.
El-Ansary, A. and Stern, L. (1972) Power measurement in distribution channels. *Journal of Marketing Research* 9, 47–52.
French, R. and Raven, B. (1959) The bases of social power. In: Cartwright, D. (ed) *Studies in Social Power*. University of Michigan Press, Ann Arbor, MI.
Fukuyama, F. (1995) *Trust: The Social Virtues and The Creation of Prosperity*. Free Press, New York.
Gaski, J. (1984) The theory of power and conflict in channels of distribution. *Journal of Marketing* 48(Summer), 9–29.
Granovetter, M. (1985) Economic action and social structure: the problem of embeddedness. *American Journal of Sociology* 91(Nov), 481–501.
Humphries, A. and Wilding, R. (2004) Long-term collaborative relationships: the impact of trust and C3 behaviour. *Journal of Marketing Management* 20(9–10), 1107–1122.
Johnson, J., Sakano, T., Cote, J. and Onzo, N. (1993) The exercise of interfirm power and its repercussions in US–Japanese channel relationships. *Journal of Marketing* 57(Apr), 1–10.
Ostrom, E. (2000) Social capital: a fad or fundamental concept? In: Dasgupta, P. and Seragilden, I. (eds) *Social Capital: A Multifaceted Perspective*. The World Bank, Washington, DC.

Chapter 16

Improving Agri-food Chain Relationships in Europe: the Role of Public Policy

Luis Miguel Albisu,[1] Maeve Henchion,[2] Philip Leat[3] and David Blandford[4]

[1] Agri-Food Research and Technology Center of Aragón (CITA), Zaragoza, Spain
[2] Ashtown Food Research Centre, Teagasc, Dublin, Ireland
[3] Scottish Agricultural College (SAC), Aberdeen, United Kingdom
[4] Pennsylvania State University, USA

Introduction

Most agri-food chains comprise an increasing number of stakeholders and efficient relationships among them have become essential to their success. The likelihood of a large number of partners being involved increases with the level of transformation (processing) and the number of services that this entails. The entire process results in a food product that frequently represents a combination of many ingredients and services, produced and performed by different firms. The interdependent firms which make up the agri-food sector currently contribute 2% of Europe's GDP. Within the EU-27, small- and medium-sized enterprises (SMEs) make up 99% of the food sector and account for 50% of turnover (European Commission, 2007a).

Consumers, at the downstream end of the chain, send signals about the acceptability of food products through their purchasing decisions. These signals have to be transmitted along the agri-food chain from retailers and food service enterprises to processors and primary producers. Producers upstream need to receive precise and clear messages in order to supply the commodities that lead to food products. Similarly, food processors and farmers need to understand evolving consumer preferences if they are to maintain their competitiveness.

Many governments seek to safeguard agri-food chain stakeholders' interests, while at the same time trying to increase economic efficiency and enhance social welfare in a sustainable manner; a preoccupation with consumer interests is becoming more prominent. This is demonstrated in the European Union when comparing the initial focus of the Common Agricultural Policy and the subsequent development of food policy. Public policy has increasingly focused on regulating different parts of the agri-food chain in recent years. Regulators are not necessarily aware of the consequences that specific rules might have for particular parts of the chain, as they usually deal with partial problems and may not have an overall vision for how the system should evolve. The Commission of the European Communities has been gathering information and analytical insights which could help in developing more comprehensive and effective instruments to achieve policy aims.

The purpose of this chapter is to analyse the role that public policies play in influencing agri-food chains and their impact on stakeholders' relationships. In doing this, recognition is given to private institutions and activities that may assist the implementation of EU policy (e.g., private quality assurance schemes which serve the needs of food safety legislation) or reduce the pressure for EU legislative involvement (e.g., codes of conduct in the retail sector which may lessen the need to prevent anti-competitive behaviour). The next section identifies the most important elements influencing agri-food relationships and areas in which policy intervention may be required with special emphasis on the EU. Relevant EU policies are identified in the following section, complemented by an assessment of their implications. The final section offers some ideas on how policy could evolve in order to support agri-food stakeholders' relationships in the future.

Elements influencing relationships among agri-food chain stakeholders

Farmers

In Europe, farming activity is carried out on many thousands of holdings. Most farms are relatively small in terms of the volume of production. As a result, farmers have limited influence over the rest of the agri-food chain, unless they produce highly specialized products for which there is limited competition, which is unusual. Some may fulfil other functions of the chain through direct marketing, an activity that has been stimulated by the wish of some farmers to have greater influence in the chain and to retain/capture more of the value-added. Cooperatives try to agglomerate volume for farmers and market what their members produce with varying levels of success. They rarely attain a dominant position, and face many challenges in meeting the need for vertical coordination (Hanf, 2009).

However, there is also a trend towards the creation of short agri-food chains which seek to connect small producers directly to consumers. Such chains mainly apply to traditional and local products and have limited market coverage. The final outputs of both short and long types of chains are food products sold in markets, where consumers frequently judge them without knowing much about the intricate processes and specific ingredients that underlie supply.

Farmers or primary producers are increasingly moving away from the use of spot markets to contractual arrangements and, in some cases towards full vertical integration, as a response to requirements in oral or written contracts with other stakeholders. Relationships with other agri-food chain stakeholders vary depending on the kind of linkages, but it is relatively uncommon for other agri-food chain stakeholders to have direct investments in farming activities.

Agri-food processors

Agri-food processing is dominated numerically by SMEs. Agri-food enterprises often sell finished products, although some undertake primary transformation and so interface with other agri-food enterprises rather than with the distributors of primary products.

Even in fresh agricultural product chains, which involve minimal processing, a large number of activities such as refrigeration, packaging, logistics, etc. are undertaken before the products of farmers and agri-food enterprises reach the market, either through the intermediary of wholesalers or retailers or, in certain circumstances, directly to the final consumer. The closer farmers are to consumers, the more direct the communication is about consumer needs and the easier it is to engage with other stakeholders about how to improve processes and products. However, proximity to consumers is not practical in many cases.

Many food brands developed by agri-food SMEs have limited market impact because of small volumes, and limited geographical distribution and consumer recognition. In some cases, reliance is placed on umbrella brands such as designations of origin, regions, countries, breeds/varieties, etc. Some agri-food firms, without strong brands, have access to large food distribution chains as local specialities and less well-known products, because they provide opportunities to distributors to offer a wider selection to consumers. This characteristic has been sufficient to establish long-term relationships among some chain stakeholders. However, there are a few large agri-food firms with leading brands and strong market penetration, and the private labels of distribution chains are increasingly significant.

Distributors

Relationships between retailers and agri-food firms depend on the kind of business in which distributors are involved. Hard discounters relate to a few firms that are able to provide large volumes. Hypermarkets have a broad product range and necessarily have to deal with many different firms, which might have international, national or regional coverage. Supermarkets, depending on the size of their establishments and geographical coverage, have a range of food products on their shelves and consequently have relationships with a diverse number of agri-food firms. Retailers that emphasize quality are probably more attentive to supplier relationships, since price might not be the main driver of business and they perceive value in developing longer-term and more stable relationships. Hence, the variety of distribution models implies a diversity of relationships.

Large retail groups are mainly looking for large-scale operations and reliable suppliers (Burt *et al.*, 2008). They have rationalized the supplier base over time to embrace a limited number of preferred suppliers in any given product category. The larger the retailer, the greater their bargaining power with the rest of the chain, but also the more difficult it may be for them to find global suppliers. Large retailers are also increasing their private label sales, requiring special agreements with suppliers. Small retailers buy from wholesale distribution centres, resulting in indirect relationships between producers and retailers and making communication among them more difficult.

Distribution outlets have to incorporate a range of food products, some of which are not produced by leading agri-food firms but usually by SMEs. Local, regional and many national food products are provided by these kinds of enterprises, which may be sourced locally or regionally rather than through central or head office purchasing. Consumer loyalty is the main strength of these food-producing SMEs, which depends on how desirable their food products are viewed to be by consumers.

Power distribution and competition along the chain

As noted, there is a tendency for big players in the agri-food chain to look for partners that are also large scale. Big distributors look for big agri-food firms and large primary producers because, among other factors, they can take full advantage of the potential offered by information and communication technologies (ICTs). SMEs feel more at ease among other SMEs and are at a competitive disadvantage with firms of significantly larger size. Market requirements and differentiated social norms mean that personal bonds and trust play a stronger role among SMEs than among large firms, which often follow strict contractual norms applied in a more impersonal way.

Agri-food enterprises do not commonly have dominant positions in agri-food chains, but there are some situations where a small number of firms might dominate supply conditions (Palpacuer and Tozanli, 2008). Agri-food firms complain about the difficulties they face in dealing with large retail groups, but farmers frequently also feel that they receive unfair treatment from other participants in the chain. Agri-food enterprises have been accused of abusing their position in the chain in certain countries and product sectors to the detriment of farmers (Zachariasse and Bunte, 2003). Power has increasingly been shifting from the middle to the downstream part of the chain. The countervailing argument is that competition among food distributors is quite fierce and that there are no dominant market positions.

The distribution sector is the most concentrated part of the food system and some large and geographically extensive chains, handling large product volumes, are present in most countries (Lang, 2003). In an increasingly international environment, distributors are better placed to know what consumers from various socio-economic groups in a given geographic area are likely to purchase, and they use this information to develop their competitive advantage. They handle large volumes of many food products, which along with their unsurpassed insights into consumer trends and behaviour, has important implications for other stakeholders in the agri-food chain.

There is a tendency to believe that distributors have extra market power that yields more profitable returns, but results on overall returns on capital for publicly quoted firms on stock markets do not show great differences from other smaller distribution chains. Walmart's consolidation of retailing in the USA, and its expansion into other geographic areas, has had an important impact on mergers in the distribution business in Europe. There has also been, over the years, greater penetration by European distribution chains in the USA. However, experience has shown that success cannot necessarily be transferred from country to country.

Stakeholders' relationships are conditioned by many of these circumstances but unbalanced bargaining power is probably the factor that affects SMEs most. This might imply disadvantageous practices, e.g., late payments. On the other hand, some SMEs are not accustomed to working under contractual arrangements established by big companies and resent entering a highly formalized and unfamiliar business environment. This requires longer-term planning and commitment as well as continuous monitoring of contract terms, thereby raising transaction costs. Dealing with large firms involves a new business philosophy which requires an investment of time and effort, and enhanced managerial capabilities.

Competition commissions, from various countries over the years, have examined whether collusion exists in the agri-food chain but have found limited evidence of anti-competitive behaviour. There are a few agri-food enterprises that are able to provide price leadership linked to differentiated products and well-known brands. There are some circumstances in which agri-food enterprises exert market leverage, but this does not apply in most situations.

In the UK, the Competition Commission (2008, p. 9), after a study lasting many years, concluded that 'competition in the UK groceries industry is effective and delivers good outcomes for consumers, but not all is well'. It identified some dominant positions and poor competition in some local markets and urban areas. The Commission recommended that 'competition' tests be applied in granting planning permission for new stores and for the extension of existing ones, as well as the development and application of a code of good business practice. It also recommended the appointment of an ombudsman to investigate claims of unfair practices.

Overall chain relationships

The structure of the agri-food sector influences, to a great extent, how agri-food chains function and perform. In most countries, primary production and agri-food processing are dominated by small holdings and enterprises that account for a large share of employment, but as a result of market concentration large companies account for the largest proportion of sales.

Communication becomes difficult because many stakeholders, who might pursue different aims, are involved without realizing that compromises are needed for good chain performance and that conflicts among stakeholders, at different levels of the chain, can have a significant negative impact on overall performance. Leadership by a dominant firm is one way to provide an overall vision for the chain, but a leader may try to meet not only explicit market demands but also use its market power to pursue its own interests in the absence of transparency (Zhao *et al.*, 2008). Overall, centralized modes of organization are gaining ground in the coordination of transactions and are leading to the substitution of private institutions for public policies (Ménard and Valceschini, 2005). Value-added is increasingly created in the downstream part of the chain, primarily in distribution and to a lesser extent the food-processing components (Commission of the European Communities, 2009a).

A clear understanding of all requirements, from primary production to final distribution, requires strong and active relationships along agri-food chains. However, normally relationships involve only two levels along the chain (farmer/processor and processor/retailer), such that those concerned try to solve problems between them. In some cases, vertical relationships can be involved, for example through cooperatives, where farmers have access to markets without going through intermediaries or through dominant firms that exert influence over the rest of the chain.

Chain relationships and consequences for product innovation

Farmers need to be aware of the competitive food products that can be produced with their raw materials. It is difficult, in general terms, to be competitive in Europe when confronted with products from other geographic areas that have more favourable production conditions, abundant natural resources and lower wages. Increasing trade liberalization will intensify competition with other parts of the world. In this environment, food product differentiation is a crucial element in the ability to compete, but this will require good relationships and understanding along the agri-food chain, plus a focus on innovation. Many SMEs do not have the resources to devote to the search for fundamental innovations, but they can seek incremental improvements. The public sector has a role to play by conducting research and encouraging knowledge transfer, as well as communicating developments in consumer preferences and promoting the adoption of new products derived from public research centres through coordination along the agri-food chain.

Competitive markets stimulate continuous food product improvement and innovation. Food products are good examples of the impact that incorporating small changes, originating from various agri-food stakeholders at different chain stage levels, can have on market share. Consumers perceive the incorporation of many changes as resulting in new products. To a large extent commercial success depends on communicating consumers' needs to stakeholders as well as the nature of food product changes to consumers. Relationships and communication among stakeholders have an increasing influence on the process of innovation. This is similar to internal communication in large firms between the production and marketing departments in creating and launching new products.

Policies related to agri-food chains

Increasingly complex and dynamic international food market relationships can, to varying degrees, contribute to a variety of market failures. Most notably are those relating to: (i) the need to ensure the provision of public goods, particularly relating to food security, safety and public health; (ii) the existence of externalities, especially the potential environmental costs of food production and processing; (iii) the presence of transaction costs and problems of information provision, which may inhibit efficient decision-making and transactions by smaller producers and by consumers; (iv) problems of market power, mainly associated with large processors and retailers; and (v) the necessity of meeting human needs (from basic nutrition to self-fulfilment) with respect to food intake. Against this background, the EU operates a range of policies aimed at ensuring a favourable business environment and a well-functioning internal market, and has schemes to allow firms to take advantage of trade opportunities. It also implements a range of policies that address broader social and environmental concerns. Policies that relate directly to the agri-food sector as well as more general policies affecting a range of sectors can affect the food industry. With respect to relationships in the agri-food chain, policies can have a number of impacts, including: facilitating access to markets, influencing the balance of power and competition, reducing price volatility and uncertainty in the chain, enhancing trust along the chain, protecting consumers' interests, improving food safety and quality, increasing the competitiveness of firms and facilitating communication efficiency and effectiveness. However, policy measures can also create barriers to entry and trade, generate higher costs, inhibit innovation, reduce competitiveness, and have unforeseen and unwanted consequences. In addition, certain groups can be affected disproportionately, e.g., SMEs, unless policies are carefully designed and implemented.

Agri-food sector policies

The Common Agricultural Policy (CAP)

The primary aim of the CAP is to support the competitiveness and sustainable development of agriculture and rural areas in the EU and to protect consumers' interests. Food security has been a fundamental objective since the signing of the Treaty of Rome in 1957. Through its impact on the supply of raw materials, and its implications for international market access, the CAP is very significant for the food industry. The policy has evolved during its 50-year history from a production-oriented support policy to embracing a more market-oriented system, in which financial support to farmers is partially or fully decoupled from production. It has also expanded to embrace food safety, environmental, plant health and animal welfare standards in line with market demands and economic realities. Decoupling has meant that farmers now need to respond to market signals if they are to engage in profitable production whilst its expanded focus embraces the concept of multi-functional agriculture.

The current CAP has changed economic relations in the chain by ensuring that relationships are based on market demand, while providing income support for farmers. It has also promoted the development of relationships, and thereby market access, outside the EU through export refunds. However, the extent to which such relationships are sustainable is questionable, given the gap between world and EU prices of many commodities. The CAP seeks to reduce price volatility, although limited success in the recent past demonstrates that there can be conflicts between raw material suppliers (i.e., farmers) and processors, particularly when dominant retailers are not willing to absorb some of the increased costs incurred within the chain as a result of higher raw material prices. None the less, recent changes in the CAP are expected to help agricultural producers to become more market-responsive, to benefit from new opportunities and simultaneously protect them from erratic short-term price movements. As the policy continues to move away from its production orientation, it will become more and more important for policy-makers to monitor the performance of agri-food enterprises and take the necessary measures that can contribute to enhancing their competitiveness.

Food safety policy

At the EU level a common basis is needed for food safety measures to ensure the free movement of safe and wholesome food in the context of the Single European Market. Inevitably, in a market of 27 member states the interpretation of food safety requirements varies between countries, so that differences in national policies can hinder free movement. Following several food crises, such as 'mad cow' disease and dioxin contamination, measures are needed to protect consumer interests and to restore consumer confidence. Measures implemented include stricter standards for food quality and food product hygiene, animal health and welfare, plant health, and reduction in the risk of contamination from external substances. EU regulations on food safety standards are geared toward quality control, process verification, labelling and traceability.

A number of key principles are enshrined in the policy including the need for a sound scientific foundation, up-to-date legislation, and the adoption of a farm-to-fork approach. Achieving the aims of food safety policy in this context requires input from a range of stakeholders including the general public, non-governmental organizations, professional associations, trading partners and international trade organizations. Furthermore, legislation must cover all aspects of the food chain: from primary production, processing, transport, and distribution through to the sale or supply of food and feed, with the legal responsibility for ensuring safety resting with the operator at each stage. Assigning responsibility for food safety to food businesses has contributed greatly to the emergence of private-based standards – an increasingly important feature of the food system environment within the EU.

The development of private standards has promoted greater chain coordination (Young and Hobbs, 2002). Retailers pursue a differentiation strategy, which for many encompasses private-label products, implying the need for coordination with other participants in the chain, and good communication flows. Traceability schemes have been put in place partly in response to a lack of trust/consumer confidence in the agri-food chain. Businesses are motivated to participate in traceability schemes, partly because downstream players often require their adoption to meet the requirements of food safety legislation, but also to facilitate access to market segments where provenance is an important criterion for consumers. Thus, information about the origin of food and its production history has become a part of the marketing strategy used by food companies and retail chains.

The aim of EU labelling policy is to provide information to help consumers make informed choices. However, there is limited evidence that consumers actually want, and will use, all the information included on labels. Indeed there is evidence of 'information overload' (McEachern and Schröder, 2004). For example, specific labelling regulations apply to beef in order to give consumers more detailed information on the beef for sale. The label must include information on where the animal was born, reared, fattened, slaughtered and processed, and an efficient communication flow is required within the chain to provide this. However, labelling also relates to quality attributes (see below) and the EC has regulations in place to provide information on the nutritional content of foods as part of a strategy to combat the growing prevalence of overweight and obese individuals. While the provision of such information may not always modify consumer behaviour, it may influence processors' behaviour as they may not want to supply products that are, for example, high in salt or saturated fats, with consequent implications for ingredient suppliers.

Agricultural product quality policy

The prevailing view in the EU is that quality is about meeting consumers' expectations on product characteristics and farming methods. Instruments are required to communicate quality attributes to the consumer in a way that results in economic benefits to the producer and others in the chain. Marketing standards and quality schemes dealing with geographical origin, traditional methods of production, or those from organic farming are the cornerstones of EU quality policy. Labels associated with these schemes can help to reduce information asymmetry between producers and consumers by allowing the communication and verification of credence characteristics. However, the success of labels can sometimes be constrained by consumers' limited knowledge of production systems (McEachern, 2008).

As common EU marketing standards have been adopted to replace various national standards, trade has been facilitated in the single market, although there are still significant differences among countries (Becker, 2009). EU requirements cover such items as definitions of products, minimum product standards, and labelling requirements. They are designed to help farmers deliver products with the quality attributes expected by consumers and to facilitate price comparisons for products of different qualities. While such standards can facilitate trade, the cost of enforcement, the fact that it is difficult to develop and apply standards for some products, the waste of edible products that do not meet standards and other issues mean that the European Commission is currently reviewing its policy in this area (Commission of the European Communities, 2008a).

Other EU measures include four specific quality schemes to develop geographical indications, organic farming, traditional specialities and products from the outermost regions of the EU. The aim of these schemes is to support rural areas by providing a vehicle for producers to obtain a price premium, while simultaneously enabling consumers to identify products with specific quality attributes. The schemes are used by producers to protect and promote their products, increase value-added and develop market power, and they have a role in facilitating entry to international premium markets where consumers are looking for quality products (Belletti *et al.*, 2009). However, the proliferation of quality schemes and labels (mandatory and voluntary, public and private) has created concern about the transparency of requirements, the credibility of claims made, and their possible effects on commercial relations. Concerns have also been raised about the sheer volume of quality messages being presented to consumers.

Evidence from Spain (Bardaji *et al.*, 2009) suggests that certification is part of the product assortment decision-making criteria used by retailers. Further anecdotal evidence (Bord Bia, personal communication) suggests that while buyers of speciality food like to see such labels on products originating from countries with a strong tradition of using this approach (e.g., Italy), they do not expect to find them on products from other countries (e.g., Ireland). Despite the fact that many of these products are mainly sold on local or regional markets (Commission of the European Communities, 2008a), labels awarded through this procedure remain virtually unrecognized by many consumers (Barjolle and Sylvander, 2000). Moreover, the added value of such labels will be limited without significant investment in a campaign or other initiatives aimed at improving consumer awareness and understanding of the labels (Carpenter and Larceneux, 2008). This means that the existence of a positive benefit/cost balance from certification systems may be questionable. Furthermore, the impact on consumer decisions is in doubt as consumer awareness and expectations vary by product and country (Barjolle and Sylvander, 2000).

The use of geographical indications can also affect horizontal relationships within the chain, as in many situations a consortium represents companies involved in such supply chains. Membership of consortia, e.g. for Olio Toscano PGI (extra-virgin olive oil) and Pecorino Toscano PDO (sheep cheese) in Italy, is voluntary but generally offers benefits such as collective marketing and investment, negotiation benefits vis-à-vis fees charged by control bodies and technical assistance. The consortia can counter-balance a fragmented supply chain and provide stronger contractual power and coordination to participants. This collective dimension has important positive effects in supporting exports by SMEs.

Certification is necessary to ensure consumer confidence for organic products and to justify any price premium they command. Until 1991 there was a plethora of national standards, which served to confuse consumers and hinder development of the sector. The European Commission sought to rectify this through a regulation that established a harmonized framework for the production, labelling and inspection of agricultural products and foodstuffs. This was expected to increase consumer confidence, ensure fair competition between producers and support the development of inter- and intra-European trade. However, the European market remains fragmented with national supermarkets tending to stock products certified nationally (Commission of the European Communities, 2008b). The European Commission is working to improve this situation and create a functioning internal market for organic products by developing a single logo. New EU legislation requires all organic products to carry an EU-wide label from July 2010, although national labels can also be used. As the EU standard closely follows the international standard adopted by Codex Alimentarius, EU organic products are recognized outside the EU.

Schemes that influence product quality often result in short food supply chains involving private initiatives, such as farmers' markets and box schemes. Such chains facilitate a direct relationship between producers or SMEs and consumers, and have a role in enhancing consumer confidence. Moreover, they can be important in the development of local and regional food sectors. Despite positive effects, these schemes can constitute a barrier to market entry by excluding other products from the market. There is a particular problem for SMEs when the need to participate in more than one scheme creates significant financial and administrative burdens. Furthermore, the amount and distribution of direct certification costs depends on certification bodies and their relationships with various actors, particularly the consortia in terms of geographical indication products (Belletti *et al.*, 2009). This can cause conflict.

Cross-sectoral policies

Innovation policy

EU innovation policy focuses on enabling European industry to position itself at the upper end of the value chain. While recognizing that price competitive segments are very important for certain groups of disadvantaged/vulnerable consumers, the maturity of the food market in most developed countries means that a major part of the EU food industry has to target the upper end of the value chain. Improving the framework for innovation has a significant influence on the ability to forge relationships among businesses throughout the EU, e.g., making access to the single market easier. Some commentators recommend a farm-to-fork policy regarding research and development of innovative products, as well as in the development of sustainable production and processing methods (European Commission, 2009b, p. 60). They argue that primary producers need to be competitive in supplying raw materials that satisfy quality, food safety, animal welfare and environmental protection requirements, whilst retailers act as the gatekeepers to consumers (and also directly influence and promote innovation) whose purchasing decisions will determine whether this will yield economic benefits. An effective innovation policy involves bringing together all stakeholders, not just those directly involved in the supply chain. While the European Union's financial support for research and innovation (primarily through the Framework Research Programmes) has been successful in building networks among EU researchers and forging links with those outside the EU, improving conditions for SME participation in research programmes is an ongoing task, particularly with regard to reducing the administrative burden. Furthermore, the Commission has proposed voluntary guidelines for improved research collaboration and knowledge transfer between public research institutions and industry. The current European Technology Platform, Food for

Life, provides a useful vehicle for an integrated approach by defining a research agenda based on input from a wide range of stakeholders.

The development of novel foods is particularly important for the food industry and innovation policy is central to its success. Regulatory bodies play a key role in the acceptance of new technologies. While regulation is designed to protect consumer health and the environment, procedures are criticized as being too time-consuming to ensure quick access to the market for innovative products, thereby hindering innovation. Restrictions with regard to genetically modified organisms (GMOs) are also considered to have consequences for competitiveness, as those who are able to use these may have access to cheaper raw materials. Forging relationships and ensuring cooperation with regulatory bodies should be an integral part of the innovation process for companies developing novel foods. Another key aspect for ensuring returns to investment in this area is improved communication by providing clear and sufficient explanations on the nature of new products to European citizens, and by providing consumers with the necessary information to enable them to make informed choices.

Industrial policy

EU industrial policy is very conscious of the need to support the development of SMEs and to ensure sustainable production and consumption.

SME policy

Given the significance of SMEs within the EU and the limitations they face, the European Commission has implemented a specific policy in their favour. It has committed to improving the business environment in which they operate by applying the 'Think Small First' principle. The Small Business Act for Europe of June 2008 sets out principles and concrete measures aimed at creating a more SME-friendly market environment. Key actions relate to increasing access to finance, reducing administrative burdens and improving SMEs' access to the internal market. The argument is that by strengthening SMEs, they will be in a better position to engage with large retailers and to enter new markets.

Another aspect of industrial policy, which is linked to environmental policy, focuses on sustainable food production and consumption. The aim is to promote sustainable development without compromising industry's long-term competitiveness. The effects of such policies, and associated consumer demand, are seen in the supply chain with many leading retailers seeking to promote an image of being more environmentally friendly. Objectives relating to the reduction of the carbon footprint and improved waste management are taking centre-stage. To achieve these objectives, all stakeholders (e.g., producers, processors, consumers, research institutions and NGOs) need to be involved. This could have both positive and negative implications for costs along the chain. The role of retailers in ensuring sustainable production and consumption has been recognized at the EU level with the establishment of a Retail Forum as part of the Action Plan for Sustainable Production and Consumption. The objective is for large retailers to commit to a series of ambitious and defined actions with clear objectives, deliverables and timelines. Such actions also offer opportunities for the development of alternative supply chains and direct relationships between consumers and producers as the proportion of consumers supporting local and seasonal production for environmental reasons grows.

Competition policy

Effective competition policy is required to ensure fair competition, support small business development and avoid the abuse of power by dominant firms in the market place. However, it must also ensure that prices are reasonable, quality standards are high, consumer choice is comprehensive and that innovation can flourish. The European Commission has wide powers to ensure that businesses and governments observe EU rules on fair competition. It can enforce these rules regardless of where companies are headquartered; the criterion is the amount of business they do within the EU. Under EU rules, it is illegal for companies to fix prices or to carve up markets between them. Furthermore, companies with a dominant position may not abuse their power to squeeze out competitors or in their dealings with smaller companies, notably suppliers. Additionally, large companies are not allowed to use their bargaining power to impose conditions on customers/suppliers which would make it difficult for them to do business with the large company's competitors. The European Commission can, and does, impose fines on companies who engage in such practices. National competition authorities are also important in this arena and may overturn decisions by the EU competition authority as in a recent example of a merger of Danish pork producers. A decision viewed as acceptable by the EU competition authority was rejected by the Danish authority as it led to a virtual monopoly at the national level.

Other policies

Faced with a mature domestic market and low growth in domestic consumption, EU trade and commercial policy is vital for the food industry to grow and prosper in the future. New opportunities in terms of markets and investments have become available in recent years as a result of globalization. Many larger food processors and retailers are investing in and exporting to developing economies, however, they face increasing competitive pressures from companies in emerging economies such as China and Brazil. A multilateral approach to trade liberalization is actively pursued by the European Commission in seeking to ensure that European food companies are not faced with unjustified obstacles in developing export markets, and equally that imports into the EU do not place European companies at a competitive disadvantage. Multilateral actions are complemented by bilateral agreements with non-EU countries in seeking to open new export markets for European food and drink products. Participating in Codex Alimentarius, and promoting the international standards developed within Codex, are important aspects of removing non-tariff barriers to trade and ensuring European food companies can compete in the global market on an equal basis. Macroeconomic policies (interest rates and exchange rates) also have an important influence on the competitiveness of the agri-food sector (CIAA, 2008; Muñoz and Sosvilla, 2008).

Regional policy is an instrument of financial solidarity and a powerful force for cohesion and economic integration. Solidarity seeks to bring tangible benefits to citizens and regions that are less well-off by seeking to decrease the economic and social discrepancies that exist between regions. Cohesion, which relates to social inclusion and solidarity within regions, underlines the principle that everyone benefits from narrowing the gaps of income and wealth between regions. The effort focuses on three objectives: convergence, competitiveness and cooperation, which are grouped together in what is now termed Cohesion Policy. The bulk of regional spending is reserved for regions with a GDP below 75% of the Union average and seeks to improve their infrastructure and develop their economic and human potential. This concerns 17 of the 27 EU countries. On the other hand, all 27 are eligible for funding to support innovation and research, sustainable development, and job training in their less advanced regions. A small amount goes to cross-border and inter-regional cooperation projects.

Initiatives undertaken as part of the ICT policy have had a significant effect on relationships and communication in the agri-food chain. For example, e-commerce provides agri-food companies with opportunities for direct sales to consumers and offers a vehicle through which to communicate with them about the heritage of the product; e-business supply chain systems support increased efficiency in product tracking and tracing as well as logistics thereby re-building consumer trust; e-invoicing supports competitiveness by making the process faster, easier and cheaper; e-government provides simplified access to public services; and radio-frequency identification (RFID) is expected to improve the efficiency of traceability systems. Support for research in this area, the development of global standards, e.g., relating to RFID and e-invoicing, and ensuring appropriate legal mechanisms are aspects of this policy that require further action.

EU biofuels policy has a direct impact on relationships in the agri-food chain as potential suppliers of raw materials to the food industry switch to being suppliers in the energy cycle. This increases the price of raw materials and can cause conflict between agri-food and energy supply chains.

Policy propositions to improve agri-food chain relationships

The review of existing policies in the previous section shows that there are a great number of regulations that affect food products and stakeholders along agri-food chains. Most of these are designed to have their greatest impact at one particular level of the agri-food chain, e.g., for farmers, processors, distributors or consumers. All have an indirect effect on the actors in the rest of the agri-food chain, but it is very rare that regulation is deliberately established to cover participants all along the chain. Traceability linked to food safety policies is probably one of the very few issues, which purposely has a vertical rather than horizontal coverage. Some policies, especially those with cross-sectoral purposes, can have an impact on all the agri-food chain stages, such as rules to improve SMEs' efficiency, but they do not have specific considerations for agri-food firms. Their location, often in rural areas, introduces restrictions and peculiarities which have to be dealt with more specifically. The Small Business Act for Europe approach should take into consideration the specificities of the agri-food sector and its implications for the rural business environment.

The way in which policies are developed and applied is clearly an outcome of responses to various interest groups and the interpretation of national and EU legislators who focus their efforts on understanding issues and determining measures to address actual and potential problems. A clear example is the CAP as applied in the agri-food sector. Originally, among its main principles, there were explicit statements that showed concern for consumers as well as farmers. However, most of the legislation has related to farming activities and the interests of farmers. Recent changes to create a more competitive agriculture mostly imply producing commodities at prices closer to world market levels, but in order to create differentiated European food products, commodities should also be differentiated, in addition to having competitive prices. Again, this is an approach that shows a more horizontal than a vertical vision.

EU policies relating to quality differentiated food products and especially those that are linked to products from particular origins are the most comprehensive, in which producer and consumer interests are both taken into consideration. Through Regulatory Councils, rules are established for the production, transformation and distribution stages. These food products are exceptional in that they are subject to integrated policies applied all along their agri-food chains. However, the significance of such products is very limited in Europe in comparison to total food sales, even in the Mediterranean countries where they have greater recognition. Nevertheless, lessons can be learned from policies in that area which could be applied to other food products.

Consumer protection, food safety concerns, quality needs, environmental requirements and many other issues increase the pressure to create new and more refined policies and regulations, which respond to citizens' and governmental concerns. Collectively these pressures are resulting in an increasing number of laws and complex requirements which are, quite often, difficult for consumers to understand and expensive for SMEs to satisfy. Nevertheless, such difficulties could have a positive effect when European food products have to compete with those from other countries. While it is difficult to optimize policies in purely economic terms, particularly when social and environmental concerns have to be considered, compliance with legislation can result in a competitive advantage and is in any case a necessary requirement for operating in the EU market.

On the other hand, there is a feeling that the increase in legislation may be excessive and that the European Union needs to simplify its procedures (European Commission, 2006). This is not an easy objective to address, but steps have been taken in that direction, and not just in the agri-food sector. A vertical vision would add new complexities to agri-food policies and the challenge is to put new ideas into practice within less complex frameworks.

Other areas where improvements can be made are through making public policies that are closer to commercial applications. The increasing number of private agri-food standards is of great concern (Henson and Reardon, 2005). A good example is the contrast between administrative rules to certify quality food products and the various quality certification schemes that private businesses apply in the market. The proliferation of rules creates confusion; many decisions about quality logos and other commercial messages in EU policies have suffered from this complexity. It seems that regulators are often removed from real competitive needs. The incorporation of more agri-food chain stakeholders in the policy process should be addressed to incorporate their points of view. There is a role for governments in providing platforms for chain actors to come together, and to fulfil the need for collective representation by those at various stages. Examples of industry initiatives in this arena that have benefited from government support include the Red Meat Industry Forum and the Cereals Industry Forum in the UK.

There have been, and still are, great concerns about power imbalances along the agri-food chain, particularly between distributors and the rest of the chain. National and EU legislation has been introduced to address this. An example of the former is the UK Grocery Supply Code of Practice. However, limited results have been achieved in avoiding the imposition of costs on suppliers for product listing in stores or promotions, or the pressures from big distribution chains to supply at prices below the cost of production, at least for certain periods of time, squeezing agri-food firms' profits and making it more difficult for them to invest in innovation. Contractual arrangements are increasing in importance and there is a feeling that consumers, as well as SMEs, should be protected from the negative effects of some of these. Little attention has been paid to improving negotiations among partners with different power levels and little consideration has been given to increasing the amount of information required to negotiate more balanced agreements. Policies should probably focus on education and training rather than trying to regulate contractual agreements, which are always difficult to control.

A key lesson is that real problems should be addressed through scientific research to generate high-quality and comprehensive impact assessments if favourable policy outcomes are to be achieved. The same applies when the European identity and its consequent problems are matched with those from other important countries and economic blocs in the world. Increased globalization requires greater awareness of how competitors are likely to react with respect to voluntary measures versus legislation, the use of genetically modified organisms, social welfare implications and other issues. Multilateral agreements, which introduce common rules for countries around the world, can play a role in this area.

There is an increasing gap between real competitive needs and actual legislation, which should be carefully scrutinized before new regulations are applied. There is potential conflict between various concerns and a need for coherence between different policy objectives and consistency among policy measures. The generation of an increasing number of regulations does not mean that better results will be achieved. The challenge is to address critical issues, through the simplest possible approach, taking into consideration the European identity in a globalizing world in which European food products have to compete.

To summarize, it can be said that competition has switched the commercial focus from horizontal to vertical concerns along agri-food chains, as chains and their leading participants have increasingly sought to out-perform competing chains. Policies, however, have traditionally been established to address horizontal rather than vertical concerns. Significant efforts have been undertaken by the High Level Group (HLG) on the Competitiveness of the Agro-Food Industry to establish an integrated approach (European Commission, 2009a, b; CIAA, 2008) as well as to measure economic competitiveness and to provide an assessment of legislation (European Commission, 2007b; Wijnands *et al.*, 2008). The group understands that the agro-food (agri-food) industry concerns the manufacturing/processing of raw agricultural products. The next step forward for policy makers should be to develop an integrated policy approach for the entire agri-food chain or according to the denomination of the HLG, the food industry, which covers the whole industry from farm to fork.

The necessity of an integrated and whole chain approach has been recognized by the European Commission in its communication on a better functioning agri-food chain in Europe (Commission of the European Communities, 2009b, c). Three cross-cutting priorities have been identified. First, the promotion of sustainable and market-oriented relationships between stakeholders, which embraces the reduction of imbalances in bargaining power and unfair trading practices through greater awareness of contractual rights and stronger action against unfair and anti-competitive practices. Second, an increase in transparency along the agri-food chain through better price monitoring at each stage, in order to encourage competition and improve resilience to price volatility. Third, fostering the integration of the European agri-food chain across Member States, through the removal of practices that fragment the internal market such as differences in labelling regulations, and encouraging the competitiveness of the chain through agricultural re-structuring and fostering greater innovation and exports by the agri-food sector. It can be expected that these will be key themes in agri-food chain policy over the next few years.

References

Bardaji, I., Iraizoz, B. and Rapun, M. (2009) Protected geographical indications and integration into the agribusiness system. *Agribusiness: An International Journal* 25(2), 198–214.

Barjolle, D. and Sylvander, B. (2000) PDO and PGI in Europe: regulation or policy: recommendations, Final Report to the European Commission, FAIR 1-CT95-0306.

Becker, T. (2009) European food quality policy: the importance of geographical indications, organic certification and food quality assurance schemes in European countries. *The Estey Centre Journal of International Law and Trade Policy* 10(1), 111–130.

Belletti, G., Burgassi, T., Manco, E., Marescotti, A., Pacciani, A. and Scaramizzi, S. (2009) The role of geographical indications in the internationalisation process of agri-food products. In: Canavari, M. (ed) *International Marketing and Trade of Quality Food Products.* Wageningen Academic Publishers, Wageningen, pp. 201–222.

Burt, S., Davies, K., Dawson, J. and Sparks, L. (2008) Categorizing patterns and processes in retail grocery internationalisation. *Journal of Retailing and Consumer Services* 15(2), 78–92.

Carpenter, M. and Larceneux, F. (2008) Label equity and the effectiveness of value-based labels: an experiment with two French Protected Geographic Indication labels. *International Journal of Consumer Studies* 32(5), 499–507.

Confederation of the Food and Drink Industries of the EU (CIAA) (2008) Policy Recommendations. Input into the High Level Group on the Competitiveness of the Agro-Food Industry.

Commission of the European Communities (2008a) Green Paper on agricultural product quality: product standards, farming requirements and quality schemes. COM (2008) 641 final. Brussels.

Commission of the European Communities (2008b) Communication from the Commission to the Council, the European Parliament, European Economic and Social Committee and the Committee of the Regions. Think Small First. A Small Business Act for Europe. COM (2008) 394 final. Brussels.

Commission of the European Communities (2009a) Communication from the Commission to the European Parliament, the Council, the European Economic and Social Committee and the Committee of the Regions. Reviewing Community Innovation Policy in a Changing World, COM (2009) 442 Final. Brussels.

Commission of the European Communities (2009b) Commission Staff Working Document (COM (2009) 591) The evolution of value-added repartition along the European food supply chain. Accompanying document to the Communication from the Commission to the European Parliament, the Council, the European Economic and Social Committee and the Committee of the Regions, A better functioning food supply chain in Europe. Brussels.

Commission of the European Communities (2009c) Communication from the Commission to the European Parliament, the Council, the European Economic and Social Committee of the Regions, A better functioning food supply chain in Europe, COM (2009) 591 Provisional. Brussels.

Competition Commission (2008) The supply of groceries in the UK market investigation. http://www.competition-commission.org.uk/rep_pub/reports/2008/538grocery.htm.

European Commission (2006) Better Regulation – simply explained. Luxembourg.

European Commission, Agriculture and Rural Development Directorate General (2007a) Note for the file. The importance and contribution of the agri-food sector to the sustainable development of rural areas. http://www.ec.europa.eu/agriculture/analysis/markets/agrifood/text_en.pdf.

European Commission, Enterprise and Industry Directorate General (2007b) Competitiveness of the European food industry. An economic and legal assessment.

European Commission, Enterprise and Industry Directorate General, Food Industry Unit, High Level Group on the Competitiveness of the Agro-Food Industry (2009a) Report on the Competitiveness of the Agro-Food Industry.

European Commission, Enterprise and Industry Directorate General, Food Industry Unit, High Level Group on the Competitiveness of the Agro-Food Industry (2009b) Roadmap of Key Initiatives.

Hanf, J. (2009) Challenges of a vertical coordinated agri-food business for cooperatives. *Journal of Co-operative Studies* 42(2), 5–13.

Henson, S. and Reardon, T. (2005) Private agri-food standards: implications for food policy and the agri-food system. *Food Policy* 30(3), 241–253.

Lang, T. (2003) Food industrialization and food power: Implications for food governance. *Development Policy Review* 21(5–6), 555–568.

McEachern, M. (2008) Guest editorial: the consumer and value based labels. *International Journal of Consumer Studies* 32, 405–406.

McEachern, M. and Schröder, M. (2004) Integrating the voice of the consumer within the value chain: a focus on value-based labelling communications in the fresh-meat sector. *Journal of Consumer Marketing* 21, 497–509.

Ménard, C. and Valceschini, E. (2005) New institutions for governing the agri-food industry. *European Review of Agricultural Economics* 32(3), 421–440.

Muñoz, C. and Sosvilla, S. (2008) Informe Económico 2007. Federación Española de Industrias de la Alimentación y Bebidas (FIAB).

Palpacuer, F. and Tozanli, S. (2008) Changing governance patterns in European food chains: the rise of a new divide between global players and regional or multi-national producers. *Transnational Corporation Journal* 17(1), 69–97.

Wijnands, J., Bremmers, H., van der Meulen, B. and Poppe, K. (2008) An economic and legal assessment of the EU food industry's competitiveness. *Agribusiness: An International Journal* 24(4), 417–439.

Young, L. and Hobbs, J. (2002) Vertical linkages in agri-food supply chains: changing roles for producers, commodity groups, and government policy. *Review of Agricultural Economics* 24, 428–441.

Zachariasse, V. and Bunte, F. (2003) How are farmers faring in the changing balance of power along the food supply chain. OECD Conference on Changing Dimensions of the Food Economy: Exploring the Policy Issues. The Hague, Netherlands, 6–7 February.

Zhao, X., Huo, B., Flynn, B. and Yeung, J. (2008) The impact of power and relationship commitment on the integration between manufacturers and customers in the supply chain. *Journal of Operation Management* 26(3), 368–388.

Chapter 17

Lessons Learned: Recommendations for Future Research on Agri-food Chain Relationships

Fabio Chaddad,[1] Christian Fischer[2] and Monika Hartmann[3]

[1] University of Missouri, Columbia, USA
[2] Massey University, Auckland, New Zealand
[3] University of Bonn, Germany

Introduction

The chapters of this volume have described theoretically (Part I) and empirically (Part II) some recent studies undertaken to better understand the nature and relevance of inter-organizational relationships in agri-food chains around the world. Moreover, Chapters 15 and 16 have discussed practical implications arising from this research for business management and policy-making, respectively.

Despite the diverse array of theories and methods used (qualitative as well as quantitative ones), a number of agri-food chains analysed (cereals, meat, vegetables and wine) and the wide geographical areas covered (Europe, USA, Australia, China and the Philippines) in this volume, there is still considerable scope for additional research. While the importance of inter-organizational relationships has been recognized it is not yet fully understood in which cases organizations in the agri-food sector should rely on traditional, open market-type transactions or rather seek to build strong relationships. In addition, there is a wide set of institutional arrangements to govern the latter (see Chapters 2 and 3). Furthermore, though approaches to define and measure sustainable inter-organizational relationships have been developed, these concepts need to be both refined and extended. Finally, despite the fact that substantial empirical evidence on the factors contributing to more sustainable inter-organizational relationships exist, it is not yet fully clear how to turn this knowledge into practical business management advice or effective policy decisions. Hence, much of what is known so far needs to be reconfirmed, more deeply and widely tested, and transformed into useful and reliable advice to practitioners and decision-makers.

Building on knowledge that has been gained up to now and that is described in this book, this chapter identifies and explores potential research opportunities in the growing field of inter-organizational relationships, focusing on governance and chain collaboration architectures, the linkage between governance and relationship sustainability, and the definition and measurement of a key construct – i.e., sustainable inter-organizational relationships. Regarding the latter issue, more detailed suggestions are proposed on how further progress could be achieved.

Governance structures and chain collaboration architectures

Dating back to the seminal work of Coase (1937) and Williamson (1979), the first generation of the transaction cost approach to governance focused on the determinants of the make-or-buy decision facing supply chain participants. A second generation of governance researchers expanded this ubiquitous question to 'to make, to buy or to ally' (e.g., Geyskens *et al.*, 2006) recognizing that relational modes of governance were increasingly prominent in the real world. This extensive body of literature has identified factors at the transaction unit of analysis that affect governance structure choice, including asset specificity, uncertainty, frequency and complexity. Geyskens *et al.* (2006) quantitatively tested the main postulates of transaction cost economics in a recent study based on a meta-analysis that considered 200 empirical papers, all of which concerned with explaining make-buy-or-ally decisions. Their findings confirm that the existence of asset specificity and volume uncertainty[1] as well as behavioural uncertainty favours hierarchical governance modes while technological uncertainty[2] promotes a choice of market governance. However, Geyskens *et al.*'s (2006) results do not support Williamson's assertion that asset specificity is of greater relevance than uncertainty as a driver for the decision between make or buy. Comparing the main determinants driving the decision between relational versus market governance the authors found that asset specificity motivates the former over the latter while the opposite holds for volume, technological and behavioural uncertainty. The meta-analysis by Geysens *et al.* provides ample evidence that transaction cost economics is a well-developed and empirically corroborated theoretical approach to the study of governance in inter-organizational relationships. Yet, the analysis shows at the same time that the diversity and complexity of inter-organizational relationships offers several opportunities to refine, extend and test the theory.

Though 'frequency' of a transaction is considered by Williamson as a determinant for the optimal organizational form of an exchange, it has received far less attention in the literature than 'asset specifity' and 'uncertainty' (Geyskens *et al.*, 2006).[3] In addition, the social dimension of 'transaction frequency' is hardly acknowledged by Williamson. However, repeated transactions (Gulati, 1995; Rooks *et al.*, 2000) as well as 'the firm's entire network of relationships' (Gulati *et al.*, 2000, p. 204) could create social embeddedness which may influence transaction costs. Chapter 3 shows that due to social relations and behavioural rules economic behaviour can considerably deviate from what is traditionally assumed in neoclassical and transaction cost economics.

---

[1] Volume uncertainty refers to the 'inability to accurately forecast the volume requirements in a relationship' (Geyskens *et al.*, 2006).

[2] Technological uncertainty refers to the 'inability to accurately forecast the technical requirements in a relationship' (Geyskens *et al.*, 2006). Such uncertainty may occur due to changes in environmental conditions (e.g., standards) or as a consequence of general technological developments.

[3] For instance, Geyskens *et al.* (2006) were not able to include 'transaction frequency' in their meta-analysis as a determinant for make-buy-or-ally decisions as too few studies had provided results regarding this determinant for governance choice. The authors stress 'Transaction frequency is deserving of greater empirical attention' (Geyskens *et al.*, 2006, p. 532).

Chapter 2 developed a continuum of generic governance structures treating inter-organizational relationships as intermediate forms between markets and hierarchies. There is a wide array of governance modes to ameliorate coordination and motivation problems in inter-organizational relationships. Cook *et al.* (2008) offer a generic typology of hybrid structures – i.e., leadership, egalitarian and cooperative hybrids – to govern relationships between organizations in agri-food chains. This is a useful starting point but much remains to be learned on how the specific mechanisms of governance work and their effects on the performance and sustainability of inter-organizational relationships. Treating inter-organizational relationships as true hybrids that combine hierarchical or organization-like attributes with market-like mechanisms and approaching governance as a system (or syndrome) of attributes appear to be fruitful avenues for future research.

One obvious avenue for future research is to describe the diversity of inter-organizational relationship governance using the conceptual framework summarized in Table 2 of Chapter 2. Specific research questions that might be included in future inter-organizational relationship studies from a governance perspective include the following. What is the logic behind different combinations of governance attributes? What purposes do they serve? Under what conditions will a particular 'syndrome of attributes' emerge and/or fit with organizational strategy and the business environment? Why do we observe different types of inter-organizational relationships (with different architectures) co-existing simultaneously and often competing in agri-food chains? A low-hanging fruit for scholars is the transaction cost analysis of producer-owned cooperatives given their increasing importance in the global agri-food system (see Chapter 11) and the relative dearth of research on the governance instruments used in these complex hybrid arrangements. Related to these queries are the issues of dynamics and stability of inter-organizational relationships and the institutional and market forces that shape their structural characteristics.

In the agri-food sector, it has been shown that farmers do not always behave in the rationality-based, selfish way that neoclassical economic theory and transaction cost economics predict. Community life requires paying attention to social ties and farmers have traditionally been at the core of rural and religious communities. This explains why farmers have been found to sell land at lower prices to relatives, close friends or other people with whom they have strong relationships, and charge premiums to those with whom relationships are weaker (Siles *et al.*, 2000). Social ties obviously need to be taken into consideration when analysing business relationships – especially in the agri-food sector.

Considering business relations as embedded in social networks and acknowledging the relevance of fairness-driven reciprocity behaviour helps to understand and explain trust and reputation-generating processes and incentives for good conduct for a large share of actors. Though preventing malfeasance of those agents that otherwise would behave selfishly can be enhanced by measures such as formal contracts, which provide extrinsic incentives such as performance contingent monetary rewards, the same measures might diminish voluntary effort contributions of reciprocal actors.[4] This phenomenon is also referred to as 'crowding out' effect (Frey, 1997). More research is needed to analyse the interdependencies between explicit performance incentives and reciprocity-based voluntary cooperation of economic

---

[4] It is assumed that reciprocal actors pursue cooperative behaviour due to intrinsic motivation. Thus, even in the absence of any monetary reward, cooperative behaviour is displayed by reciprocal actors as they are motivated by factors such as fairness-driven reciprocity.

agents, thus linking the insights of behavioural economics and social structure theory with those of the transaction cost approach.

In addition, little empirical research so far has analysed fairness concerns directly related to inter-organizational relationships. This holds especially with respect to field studies, making this an important area for further research. Moreover, social structures and the relevance of social norms likely differ between sectors and countries. Comparing the relevance of fairness concerns across countries and sectors is an additional interesting area for studies yet to come.

Governance mode and sustainability of inter-organizational relationship

More in-depth analysis is needed to examine the relationship between governance type choice and the sustainability of inter-organizational relationships. Examining survey data from agri-food chains in several EU countries, Fischer *et al.* (2009) explore whether different governance mechanisms (relational versus explicit contracts) influence relationship sustainability. The results indicate that relationship sustainability is largely independent of the adopted contract type. Based on a structural equation model, the authors also analysed the main determinants of sustainable inter-organizational relationships (SIRs). While 'effective communication' and the 'existence of personal bonds' have a positive and significant influence on SIRs – independent of contract type – the determinants 'key people leaving' and 'unequal power distribution' revealed a significant negative impact on SIRs only in the case of relational contracts (Fischer *et al.*, 2009). Further studies examining these relationships in other countries and chains would be interesting extensions of this work.

Another relatively unexplored research topic is the role of the institutional environment – i.e., the rules of the game (North, 1990) – on inter-organizational governance and sustainability. For example, are there any differences between countries on how transactions along agri-food chains are governed? Do these differences in the institutional environment affect the sustainability and effectiveness of inter-organizational relationships? Answering these questions would entail expanding the Fischer *et al.* (2009) EU survey data to other parts of the world – including developing countries – but the pay-offs would probably offset the costs.

Motivation of the partners in an inter-organizational relationship is assumed to be a crucial determinant of SIR. Two types of motivation can be distinguished: extrinsic and intrinsic. With extrinsic motivation the activity provides indirect satisfaction (e.g., through monetary compensation), whereas intrinsic provides immediate satisfaction (e.g., through the fulfilment of own needs). According to the theory of human motivation, extrinsic and intrinsic motivations are interdependent (Frey, 1997). Research conducted in this area is so far primarily oriented towards intra-firm relationships (e.g., Osterloh *et al.*, 2002). An extension of these studies with a focus on the role of motivation in inter-organizational relationships would be of significant value.

As discussed in Chapter 3 of this book, firms can obtain joint benefits if they are motivated to invest in intangible 'relationship-specific pool resources', i.e., freely accessible and reserved to exchange or network partners (e.g., tacit knowledge creation and transmission) (Osterloh *et al.*, 2002). Whenever the contributions to these inter-firm-specific resources of each partner are not easily observable and thus not verifiable, free-riding can take place. There is a trade-off between self-interest where each firm can reap a higher individual benefit by not contributing and joint benefits where all actors involved in a relationship are better off if everybody is willing to invest in the creation of those 'relation-specific pool resources' (Osterloh *et al.*, 2002). Thus, actors who are primarily extrinsically motivated have little incentive to contribute to the joint effort and might free-ride on partners' efforts. As a consequence, 'relationship-specific pool resources' are expected to be under-supplied. A social dilemma then arises for inter-organizational relationships. Following Osterloh *et al.* (2002), solving social dilemmas is a problem of strategically managing the motivation, rather than precisely measuring and controlling it. Practically, if rewards and commands (extrinsic motivation) are perceived as supportive (i.e., appreciating members' contribution) and not as controlling, then intrinsic motivation tends to be strengthened. Otherwise, even if considerable intrinsic motivation exists in the first place, based on mutual acknowledgement of one's engagement and contribution, it can be 'crowded out' by the external control (Osterloh *et al.*, 2002).

Definition and measurement of inter-organizational relationships

A further issue that would need more academic research concerns the exact nature of the concept of SIRs. In Chapter 4 of this volume an attempt has been made to review the recent literature on this topic and to propose a definition of SIRs. Subsequent empirical work (e.g., Chapter 7; Fischer *et al.*, 2008, 2009; Reynolds *et al.*, 2009) successfully used the SIR construct. However, some of the measures of relationship stability proposed in Chapter 4 did not perform satisfactory in the empirical estimations, indicating that the definition and/or the measurement of the SIR sub-constructs still need to be further refined. Building on more recent developments in construct and measurement theory, in what follows we offer some suggestions of how further progress in this matter could be achieved.

SIRs – a Type I or Type II construct?

Multidimensional constructs, in theory, can be of three different types. These types differ according to two criteria: (i) 'relational level'; and (ii) 'relational form' (Law *et al.*, 1998). 'Relational level' refers to whether or not a construct is on the same level of abstraction as its dimensions (indicators). 'Relational form' is concerned with the issue whether or not individual dimensions can be algebraically combined to form an overall representation of the construct.

Type I (or 'latent') constructs exist only at a deeper and more embedded level as their dimensions. In other words, construct and dimensions essentially represent the same but each dimension constitutes a different way of expressing (or *reflecting*) the construct, which in turn encompasses all these different dimensions, thus being at a deeper level of abstraction. A particular dimension (indicator or measure) is non-essential to define the construct. That is, leaving one dimension away or replacing it by another one does not change the fundamental meaning of a construct. In statistical models, the latent construct is extracted from the common (i.e., shared) information (variance) contained in its dimensions (measures).

Type II ('aggregate') or Type III ('profile') constructs are at the same level of abstraction as their dimensions. That is, such constructs are *formed* by their dimensions and each dimension is essential to exactly define a construct. The difference between Type II and Type III constructs relates to their 'relational form', i.e., whether the individual dimensions can be algebraically combined to form an overall representation of the construct, or not. If they can, the construct is called Type II. In such cases all dimensions are of an ordinal nature. That is, for each dimension the attributes 'high' and 'low' can be assigned and have meaning (e.g., job satisfaction). 'Profile' multidimensional constructs are of a nominal nature and the different states which they may assume cannot meaningfully be ranked (e.g., personality).

Deciding on a particular type for a construct is a matter of theoretical justification. This is in particular true for distinguishing Type I from Type II/III constructs (Edwards and Bagozzi, 2000). In the past, there has been a general tendency to misclassify constructs (Jarvis *et al.*, 2003) and to assume more often than theoretically justified Type I models for reasons of ease of data collection and statistical analysis. Moreover, the empirical consequences of using an incorrect construct type may not be severe if dimensions (measures) can be assumed to be highly correlated with each other (Diamantopoulos and Winklhofer, 2001).

As discussed in Chapter 4 and applied in Chapters 7 and 8, the SIR construct was assumed to be of Type I. This is line with previous research, in particular involving the relationship quality construct. However, given that discussions have been intensified in the literature (e.g., Edwards and Bagozzi, 2000; Diamantopoulos and Winklhofer, 2001; Jarvis *et al.*, 2003), in the following it is explored whether a Type II classification of the SIR construct would also be justified, or potentially even more defensible.

The SIR construct presented in Chapter 4 is of 'third order', formed by the second-order relationship quality (RQ) and relationship stability (RS) sub-constructs which themselves are formed by first-order constructs (commitment, satisfaction, etc.). The reason for assuming a Type II SIR construct is, first, that it is made up of similar but not identical dimensions (RQ and RS), both of which are essential to fully define the SIR construct. As discussed in Chapter 4, RQ may be seen to represent static aspects of SIRs while RS refers to dynamic ones. Given that the two second-order constructs may not *reflect* the same concept but rather *form* the SIR construct, a Type II classification could be justified. Second, both second-order constructs are of ordinal character and the attributes 'high' and 'low' are meaningful for both RQ and RS, as is talking of 'highly sustainable inter-organizational relationships'. As to the second-order constructs, one could argue in the same way and classify them as Type II. First, all dimensions of RQ and RS are essential to define the respective constructs. That is, all dimensions are complements rather than substitutes. Second, commitment, satisfaction, trust, mutual dependence, conflict resolution capacity and positive collaboration history are also all of ordinal nature and can be ranked on some sort of intensity scale.

The measurement of the SIR construct – reflective or formative specification?

As to the measurement of the suggested SIR construct two common research approaches are considered: data-rich quantitative business surveys and data-sparse quantitative case studies. In the former approach, statistical analysis of the obtained data is possible which allows for the estimation of the significance and magnitude of the individual components (dimensions) within the SIR construct. This can be done overall and for relevant sub-groups and thus, for example, it is possible to determine which component of the SIR construct is more important in one examined group relative to another. In the case study approach, such weight estimation is in general not possible but alternative methods such as index calculations for arriving at SIR scores can be applied, for instance, in order to compare them across study units.

Statistical models

When a large database is available, the quantitative description of a non-directly observable, latent construct can be achieved in two ways. First, the construct can be measured indirectly by means of observable indicator variables. A usual method to do this is the use of latent variable structural equation modelling. The standard case is that latent constructs are extracted as common factors from at least one available indicator variable (a single measurement 'item'). However, the use of multiple items is considered as being more reliable since in this way the common variance ('communality', i.e., a measure of the unobservable construct) can be clearly separated from random measurement errors, which are assumed to be present in each item. Second, a latent construct can be indirectly estimated as an intervening variable within a larger statistical model, also usually a structural equation model. Both approaches can be applied for the suggested SIR construct.

In structural equation modelling, measurement models can be specified in two different ways which relate to the theoretical nature of the construct under consideration discussed above. While the fundamental distinction between these two specification approaches has been known for a long time (see e.g., Bollen, 1989), the discussion on this topic has intensified in the literature over the last decade (e.g., Edwards and Bagozzi, 2000; Diamantopoulos and Winklhofer, 2001; Jarvis *et al.*, 2003).

The first way to specify a measurement model for an underlying construct is the use of reflective measures (or 'effect' indicators) which make up a classical measurement scale. In structural equation modelling depictions, reflective specification is represented by single-headed arrows pointing from the construct (represented as a circle or an oval shape) to the indicators (represented as squares). Reflective specification relates to the above mentioned Type I construct, with the indicators being measured variables and not latent constructs. The underlying statistical method to quantify the construct is common factor analysis.

The other way to specify measurement models is by means of formative (or 'cause') indicators. Here the single-headed arrows point from the indicators to the construct which, in this case, is by nature more an index than a scale (Diamantopoulos and Winklhofer, 2001). This measurement model relates to the Type II (the aggregate) model, discussed above, with the same distinction that indicators are measured variables rather than latent ones. The problem is that such a measurement model in itself is not identified and thus cannot be estimated (i.e., the construct cannot be quantified). The reason for this is that in formative measurement models, the underlying statistical model is regression analysis, which requires the existence of (data on) a dependent variable (the construct) to estimate the contribution of the individual indicators forming it. Since the dependent variable (the construct) is not directly observable and thus not measured, the regression cannot be conducted.

One way to overcome this problem is to transform the formative measurement model into a multiple-indicators-multiple-causes (MIMIC) model by adding at least one reflective indicator to the construct (Diamantopoulos and Winklhofer, 2001). This indicator (or the extracted shared variance in the case of several added reflective measures) is then used as the dependent variable which now makes possible the estimation of the formative part of the MIMIC model. A fully specified measurement model of the SIR model is depicted in Fig. 1. Both the third-order SIR construct and the two underlying second-order RQ and RS constructs are specified as MIMIC models. As such they are all individually and in themselves fully identified. While for each of these three constructs one reflective indicator would be sufficient to make them fully identified, two are used in Fig. 1.[5] As to the RQ construct, earlier studies (Bigne and Blesa, 2003; Rossomme, 2003; Selnes, 1998) suggest that significant (and positive) correlations (the double-headed arrows) exist between its dimensions commitment, satisfaction and trust. For the RS construct, no clear previous results exist. However, intuition suggests that conflict resolution capacity and a positive collaboration history are positive correlated, since if conflicts are not resolved in a satisfactory way, a relationship will break and the collaboration history will be short and not positive.

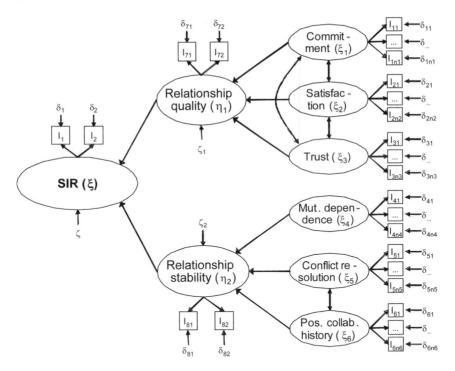

Fig. 1. A formative measurement model of SIRs.

[5] Note, in practice, the number of items which can be included into a survey instrument is limited because there is a maximum number of questions which respondents (or interviewees) are willing to answer. Consequently, construct measurement is a compromise between theoretical requirements and practical feasibility aspects.

In regard to the included error terms, all depicted δ_{ij} represent the conventional measurement errors usually specified in structural equation models. However, given that the three higher-order constructs are specified with formative dimensions, the included $\zeta_{(i)}$ terms have a different meaning. Following Diamantopoulos (2006), the $\zeta_{(i)}$ terms reflect the influence of all non-included (omitted) dimensions of a formative construct, provided that its reflective indicators (or better, their communality) represent the construct's full and true meaning. Hence, it is a measurement error at the level of the construct and not at the level of its indicators. Nevertheless, since in the SIR case the RQ and RS constructs serve as measures for the third-order SIR constructs, the $\zeta_{(i)}$ terms may be interpreted as conventional indicator measurement errors, with the difference that the indicators are not directly but indirectly measured.

Another approach to estimate the SIR construct in a structural equation modelling framework would be to use a reduced measurement model specification in which the second-level constructs (RQ and RS) are modelled as intervening variables. This can only be done within a larger statistical model, but it is possible for the SIR construct. In practical terms, this would imply the removal of the reflective measures I_{71}, I_{72}, I_{81}, I_{82} from the endogenous constructs η_1 and η_2. In addition, in order to make the SIR construct fully identified, the errors terms ζ, ζ_1 and ζ_2 also need to be removed (which is permissible, see Diamantopoulos, 2006), and two additional constraints need to be added (e.g., constraining two of the paths pointing to or pointing away from η_1 or η_2 to be equal to 1). Technically, the two endogenous constructs η_1 and η_2 represent then the summary influences of the three each underlying measured variables on SIR. That is, variance (or communality) of the SIR reflective measures is related to summary influences η_1 and η_2 in a way to optimize the overall model fit. Since this model is more parsimonious than the one depicted in Fig. 1, its main advantage is that fewer data need to be collected (i.e., respondents have to answer fewer questions). On the other hand, asking respondents a few additional questions to directly assess RQ and RS would offer the advantage to easily validate the consistency of the obtained answers. In both SIR specifications (the full and the reduced one) the path coefficients should approximately be the same if all used items accurately measure what they are supposed to measure, and construct errors ζ, ζ_1 and ζ_2 should be small. As a consequence, it may be advisable to integrate the additional second-level items, to estimate both versions of the model, and to carefully compare and to interpret the obtained path coefficients. In fact, this could prove to be a useful way to technically judge the measurement reliability of the proposed construct, an issue for which so far no standard procedures exist in formative measurement models (Diamantopoulos and Winklhofer, 2001).

The proposed SIR construct can be used as both a dependent and independent variable within a larger structural equation model. For example, analysts may be interested in examining which factors determine sustainable inter-organizational relationships in certain supply chains, or at different chain stages (Chapter 7). As an independent variable, the SIR construct may be used for quantifying the impact of sustainable relationships on business outcome (e.g., chain profitability, innovation activity, output growth, as done in Chapters 13 and 14).

Index construction

The SIR construct discussed above may also be used apart from data-intensive structural equation modeling,[6] e.g., in quantitative value/supply chain case studies. In those studies, analysts could be interested in deriving an overall SIR score on which different chains can be compared or benchmarked.

In quantitative case studies, the problems associated with measurement errors and valid estimations of structural interdependencies are probably not of primary importance. Rather, some crude but easy-to-handle measures to quantify the phenomena under investigation are needed. Therefore, it is probably more appropriate to measure the SIR-underlying first-level constructs (i.e., commitment, satisfaction, etc.) by means of a single question and using some sort of simple rating scale (e.g., 1 to 5, or 1 to 7).

Assuming first that all first-order SIR components carry equal weight (i.e., the weights ω_k, $k \in [1, 6]$, are all 1), the second-order constructs RQ and RS can be calculated by adding up the obtained scores from the first-order indicator variables. The obtained summation result can be standardized in order to lie in a pre-defined and easily interpretable range. Thus, the relationship quality index (RQI) for an individual study object i consists of the responses to the variables commitment (x_1), satisfaction (x_2) and trust (x_3). RQI_{max} would be equal to 21 and RQI_{min} equal to 3 if all variables are measured on a 1 to 7 scale:

$$RQI_i = \left(\frac{(\omega_1 \cdot x_{1i} + \omega_2 \cdot x_{2i} + \omega_3 \cdot x_{3i}) - RQI_{min}}{RQI_{max} - RQI_{min}} \right) \cdot 50 \tag{1}$$

The relationship stability index (RSI) can be defined in a very similar way, building on the variables' mutual dependence (x_4), conflict resolution capacity (x_5), and positive collaboration history (x_6):

$$RSI_i = \left(\frac{(\omega_4 \cdot x_{4i} + \omega_5 \cdot x_{5i} + \omega_6 \cdot x_{6i}) - RSI_{min}}{RSI_{max} - RSI_{min}} \right) \cdot 50 \tag{2}$$

Both indices have a range of [0; 50]. The summation of these components would yield the sustainable inter-organizational relationship index ($SIRI$), ranging from zero to 100:

$$SIRI_i = RQI_i + RSI_i \tag{3}$$

The assumption of equal weights for the individual first-order components of the SIR construct may not be reasonable in all analytical situations. While this is a commonly chosen approach in many studies (e.g., Simatupang and Sridharan, 2005), another option would be to explicitly ask case study participants about the relative importance of the individual indicator variables in their relationships, where these variables may be named

---

[6] A common recommendation for structural equation modelling is to have at least 15 cases per measured variable (Stevens, 1999). This would imply needing at least 300 (reduced SIR model) to 360 (full model) cases in order to be able to reliably estimate the path coefficients for the suggested SIR construct. Incorporating the SIR construct in larger structural equation models would of course require even larger sample sizes.

$w_1,...,w_6$. If there are responses from N participants, the ω_k may be computed as mean scores:[7]

$$\omega_k = \frac{\sum_{i=1}^{N} w_{ki}}{N}.$$

Note, however, that unless a similar explicit weight-determining procedure is adopted for the second-order constructs RQ and RS, these will still be equally weighted in the *SIRI* as expressed in equation (3). Different weighting approaches, such as using individual weights for each relationship, are also possible and further research may try to identify the solutions which are best suited for individual analysis situations.

Looking ahead, more generally, there are various avenues which future research involving the SIR construct may take. First, construct definitions and specifications are subject to constant discussion and adaptation. The suggested SIR construct specification may need further theoretical scrutinizing. In the literature, terms like 'relationship strength' (Donaldson and O'Toole, 2000), 'relationship intensity' (Santoro, 2000) or 'relationship closeness' (Srivastava and Singh, 2010) are used and at this stage it is not entirely clear how these terms relate to the SIR concept. Hence, there is scope for both expanding and refining the SIR definition and specification. Second, it would be interesting to know how structural estimation results (i.e., β weights) differ if alternative SIR measurement models are used. General simulation study results exist (Jarvis *et al.*, 2003) which have investigated potential biases from construct misspecification and which suggest that such biases can be substantial, although they do not necessarily have to be large (Diamantopoulos and Winklhofer, 2001). However, no specific studies for the SIR construct exist so far and it would be useful to see the sensitivity of structural model estimates as a result of different SIR specifications. Third, if large biases were found, new empirical work would need to be conducted for validating, or replacing, the structural SIR estimates in existing studies.

Conclusion

This chapter has presented potential research opportunities in the growing field of inter-organizational relationships that could build on the theories and empirical evidence discussed in this volume. Several research opportunities were identified regarding governance and chain collaboration architectures, the linkage between relationship governance and sustainability, and the definition and measurement of sustainable inter-organizational relationships. Extending, refining and combining existing theories and using additional sources of data and new research methods to confirm or disprove empirical evidence would be of significant value to researchers and practitioners interested in analysing and managing sustainable inter-organizational relationships in the global agri-food system.

---

[7] Depending on how the w_k are measured (e.g., on a 1 to 7 scale), the obtained ω_k need to be normalized so that their sums range from 0 to 1 in both equations (1) and (2). Accordingly, the RQI_{max}, RQI_{min}, RSI_{max} and RSI_{min} need to be adapted, which would then be 1 and 7, respectively in case the x_k are measured on a 1 to 7 scale.

References

Bigne, E. and Blesa, A. (2003) Market orientation, trust and satisfaction in dyadic relationships: a manufacturer–retailer analysis. *International Journal of Retail & Distribution Management* 31(11), 574–590.

Bollen, K. (1989) *Structural Equations with Latent Variables*. Wiley, New York.

Coase, R. (1937) The nature of the firm. *Economica* 4, 386–405.

Cook, M., Klein, P. and Iliopoulos, C. (2008) Contracting and organization in food and agriculture. In: Brousseau, E. and Glachant, J. (eds) *New Institutional Economics: A Guidebook*. Cambridge University Press, Cambridge, pp. 292–304.

Diamantopoulos, A. (2006) The error term in formative measurement models: interpretation and modelling implications. *Journal of Modelling in Management* 1(1), 7–17.

Diamantopoulos, A. and Winklhofer, H. (2001) Index construction with formative indicators: an alternative to scale development. *Journal of Marketing Research* 38(May), 269–277.

Donaldson, B. and O'Toole, T. (2000) Classifying relationship structures: relationship strength in industrial markets. *Journal of Business & Industrial Marketing* 15(7), 491–506.

Edwards, J. and Bagozzi, R. (2000) On the nature and direction of relationships between constructs and measures. *Psychological Methods* 5(2), 155–174.

Fischer, C., Hartmann, M., Bavorova, M., Hockmann, H., Suvanto, H., Viitaharju, L., Leat, P., Revoredo-Giha, C., Henchion, M., McGee, C., Dybowski, G. and Kobuszynska, M. (2008) Business relationships and B2B communication in selected European agri-food chains – first empirical evidence. *International Food and Agribusiness Management Review* 11(2), 73–99.

Fischer, C., Hartmann, M., Reynolds, R., Leat, P., Revoredo-Giha, C., Henchion, M., Albisu, L. and Gracia, A. (2009) Factors influencing contractual choice and sustainable relationships in selected agri-food supply chains. *European Review of Agricultural Economics* 36(4), 541–569.

Frey, B. (1997) *Not Just for the Money: An Economic Theory of Personal Motivation*. Edward Elgar, Cheltenham/Brookfield.

Geyskens, I., Steenkamp, J. and Kumar, N. (2006) Make, buy or ally: a transaction cost theory meta-analysis. *Academy of Management Journal* 49(3), 519–543.

Gulati, R. (1995) Does familiarity breed trust? The implications of repeated ties for contractual choice in alliances. *Academy of Management Journal* 38(1), 85–112.

Gulati, R., Nohria, N. and Zaheer, A. (2000) Strategic networks. *Strategic Management Journal* 21(3), 203–215.

Jarvis, C., Mackenzie, S. and Podsakoff, P. (2003) A critical review of construct indicators and measurement model misspecification in marketing and consumer research. *Journal of Consumer Research* 30(Sept), 199–216.

Law, K., Wong, C.-S. and Mobley, W. (1998) Towards a taxonomy of multidimensional constructs. *Academy of Management Review* 23(4), 741–755.

North, D. (1990) *Institutions, Institutional Change, and Economic Performance*. Cambridge University Press, Cambridge.

Osterloh, M., Frost, J. and Frey, B. (2002) The dynamics of motivation in new organisational forms. *International Journal of the Economics of Business* 9(1), 61–77.

Reynolds, N., Fischer, C. and Hartmann, M. (2009) Determinants of sustainable business relationships in selected German agri-food chains. *British Food Journal* 111(8), 776–793.

Rooks, G., Raub, W., Selten, R. and Tazelaar, F. (2000) How inter-firm co-operation depends on social embeddedness: a vignette study. *Acta Sociologica* 43, 123–137.

Rossomme, J. (2003) Customer satisfaction measurement in a business-to-business context: a conceptual framework. *Journal of Business & Industrial Marketing* 18(2), 179–195.

Santoro, M. (2000) Success breeds success: the linkage between relationship intensity and tangible outcomes in industry–university collaborative ventures. *The Journal of High Technology Management Research* 11(2), 255–273.

Selnes, F. (1998) Antecedents and consequences of trust and satisfaction in buyer–seller relationships. *European Journal of Marketing* 32(3/4), 305–322.

Siles, M., Robison, L., Johnson, B., Lynne, G. and Beveridge, D. (2000) Farmland exchanges: selection of trading partners, terms of trade, and social capital. *Journal of American Sociological Farm Managers and Rural Appraisers* 63(1), 127–140.

Simatupang, T. and Sridharan, R. (2005) The collaboration index: a measure for supply chain collaboration. *International Journal of Physical Distribution & Logistics* 35(1), 44–62.

Srivastava, V. and Singh, T. (2010) Value creation through relationship closeness. *Journal of Strategic Marketing* 18(1), 3–17.

Stevens, J. (1999) *Applied Multivariate Statistics for the Social Sciences*. Lawrence Erlbaum Associates, New Jersey.

Williamson, O. (1979) Transaction-cost economics: the governance of contractual relations. *Journal of Law and Economics* 22(2), 233–261.

Index

abattoirs, 152
adaptation to disturbances, 55
advantage
    collaborative, 15, 75
    competitive, 15, 17, 64, 65, 66, 75,
        150, 165, 192
adverse selection, 62, 64
Ag Processing, Inc., 183
AgFirst, 183
agility, 41
AgriBank, 183
agricultural
    cooperatives. See *cooperatives*
    development, 74
agri-food supply chains
    barley-to-beer, 17, 94, 98, 135–49
    barley-to-whisky, 17, 135–49
    cattle-to-beef, 17, 95, 98, 150–63
    cereals, 16, 119–34
    cereal-to-bread, 97
    grapes-to-wine, 19, 220
    meat, 16, 119–34
    pigs-to-cured ham, 96
    pigs-to-pigmeat, 96
    pigs-to-sausage, 94
    vegetables, 191–205
    wheat-to-bread, 18, 19, 164–76, 206–
        19
agri-food systems, 18, 21, 45, 58, 119,
    177, 269
agri-food trade, 26
Agrilliance, 187
Ahold, 28
alignment of business goals. See *goals*
alliances, 18, 47, 48, 101
    cross-industry, 47, 49
    equity, 47
    horizontal, 47, 49
    motivations for formation, 51
    non-equity, 47
    strategic, 45, 49, 187
    vertical, 47

altruism, 67
Aragon, 19
Arkansas Dairy Cooperative Association,
    187
Asia, 31
asset specificity, 48, 166, 268
assets
    relationship-specific, 20, 65
    transaction-specific, 194, 246
Australia, 16, 19, 105, 138, 208, 220–34
autonomy
    loss of, 157

bakeries, 168
BÄKO, 172, 174
barley, 139
beef cattle, 151
behaviour, 14, 18, 61–73, 67, 106, 192,
    240, 269
    anti-competitive, 254
behavioural economics, 14, 61–73, 270
benefit. See commercial benefit
best practice, 20, 237–49
bonds
    personal, 11, 17, 19, 121, 126, 141,
        144, 208, 253
    social, 111
Brazil, 261
BRC Global Standard, 36
bread, 207
British Retail Consortium Global
    Standard for Food Safety, 36
business relationships. See *relationships*
business systems, 14, 40
business ventures, 242, 244
    non-cooperative, 18
buy, make or collaborate, 12, 53, 75, 268

C3 behaviour, 243
Camgrain, 143, 144
capacity-building, 162, 247

capital intensity, 18
Cargill, 47, 187
Carrefour, 28, 37
cartels, 38
case studies, 143
centralization
   degree of, 55
CF Industries, 188
chain captain, 52, 177
chain performance. See *performance*
China, 19, 29, 30, 51, 191–205, 207,
   246, 261
CHS, Inc., 183, 187
cluster of firms, 46
Coase, Ronald, 45, 53, 135, 268
CoBank, 183
Codex Alimentarius, 261
cohesion of supply chains, 17
collaboration, 11, 14, 20, 21, 38, 46, 51,
   92, 110, 191, 243
   architectures, 21, 57, 268
collusion, 38
commercial benefit, 17, 146, 147, 221,
   244
commitment, 19, 77, 79, 92, 108, 111,
   137, 162, 166
   attitudinal, 79
   manifest (behavioural), 79
Common Agricultural Policy (CAP), 20,
   95, 151, 158, 250, 256
communication, 108, 116, 119, 137, 144,
   147, 160, 161, 195, 208, 224, 230, 254
   effective, 17, 19, 20, 121, 126, 243,
      248
   frequency, 141, 243
   means of, 141
   quality, 17, 19, 141
   two-way, 140
competition, 14, 38, 50, 53, 69, 91, 106,
   141, 150, 156, 241, 253, 261
competitiveness, 13, 15, 17, 18, 19, 20,
   75, 119, 136, 164, 206–19, 239, 255,
   260
concentration, 38

conflict, 52, 108
   resolution, 19, 53, 76, 225
   resolution capacity, 82, 141
consolidation, 45
construct, 77, 122, 141, 166, 209, 222,
   271
   measurement, 272
   Type I ('latent'), 271
   Type II ('aggregate'), 271
   Type III ('profile'), 271
consumer, 13, 29, 32, 80, 248, 250, 252,
   255, 257
   awareness, 52
   demand, 29
   expectations, 91
   requests, 12, 32
   satisfaction, 248
contracts, 20, 35, 61, 63, 69, 81, 91, 98,
   140, 177, 191–205, 244
   equity-based, 50
   explicit, 15, 63
   fair, 247
   formal, 54, 101
   incomplete, 15, 52, 69
   informal (verbal), 101
   non-equity, 50
   relational, 45
   vertical, 28
   written, 101
control
   mechanisms, 49, 54, 56
convergence in demand, 32
cooperation, 17, 18, 19, 38, 52, 110, 165,
   222, 230
cooperatives, 45, 52, 55, 56, 97, 101,
   177–90, 254, 269
   agricultural, 18
   marketing, 57
   processing, 57
coordination, 17, 38, 39, 49, 52, 61, 115,
   135, 151, 194
   mechanisms, 54, 56
corporate social responsibility (CSR), 33
corruption, 192

costs
  agency, 91
  exchange, 91
  opportunity, 64
  predictability, 98
  switching, 82, 95, 96, 155, 164, 166,
    167, 173
culture, 249
  business, 158
customer relationship management
  (CRM), 248
customer retention, 159

DairiConcepts, 51
Dairy Farmers of America (DFA), 51,
  183, 187, 188
Dairy Partners Americas (DPA), 48
dairy sector, 28
Dairymen's Marketing Cooperative, 187
Danone, 27
Dean Foods, 187
Denmark, 208, 261
dependence, 110
developing countries, 245
diets, 30
discrete choice analysis, 156
diversification, 50
due diligence, 244

e-business, 262
economic
  agents, 45, 62, 68
  games, 67
  organization, 46, 135, 177
effectiveness, 17, 119, 135
efficiency, 11, 15, 17, 69, 71, 119, 135,
  250
embeddedness, 15, 182
  local, 141
  of market exchange, 62
  social, 63, 64, 137, 240, 268
emerging markets (EMs), 25, 30
entrepreneurial activities, 13, 40
equal power distribution, 16, 19, 122,
  126, 208

European Union (EU), 30, 38, 40, 250–
  66
  countries, 16, 92, 120, 208
  markets, 152
  policies, 20, 250–66
exchange hazards, 63
expert interviews, 92

factor analysis, 196, 213, 227
fairness, 15, 20, 67, 68, 69, 77, 112, 269
farmers, 14, 17, 18, 19, 20, 39, 45, 52,
  55, 97, 152, 192, 245, 251
Farmland, 188
Finland, 16, 120, 126
flexibility, 49, 158
Florida's Natural, 183
Fonterra, 48, 51, 188
food
  policy, 250
  quality, 15, 18, 25, 31, 33, 52, 91, 105,
    114, 119, 191, 255
  safety, 15, 25, 31, 33, 91, 97, 191, 255
  scares, 35
food service sector, 28, 29
forbearance, 111
foreign direct investments (FDIs), 27
formalization
  degree of, 55
franchising, 61, 101
fraud, 64
friendship, 17, 77, 209, 237
  personal, 16, 113, 116, 144
fruit, 105
FS Industries, 188

Galizzi, Giovanni, 11
genetically modified organisms (GMOs),
  32, 260
geographical indications (GIs), 258
Germany, 16, 18, 92, 119, 120, 126,
  164–76, 208
Ghana, 109
Global Berry Farms, 32
Global Food Safety Initiative (GFSI), 36
GlobalGAP, 36
globalization, 13, 25, 26, 261

goals
  alignment of, 17, 20, 242
  organization-specific, 52
good agricultural practice (GAP), 35,
  246
governance, 21, 48, 50
  characteristics, 14
  contractual, 19
  devices, 64
  forms, 63, 151
  informal, 193
  mechanisms, 14
  multidimensional nature of, 53
  relational, 19
  structures, 12, 45–60, 61, 101, 268
government policies, 20, 250–66
Gowlett Grain, 143, 146
Granovetter, Mark, 62, 137, 182, 241
grape and wine industry, 19
Greencore Malt, 143, 146
Greene King, 143, 146
Grocery Supply Code of Practice, 263
growers, 16, 20, 105, 220, 245
Growmark, 183
guanxi, 19, 191–205, 240
  networks, 19, 192

hazard analysis and critical control point
  (HACCP), 35
health, 32
hierarchy, 14, 45, 48, 50, 56, 61, 101
  macro-, 183
hold-up problems, 28, 62, 64
homo
  economicus, 67
  reciprocity, 67
horizontal
  business systems, 40
  integration, 50
  ties, 181
Hortifrut, 32
human skills, 40
hybrid, 14, 48, 52, 56, 61, 101
  arrangements, 12, 45
  forms, 54, 194
  relationships, 77

incentives, 55
independence, 52, 157, 175
index
  of relationship sustainability, 123, 276
India, 30
Indonesia, 29, 30
information, 160
  access to, 113
  asymmetries, 80, 106, 119
  quality, 141, 243
  sharing, 20, 102, 106, 115, 137, 195,
    241, 247
information and communication
  technologies (ICTs), 13, 25, 39, 253,
  262
innovation, 159, 255, 259
integrity, 116, 244
interactions
  professional, 75
  socio-technical, 66
International Food Standard (IFS), 34
investments, 31
  relationship-specific, 20, 52, 57, 65,
    76, 77, 109, 155, 173, 239, 243, 246,
    248
  symmetric, 63
  transaction-specific, 96, 192, 194, 197,
    248
investors, 18
Ireland, 16, 17, 92, 95, 96, 120, 126,
  150–63
Italy, 258

Japan, 246
joint ventures, 18, 45, 50, 187
JR Simplot Co., 188

Keenans, 154
Kepak, 154
key people, 64
  leaving an organization, 16, 122, 126
KK Club, 154
knowledge
  sharing, 65, 74
Korea, 246

Label Rouge chickens, 34
labour market, 71
Land O'Lakes, 183, 187, 188
Latin America, 31
leadership, 138, 254
less developed countries (LDCs), 25, 30
livestock auction marts, 153
logistics, 29, 137, 138, 252
Lone Star Milk Producers, 187
long-term orientation, 157
loyalty, 15
    card schemes, 248

make-or-buy decision. See *buy, make or
    collaborate*
market, 14, 45, 48, 50, 55, 61, 101, 135,
    150
    agents, 16
    transactions, 12, 50, 77, 98
market power. See *power*
marketing
    agencies in common (MACs), 186
    cooperative, 57
    direct, 251
    of farm commodities, 18
    performance, 19
Maryland & Virginia Milk Producers
    Cooperative Association, 187
Ménard, Claude, 54, 101, 177
Mexico, 28
MFA, Inc., 183
Michigan Blueberry Growers Marketing,
    32
Mitsui & Co., 187
MoArk LLC, 188
Monsanto, 47
moral hazard, 15, 62, 64
Morocco, 29
motivation, 62, 270
    extrinsic, 270
    intrinsic, 270
mutual dependence, 82, 167

Naturipe Berry Growers of California, 32
negotiations, 191, 202

neoclassical
    assumptions, 15, 66
    economics, 61
Nestlé, 27, 48
netchains, 18, 177–90
Netherlands, 119
networks, 45, 46, 52, 91, 177, 188
    of firms, 101
    guanxi, 192
    of relations, 64
new institutional economics, 136
New Zealand, 155
non-governmental organizations
    (NGOs), 33
norms, 240. See also *social norms*

Ocean Spray Cranberries, 47
opportunism. See opportunistic
    behaviour
opportunistic behaviour, 16, 52, 61, 64,
    67, 76, 81, 91, 107, 110, 111, 115,
    182, 192, 222, 239, 246
organic food products, 259
ownership, 161
    of farm assets, 55, 57, 181

partial least squares (PLS) estimation,
    196
partner selection, 242
partnerships, 39, 45, 55, 74, 98, 155, 237
PepsiCo, 47, 51
performance, 17, 19, 61, 137, 146, 147,
    151, 159, 192, 269
    ambiguity, 106
    chain, 13, 91
    contract, 109
    of agri-food enterprises, 256
    of relationship, 167
    satisfaction, 223
Philippines, 16, 105
Poland, 16, 28, 120, 126, 210
positive collaboration history, 83, 141

power, 16, 108, 110, 115
  asymmetries, 20, 101, 119, 160, 162,
    224, 240, 249
  bargaining, 16, 102
  distribution, 253
  market, 17, 38, 113
  negotiation, 218
prices, 16, 18, 38, 55, 105, 113, 114,
  139, 146, 173, 177, 220, 238
  fair, 20, 153, 238, 245
  pressure, 101
  transparency, 95
principal component analysis (PCA),
  112, 122
principal–agent
  literature, 53
  theory, 69
private labels, 252
problem resolution, 17, 20, 144, 243
process quality, 32
processing cooperative. See *cooperatives*
processors, 19, 39, 52, 55, 96, 97, 152,
  239, 247, 251
produce, 16, 105, 114
producer clubs, 154, 162
product differentiation, 37
production and distribution systems, 11,
  12
professional regard, 17, 144
profit, 18, 106, 110, 114, 220
  margin, 16, 97
  motives, 11
psychology, 66

quality certification schemes, 156, 159

radio-frequency identification (RFID),
  262
reciprocity, 15, 66, 67, 269
regression analysis, 112

relationship
  age, 141
  costs, 15, 18, 165
  length, 64, 164, 174
  management, 237–49
  managers, 243
  marketing, 165, 166
  performance, 167
  phases, 167
  profit, 164
  quality, 13, 18, 19, 78, 79–81, 122,
    137, 141, 147, 171, 206–19
  satisfaction, 11, 19, 192, 195
  stability, 13, 78, 81–83, 122, 141, 171,
    271
  sustainability, 13, 16, 17, 18, 75, 119–
    34, 121, 136, 141, 208
  types, 151, 153, 156, 165, 170
  value, 13, 19, 115, 220–34
relationships
  arm's-length, 50, 76
  behavioural dimensions, 18
  business, 75, 77
  channel, 110
  contractual, 17, 144
  dyadic, 46, 75
  economic dimensions, 18
  history of, 83
  inter-group, 77
  inter-personal, 77, 116, 237, 244
  life-cycle models, 83
  long-term, 70, 75, 101, 110, 161, 165,
    171, 192, 206, 222, 247, 252
  non-arm's-length, 150, 165
  personal, 16, 20, 66, 102, 108, 113,
    119, 182, 192, 240, 246
  safeguards, 54
  social, 15
  sustainable, 74, 78, 119–34, 150, 164–
    76
  termination of, 18, 166, 174
  trading, 224
  typology, 50
  unstable, 53
  vertical, 38

relationships
  buyer–seller, 19, 74, 167, 192
  buyer–supplier, 46, 51, 52, 180, 220–
    34
  customer–supplier, 221, 224
  farmer–processor, 18, 52, 120, 129,
    156
  farmer–retailer, 52
  processor–retailer, 18, 95, 120, 129
  producer–processor, 95
  retailer–processor, 155
  supplier–customer, 151
relationship-specific assets. See *assets*
reliability, 15, 66
rents
  relational, 65, 150, 165, 168
  sharing, 52
repeated market transactions, 12, 50, 156
reputation, 15, 16, 20, 64, 69, 109, 113,
  240, 244, 269
resource
  allocation, 20, 243
  pooling, 52
resource-based view (RBV), 49
resources, 66, 91, 109, 239, 248, 254
  financial, 191
  relationship-specific pool, 271
Retail Consortium Global Standard for
  Food Safety (BRC), 34
retailers, 17, 19, 29, 39, 52, 96, 105, 140,
  152, 248, 252
rewards, 15, 16, 20, 77, 107, 138, 156,
  160, 244, 269
  distribution of, 17, 20, 244
Riceland Foods, 188
risks, 34, 92, 106, 160, 194
  management of, 25
  of crop failure, 239
  relational, 15, 76

Safe Quality Food (SQF), 34
safeguards, 64, 76, 80
  contractual, 63
  mechanisms, 194

satisfaction, 19, 77, 79, 107, 114, 137,
  144, 159, 244
  business, 19
  economic-rational, 80
  social-emotional, 80
Shoei Foods, 47
small- and medium-sized enterprises
  (SMEs), 16, 19, 20, 129, 206, 250–66
social
  bonds, 111
  capital, 66
  interaction, 246
  norms, 15, 64, 71, 92
  psychology, 54
  sanctions, 64
  ties, 15, 64, 71, 75, 269
  welfare, 250
sociology, 54, 62
South America, 151
Southern Marketing Agency, Inc.
  (SMA), 186
Spain, 16, 19, 92, 96, 97, 120, 127, 206–
  19, 258
spillover effects, 28
spot market. See *market*
standards, 13, 32, 33, 34
  marketing, 257
  private, 257
Starbucks, 51
strategic
  alliances. See *alliances*
structural equation model (SEM), 16, 17,
  19, 121, 136, 210, 227
Suiza Foods, 188
Sunsweet Growers, 47
supermarkets, 14, 29, 41, 140, 155, 191,
  203, 252, 259
supply chain
  analysis (SCA), 46, 179
  management, 136, 207
supply continuity, 18
sustainability, 17, 21, 77, 164, 260
sustainable inter-organizational
  relationships (SIRs), 13, 15, 21, 75,
  83, 271–77

synergies, 49, 66

Teagasc, 153
technology, 20, 39, 157, 191, 247, 248
Tesco, 28, 37
theory
  equity, 77, 80
  of social structure, 14, 61–73, 270
  social exchange, 77
  socio-economic, 61
  transaction costs. See *transaction cost
    economics (TCE)*
third-party enforcement, 63, 69, 70
ties. See social ties
traceability, 25, 35, 39, 96, 173, 256
tracking and tracing. See *traceability*
traditional and local food products
  (specialities), 251, 252, 258
transaction cost economics (TCE), 12,
  14, 45–60, 61, 80, 246
transaction costs, 19, 102, 136, 193
  minimization of, 12, 181
transaction frequency, 268
transactions, 15, 45, 48, 55, 61, 63, 81,
  91, 96, 106, 150–63
transition economies, 113, 240
transparency, 16, 39, 115, 138
  lack of, 101, 106, 254
trust, 15, 16, 17, 18, 19, 20, 62, 66, 76,
  80, 91–104, 105–18, 119, 137, 144,
  162, 165, 166, 168, 182, 223, 239,
  244, 253, 269
  inter-personal, 193, 194, 197
  personal, 172
  types of, 92

trustworthiness, 64, 107, 242

uncertainty, 15, 53, 114, 193, 194, 255,
  268
  avoidance, 141
unequal power distribution. See *power
  assymmetries*
Unilever, 27
United Kingdom (UK), 16, 17, 36, 92,
  98, 120, 126, 135–49, 254
United States of America (USA), 18, 27,
  40, 52, 177–90, 253
utility maximization, 61, 67

value chains, 13, 26, 41
value creation, 45, 46, 222
vegetables, 19, 105, 191
Venturini, Luciano, 11
vertical
  business systems, 14, 40
  collaboration, 75
  cooperation, 48
  coordination, 28, 45, 169, 251
  integration, 50, 101, 135, 161, 194
  ties, 179, 181
vulnerabilities, 62, 106

Walmart, 28, 253
wholesalers, 105
Williamson, Oliver, 48, 56, 165, 194,
  268
wine, 220
World Trade Organization (WTO), 241